A New History of Life

Selected Books by Peter Ward

Rare Earth: Why Complex Life Is Uncommon in the Universe
(with Donald Brownlee)
The Life and Death of Planet Earth (with Donald Brownlee)
Gorgon: Paleontology, Obsession, and the Greatest Catastrophe in Earth's History
Life as We Do Not Know It: The NASA Search for (and Synthesis of) Alien Life
Out of Thin Air: Dinosaurs, Birds, and Earth's Ancient Atmosphere
Under a Green Sky: Global Warming, the Mass Extinctions of the Past, and What They Can Tell Us About Our Future
The Medea Hypothesis: Is Life on Earth Ultimately Self-Destructive?
The Flooded Earth: Our Future in a World Without Ice Caps

A NEW HISTORY OF LIFE

THE RADICAL NEW DISCOVERIES ABOUT THE ORIGINS AND EVOLUTION OF LIFE ON EARTH

Peter Ward and Joe Kirschvink

B L O O M S B U R Y
LONDON • NEW DELHI • NEW YORK • SYDNEY

Bloomsbury Publishing
An imprint of Bloomsbury Publishing Plc

50 Bedford Square
London
WC1B 3DP
UK

1385 Broadway
New York
NY 10018
USA

www.bloomsbury.com

BLOOMSBURY and the Diana logo are trademarks of Bloomsbury Publishing Plc

First published in Great Britain 2015

British Library Cataloguing-in-Publication Data
A catalogue record for this book is available from the British Library.

ISBN: HB: 978-1-4088-3552-4
TPB: 978-1-4088-4465-6
PB: 978-1-4088-5558-4
ePub: 978-1-4088-4280-5

2 4 6 8 10 9 7 5 3 1

Typeset by RefineCatch Ltd, Bungay, Suffolk
Printed and bound in Great Britain by CPI Group (UK) Ltd, Croydon CR0 4YY

From PW: For Dr. Howard Leonard and Dr. Peter Shalit, life historians, and to the great Robert Berner, Yale University

From JK: To the memories of Dr. Eugene M. Shoemaker, Dr. Heinz A. Lowenstam, and Dr. Clair C. Patterson of Caltech. They and many others left fingerprints all over my brain.

Contents

Introduction

HISTORY in almost any form is the academic subject perhaps most hated by school kids. One of the most thoughtful examinations was by James Loewen, in *Lies My Teacher Told Me*.[1] His conclusion can be summarized in one word: "irrelevance." Loewen wrote, "The stories that history textbooks tell are predictable; every problem has already been solved or is about to be solved . . . Authors almost never use the present to illuminate the past—the present is not a source of information for writers of history texts."

Loewen's message is quite clear. As American history is now taught in high schools, the past and present are disconnected, such that history has no effect on, or relevance to, our day-to-day lives. Yet that conclusion is so untrue, especially for the history of life, so ancient it is written in rocks, molecules, models, and on the DNA strands found within our every cell. Its relevance is that it gives us place and context. The history of life also just might save us from near-term extinction, if we take note of it and heed its warnings.

In the early 1960s the great American writer James Baldwin wrote: "People are trapped in history, and history is trapped in them."[2] He was speaking to race when he penned those words. But the statement is equally true if the word "people" is replaced with "all of life on Earth, present and past," for each strand of DNA, in each of our cells, is an ancient record of biological history, written in simple code and passed down from generation to generation. One could say that DNA is nothing *but* history, one with a physical manifestation that was slowly melded and accumulated over countless eons by the most pitiless of all phenomena—natural selection. DNA is a history that *is* inside us—and yet one that is our master as well, the blueprint for our bodies, and dictator of what we *will* pass on to our children, gifts that can be blessings—or deadly time bombs. We are indeed trapped in this particular vehicle of history as much as it is trapped in us.

The history of life gives us the answers to so many of the perplexing questions confronting us all: How did we humans come to inhabit this thin, late-occurring, and very marginal twig on the giant tree of life? What wars did our species have to pass through? What calamities mark the human branch of life's 4-billion-year-old tree? The past can help us understand our place among the twenty or more million species now living—and the untold billions now extinct. When a species is no longer extant, what is also destroyed is the as yet unrealized future evolution of untold species.

In the pages to come we will look at the long road to our present and the distant trials our long-ago ancestors had to pass through: fire, ice, hammer blows from space, poison gas, the fangs of predators, pitiless competition, lethal radiation, starvation, enormous changes of habitat, and episodes of war and conquest amid a relentless colonization of every habitable corner of this planet—each an episode that left its mark in the total sum of DNA now extant. Each crisis and conquest was a forge that changed genomes by adding or subtracting all manner of genes, each of us the descendant of survivors tempered by catastrophe and quenched by time.

There is a second and perhaps even greater reason to pay attention to the history of life encapsulated in this quote from Norman Cousins: "History is a vast Early Warning System."[3] This wisdom dates back to near the end of the Cold War. Newer generations of humans have little sense of what it was like to grow up in the 1950s and 1960s, when weekly noontime siren tests told us children of that dark time that Armageddon was but one terrifying siren scream away, and that every faint sound of a late-night jet might be the start of the end.

Human warfare has repeatedly, and seemingly incessantly, exacted a hideous toll on humanity—physically, economically, and emotionally. In many ways the history of life has undeniable similarities to human conflict and warfare. The coevolutionary development of offensive weapons by predators (better claws, teeth, gas attack, even poison-tipped barbs to catch and kill food species) caused equally rapid countermeasures in the predators' prey, including better body armor, speed, hiding ability, and sometimes defensive weapons as well—all of which is technically called the biological arms race. Many of the great events of evolution cannot be repeated; evolution has had a long period to fill the biosphere with highly competitive and efficient organisms, making it unlikely to repeat, for instance, the Cambrian explosion when all of the basic body plans of animals came into being. But what *can* be repeated are things antipodal to living and diversifying, such as extinction, or extinction writ larger—the dreadful past catastrophes of deep time, the mass extinctions.

With every molecule of carbon dioxide we pump into the atmosphere we are ignoring the early warning sirens that rapid rises in carbon dioxide are the commonality between more than ten mass extinctions of the deep past and what is happening today. Those extinctions were caused not by asteroid impact, but from rapid increases in volcanically produced atmospheric greenhouse gases and the global warming they produced. A terrifying new paradigm of mass extinction has arisen this century: "greenhouse mass extinctions," a name overtly chosen to describe the cause of the vast majority of species killed off by mass extinctions in the past.[4]

Evidence of when, where, and how these greenhouse extinctions took place now blares at us from a wide variety of data. To those hearing these siren calls, the danger seems real enough. Yet too many have ignored or missed vast moralities from the past, and what is a possible future. The history of life provides an early warning system that tells us we *must* reduce human-caused greenhouse gas emissions, but it is *human* history that tells us that we probably will not heed the warnings and reverse the damage until a succession of climate-induced mass human mortalities gives us no choice.

Scientific information from so-called deep time is the single most ignored aspect of the climate change debate. George Santayana wrote the most oft-repeated aphorism about history, one so commonly used that it is trite: "Those who ignore history are condemned to repeat it."[5] With regard to a clear history of mass extinction brought about by atmospheric carbon dioxide levels rapidly approaching in the near future, however, we should especially take heed of the most important word in Santayana's prophecy: the word "condemned."

WHAT IS NEW IN THIS "NEW HISTORY OF LIFE"

No single book could ever do justice to the history of life. Choices had to be made, and those choices were largely dictated by our directive around the word "new." The last "complete" single-volume history of life was written in the mid-1990s: the exquisite and bestselling *Life: A Natural History of the First Four Billion Years of Life on Earth*,[6] by the British paleontologist and science writer Richard Fortey. His "takes" were marvelous and the book remains a joy to read, or in our case, to reread nearly two decades after its publication. But because science advances so quickly, there is much that was not known then compared to now. There are even two new scientific disciplines that barely existed in the mid-1990s: astrobiology and geobiology. The advances in instrumentation have led to entirely new understanding, while outcrops of strata containing fossils from times or taxa previously unknown have come to light. Even the sociology of how science is done has changed, for now the most important scientific breakthroughs are acknowledged to take place between the boundaries of the previously august and familiar disciplines of geology, astronomy, paleontology, chemistry, genetics, physics, zoology, and botany each symbolically separated into their own buildings on most university campuses, each with not only a faculty with its own rules and boundaries, but entire fields with their own vocabularies and favored methods of disseminating information coming from research.

We have used three themes in the pages to come as lodestones for the history we have chosen to feature. First, we posit that the history of life has been more affected by catastrophe than the sum of all other forces, including the slow, gradual evolution first recognized by Charles Darwin—based on his training by the dominant teachers of uniformitarianism. The guiding principle of geology for more than two centuries—the Principle of Uniformitarianism—was first developed by James Hutton and Charles Lyell in the late 1700s.[7] It was taught to and ultimately became the prime scientific influence on generations of young naturalists, including Charles Darwin.[8] The discovery of the dinosaur-killing asteroid that hit our planet 65 million years ago was the start of this paradigm shift toward one that has been sometimes called neocatastrophism,[9] a take on catastrophism, the paradigm that preceded uniformitarianism.

As we will show in this book, uniformitarianism, as it applies to ancient worlds as well as the mode and tempo of evolution, is outmoded and largely refuted. The modern world is *not* the best tool for explaining many times and events in the deep past that indeed were sudden—not gradual. For example, there are no modern examples that can explain the "snowball Earths," or the "great oxygenation event," or the sulfur-rich "Canfield oceans" that lasted for more than a billion years and in all that time impeded the first evolution of the animal grade of complexity. Even the dinosaur killing K-T Cretaceous Period–Tertiary Period mass extinction (now termed the K-Pg, or Cretaceous-Paleogene Period, but we hope our colleagues forgive us if we stick to the better sounding and better known K-T) has no parallel today; nor the type of atmosphere and ocean that allowed life to form on Earth; or an atmosphere with carbon dioxide levels so high that there is not a scrap of ice anywhere on the planet. The present is not a key to most of the past; in fact, it is barely a key to the Pleistocene. Making it so has limited us in our vision and understanding.

Second, while we may be carbon-based life, composed of "long-chain" carbon molecules (carbon atoms strung together to form proteins), it is the influence of three different kinds of molecules, simple molecules that exist as simple gases, that have had the greatest influence on the history of life: oxygen, carbon dioxide, and hydrogen sulfide. Sulfur, in fact, may have been the single most important of all elements in dictating the nature and history of life on this planet.

Finally, while the history of life may be populated by species, it has been the evolution of ecosystems that has been the most influential factor in arriving at the modern-day assemblage of life. Coral reefs, tropical forests, deep-sea "vent" faunas, and many more—each can be viewed as a play with differing actors but the same script over eons of time. Yet we know that on occasion in the deep past entirely new ecosystems appear, populated by new kinds of life. The appearance of life that can fly, for instance, or life that can swim or walk—each was a major shift in evolutionary innovation that changed the world, and in each case helped create a new kind of ecosystem.

WHAT WE BRING

An author's background affects the biases inherent in any written history. Peter Ward has been a paleobiologist since 1973, and has published extensively on modern and ancient cephalopods as well as on mass extinctions of vertebrates and invertebrates. Joseph Kirschvink is a geophysical biologist who began his work on the Precambrian-Cambrian transition, but then expanded to looking at older times (the great oxidation event) as well as being the discoverer of the snowball Earths—major parts of life's history. Together we have subsequently worked on the Devonian, Permian, Triassic-Jurassic, and Cretaceous-Tertiary (this time interval recently renamed the Paleogene Period) mass extinctions.

We have worked together in the field since the mid-1990s. These trips included study of the Permian mass extinction in South Africa, from 1997 to 2001; the study of Upper Cretaceous ammonites in Baja California, California, and the Vancouver Island region; the study of the Triassic-Jurassic mass extinction in the Queen Charlotte Islands; the study of the K-T mass extinction in Tunisia, Vancouver Island, California, Mexico, and Antarctica; and the study of the Devonian mass extinction in Western Australia.

The voice we bring to this book is meant to be a seamless duet, but there are passages where one or the other of us self-identifies because of the nearness of the topics to some particular interest we have, or because we were integral in the history of some aspect of the science being reported.

NAMES AND TERMS

Earlier we noted that the number of species on Earth is in the millions. Most who study life will acknowledge that the current number of formally defined species (which requires a name for both genus and species) is probably less than 10 percent of the actual number of currently living species.[10] But how many have there been in the past? Billions, certainly. That makes the writing of a history of them a daunting process. Paleontology, biology, and geology all have entire vocabularies of highly specific jargon, and it is our job to use the English language in an understandable way to make sense of so much of the multisyllabic jargon—or, in the case of NASA, decipher their endless acronyms. Perhaps even more daunting, by necessity we will have to introduce many of the Latin names for the many creatures great and small that produced and daily continue life's history on Earth.

Finally, a full acknowledgment of the large number of people helping us in our journey to write this book will come at the end of the text. But Ward would like to specifically shout out to two scientist-writers who have profoundly influenced him: Robert Berner, whose work on oxygen and carbon dioxide is absolutely integral to the work written here, and Nick Lane, a prolific scientist and writer whose books are pinnacles of clarity and insight, whose work profoundly influenced at least one of the coauthors, and whose books remain groundbreaking and current.[11]

CHAPTER I

Telling Time

UNTIL recently, the history of life had an arcane time scale, measured not in years, but in the relative positions of rocks scattered about the Earth's crust. In this chapter we will look at the geological time scale, the tool used in discovering the relative sequence of life's history on Earth.

The geological time scale is a rickety old contraption, held together by nineteenth-century rules and current European formality. The newer generations of geologists do not love the hoary and very stuffy series of conventions involved with the time scale, and still required by an increasingly aged set of geologists who were trained in the old tradition. To this day, any change has to be approved by committees;[1] all time units have to be associated with a "type section"—a real stack of sedimentary rocks chosen to best represent a given time interval. The type section is supposed to be readily accessible and must be undisturbed by tectonism, heating, and "structure" complexity (such as faults, folds, and other tricky mashing of the originally horizontal sedimentary beds). The section should not be upside down (which happened more often than one would think), should have lots of fossils (both macro and micro), and should also have beds, fossils, or minerals that can be dated with "absolute" ages (a date in actual years) through some combination of radiometric age dating, magnetostratigraphy, or some form of isotope age dating (such as carbon or strontium isotope stratigraphy).

The time scale is complicated and often useless in the sense that when someone says a rock is Jurassic in age, they are in reality saying the rock in question is of the same age as the designated type section for the Jurassic, which was in the Jura Mountains in Europe. But it is what we Earth and life historians have to work with to discover the age of rocks by their fossils, as well as to communicate their actual age to others. Although more modern tools than dating events and species based on their relative position in piles of sedimentary rock are sometimes available[2]—including the determination of a fossil's actual age through the use of isotopic dating, such as the well-known use of carbon 14 or other kinds of "radiometric" dating using the known rates of decay of various elements contained in the rock—in fact very few fossils are found in beds or are made up of materials allowing this kind of absolute age dating. Usually it is fossil content only that is available, yet from this the rock must be dated.

The geological time scale remains not only the major tool in dating all *rocks* on Earth (categorized by their age, rather than on their lithological characteristics), but also the means by which *events* in the history of life are dated. Using intricate names and seemingly random and dissimilar intervals of time, the time scale remains a thoroughly nineteenth-century tool, and more often than not is an impediment not so much because of the manner in which it was developed, but by the rigid and bureaucratic fashion in which it was formalized and codified into what we have today. Only in the last decade have new geological "periods" been put in place. The formation and common usage of these two new periods are central to our new understanding of the history of life: the Cryogenian period, from 850 to 635 million years ago, followed immediately by the Ediacaran period, from 635 to 542 million years ago.

ARRIVING AT A 2015 TIME SCALE

The first half of the eighteenth century was both the time when the field of geology was born and the time when the geological time scale as we know it now was put in place. During this time, the various eras, epochs, and periods were defined and in so doing replaced a more ancient system.[3] Prior to 1800, each kind of rock observed on the Earth was thought to be of one specific age. The hard igneous and metamorphic rocks, the core of all mountains and volcanoes, were presumed to be the oldest rocks on Earth. The sedimentary rocks were younger, the result of a series of world-covering floods. This principle—called neptunism—held sway, and even developed to the point that specific kinds of sedimentary rocks themselves were thought to have specific ages. The omnipresent white chalks that stake out the northern limits of the European subcontinent and then continue into Asia were considered of a single age, different from the sandstones, and different again from finer mudstones and shale. But in 1805 a discovery was made that changed everything. William "Strata" Smith[4] was the first to recognize that it was not the order of lithological types that determined their age, but the order of fossils within the rocks themselves that could be used to date and then correlate strata to distant locales. He showed that various rock types could have many different ages—and that the same succession of fossil types could be found in far separated regions.

The principle of faunal succession opened the door to the formation of the time scale in its modern sense.[5] Life was the key, life preserved in fossils, and the relative difference of fossil content could be used to distinguish a succession of rocks on the surface of the Earth. The largest division was of older rocks without fossils, beneath rocks where fossils were commonly present. The oldest *fossil*-bearing unit of time was named the Cambrian, after a tribe from Wales, and thus all the rocks older than this came to be known as the Precambrian. From the Cambrian onward, the fossil-bearing rocks came to be known as the Phanerozoic or "time of invisible life." The Proterozoic era, the last before animals evolved, succeeds the older Archean and Hadean eras.

Very quickly the periods of the Phanerozoic were defined, all based on fossil content. Within decades of true scientific collection, curation, and "bookkeeping" of fossils (a compilation of the first and last occurrences of particular fossil groups in the record), it was seen that the Phanerozoic was divisible into three major intervals of time and accumulations of rock. The oldest was named Paleozoic (or old life) era, the middle the Mesozoic era, and the most recent the Cenozoic era.

Even before these eras were put in place, most of the period names still used today were in place. In successive order, the Cambrian, Ordovician, Silurian, Devonian, Carboniferous (this is the European usage; the Carboniferous is subdivided into the Mississippian and Pennsylvanian periods in North America), and Permian comprised the

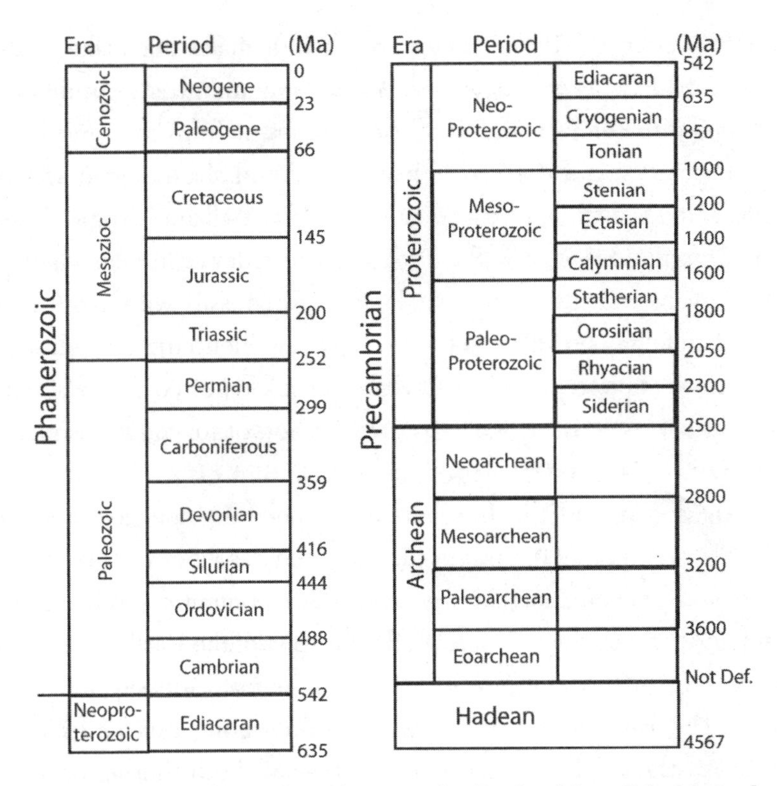

The current version of the Geological Time Scale. (Updated from Felix M. Gradstein et al., "A New Geologic Time Scale, with Special Reference to Precambrian and Neogene," *Episodes* 27, no. 2 (2004): 83–100)

Paleozoic era; the Triassic, Jurassic, and Cretaceous comprised the Mesozoic; and the Paleogene and Neogene (formerly Tertiary), and Quaternary periods comprised the Cenozoic.

By 1850 the periods were in place and new ones were rarely accepted (although many late nineteenth-century geologists tried to get the glory of defining a whole new period, which by then could only take place by cannibalizing already existing units). Only one such attempt actually succeeded, and this was by an Englishman named Charles Lapworth,[6] who carved out an Ordovician period by successfully claiming that some underlying Cambrian and overlying Silurian rocks deserved to be their own geological period, and he managed to persuade enough of the rest of geology to make it so in 1879. By that time the two English bulldogs who had pioneered the naming of periods—Adam Sedgwick for the Cambrian and Roderick Murchison for the Silurian and Permian periods—had died, leaving an ownership vacuum that Lapworth exploited. All of these men had gigantic egos, and fought ferociously for "their" time periods.

The most important real change to the geological time scale in terms of the history of life came with the addition of the Cryogenian and Ediacaran periods, during the Proterozoic era, and the time when life was readying the advent of animals. But long before the evolution not only of animals, but life itself, the Earth had to undergo significant changes to support life. The Cryogenian period (from Greek "cold" and "birth") lasted from 850 to 635 million years ago, and it was ratified by the ruling body on geological names, the International Commission on Stratigraphy and the International Union of Geological Sciences (IUGS), in 1990.[7] It forms the second geologic period of the Neoproterozoic era, and is followed by the Ediacaran period, also new compared to the other periods. Both of these time intervals are seminal times in the history of life, as we will see in greater detail in chapters to come. The Ediacaran period was named after the Ediacara Hills of South Australia—the last geological period of the Neoproterozoic era and of the Proterozoic eon, immediately preceding the Cambrian period, the first period of the Paleozoic era and of the Phanerozoic eon. The Ediacaran period's status as an official geological period was ratified in 2004 by the International Union of Geological Sciences (IUGS).[8]

The geological time scale as constructed is a mishmash of nineteenth- through twenty-first-century science. It is analogous in this to the biological sciences dealing with the classification of organisms, as both are based on historical claims, observations, and precedence of terms and definitions, which often collide with new means of definitions—of both time and species, in the latter's case. Just as DNA analyses have radically changed our view of evolution, so have new methods of dating rocks collided with the old "relative" time scale based on the superpositional relationships of rocks and their fossils. Quite often the collisions are monumental. We wonder what the geological time scale will look like a century from now, especially since modern universities no longer train and produce specialists capable of the high standard of fossil identification necessary to really define geological

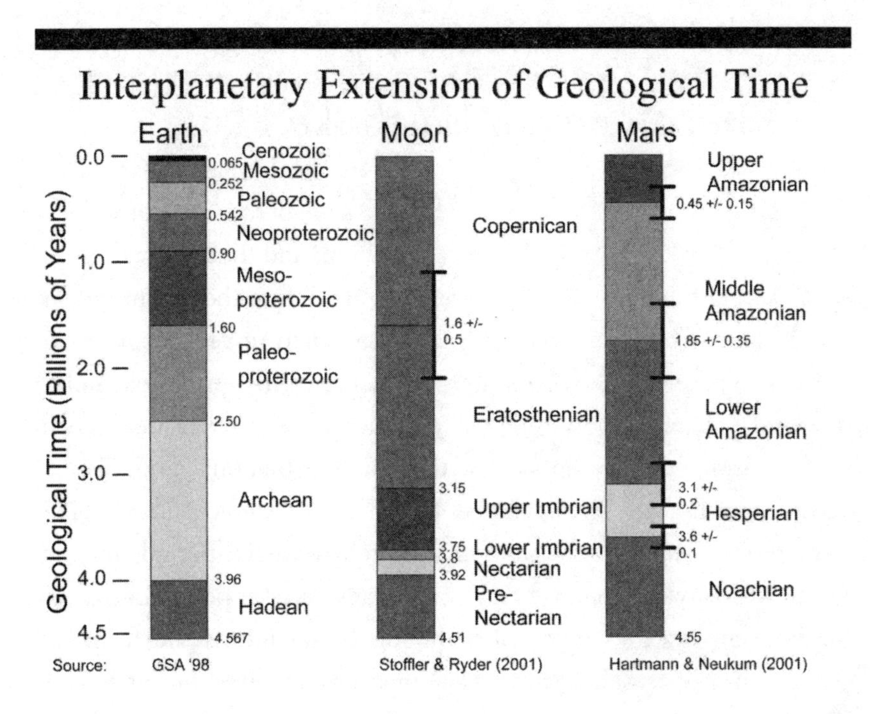

time. This would not matter if some new *Star Trek* kind of tool allowed all rocks to be dated with the flip of a switch or scan. Sadly, that will probably never be the case. We are encased in history in both the rocks and their historical dating methods and definitions. This geological time scale has even been extended to other planets and moons, based on the number of impact craters per unit area, and each body has its unique set of geological terms that we must also learn.

CHAPTER II

Becoming an Earthlike Planet: 4.6–4.5 GA

WE no longer believe, as even the most enlightened of Renaissance thinkers did, that the Earth is the center of the universe, center of the solar system, sole place where life exists in the universe, inhabited by intelligent creatures who are the image of a vast, creating god. We know now that the Earth is but one planet of many—and that its life might be similarly quite unremarkable. The most recent example of this is our search for Earthlike planets, or ELPs. More and more are being found each year,[1] a set of discoveries that is changing the conversation about the frequency of life in the cosmos. But does being "Earthlike" mean having life? Let us look at what our planet went through in its early evolution to the point that it became habitable and eventually inhabited by life.

Between the 1990s and the present, two very specific and paradigm-changing transformations swept the studies that together provide us with a history of life on our planet. Prior to about this time, very little attention was paid by Earth historians to our Earth as a planet that is just one among many. And in like fashion, little attention was paid to its life as being only one kind of life that should be present in the vastness of the cosmos. Yet the discovery of planets orbiting other stars utterly changed the status quo, both scientifically and societally.[2] These findings were a great jolt that transcended the primary fields interested in planets beyond Earth—astronomy and certain specialized branches of geology concerned with what are now called exoplanets—to biological fields, and even into religion. Geoff Marcy, one of the first of the exoplanet discoverers, recounts that one of the first phone calls he received after the momentous discovery of exoplanets came from the Vatican. The Catholic Church, wise in the ways of astronomy, wanted to know if this planet could support life, with all of the religious implications that entailed.

The very first exoplanet was found in 1992 (a planet orbiting a pulsar),[3] followed by the 1995 discovery of a planet orbiting a "main sequence" star, the kind that would be far kinder for the evolution of life than pulsars, which have the nasty habit of emitting great bursts of life-sterilizing energy periodically onto any orbiting planet.

Only a year after this second exoplanet find, another and quite different astronomical discovery further electrified the scientific, political, and public worlds—the report of a meteorite from Mars[4] that was inferred by NASA scientists to contain possible fingerprints of life (and perhaps even fossil microbes). Together, these findings helped launch the new field called astrobiology.

Huge sums of money were targeted at history of life subjects and problems that prior to this had been poorly funded and little investigated, such as the origin and nature of the first life on Earth. This great change began in the last half of the 1990s, and by the new century it was one of the most stimulating fields of science. It transformed science and continues to transform the topic of this book: a history of

While the phrase "Earth-like planet" (or ELP) is now commonly used, we should consider just which Earth we are talking about: the Earth of early in its history—upper left, a complete "waterworld"—or that of the far lower right, some billions of years from now when the oceans have been lost to space.

life on Earth, and our understanding of the potential for life on other planets and the histories of "other" life.

That our planet is one of many potentially habitable planets, and that our life is but one of many possible chemical recipes, are now givens to many astrobiologists. But the many needs of complex organisms equivalent to our Earth's current animals and higher plants are not trivial. Our kind of life is probably not unique (at least in terms of complexity). But one of us (Ward) has argued that the word "rare" is appropriate, and hence his Rare Earth Hypothesis[5]: that while microbial life may be common in the universe, the systems and especially time of planetary environmental stability allowing evolution to eventually produce animal equivalents might be rare indeed.

WHAT IS AN "EARTHLIKE PLANET"?

Perhaps it is terrestrial chauvinism, or perhaps it is true that only life such as our own is possible in the universe. But the search for exoplanets has, at its core, the central goal of finding other "Earths." The question becomes to define just what an Earthlike planet really is. We all have a conception of our planet in the present day: dominated by oceans, a green and blue place, and our place. But as we go back in time and forward in time, we find that the Earth was and absolutely will be a place very different from the planet we now call home. Earthlike is really a time as well as a "place" definition, it turns out.

There are various definitions that are current in astronomy and astrobiology, the two fields most concerned with defining just what kind of planet we live on. At its most inclusive, an Earthlike planet has a rocky surface and higher-density core. In its most restricted sense, it should share important necessities of "life as we know it," including moderate temperatures and an atmosphere that allows liquid water to form on the surface. "Earthlike planet" is often used to indicate a planet resembling modern Earth, but we know that the Earth has changed greatly during the past 4.567 billion years since it formed. During parts of its history, our own Earthlike planet could not have supported life at all, and for over half of its history complex life such as animals and higher plants was impossible. The Earth was wet for virtually all of its history. Within 100 million years of the moon-forming event, where a Mars-sized protoplanet slammed into a still-accreting Earth-sized body, there was liquid water. Coincidence? Or simply a result of the great rain of water-heavy comets smashing onto the Earth's surface and creating an extraterrestrial deluge?

The evidence is found in tiny sand grains of the mineral zircon[6] radiometrically dated to as old as 4.4 billion years ago. They have the isotopic fingerprint of ocean water being sucked down into the mantle via a plate-tectonic-style subduction process. Even though our sun was far less energetic in earliest Earth history, there were enough greenhouse gases in the atmosphere to keep our planet warm. But even more important than heat from the sun, the volcanic activity on early Earth may have been ten times what it is now—and consequently a great deal of heat was streaming out of the Earth and warming its oceans and land. Some astrobiologists now think that life on Earth could not start until planetary heat cooled far lower than it was in the first billion years of Earth history, which is one of many reasons to think that Earth life could possibly have started on another planet, such as Mars. But there was another Earthlike planet early in our solar system history: Venus.

Early in its history[7] Venus should have been in the sun's habitable zone, although it now has a surface temperature of nearly 900°F (500°C) due to a runaway greenhouse effect that surely sterilized its surface (although some think there may be microbial life in its atmosphere, this seems to us to be a pretty slim chance). In contrast, the geological record of Mars shows clearly that it once had flowing water, even in major rivers and streams that could round pebbles and form alluvial fans.[8] Now the water is lost, frozen, or just a faint vapor in the near vacuum of its atmosphere. Presumably its lower mass prohibited the plate tectonic processes essential for crustal recycling, which lowered the thermal gradients in its metallic core that are needed to generate an atmosphere-protecting magnetic field, and the greater distance from the sun allowed it to slip more easily into a permanent "snowball Earth" condition. If life ever existed on Mars, it might still exist in the subsurface, powered by the slight geochemical energy of radioactive decay.

Prior to about 4.6 billion years ago[9] (from this point on, GA refers to billion years ago) the proto-Earth formed from the coalescence of variously sized "planetesimals," or small bodies of rock and frozen gases that condensed in the plane of the ecliptic, the flat region of space in which all our planets orbit. At 4.567 GA (rather precisely dated, and numerically easy to remember), a Mars-sized object appears to have slammed into this body, causing the nickel-iron cores of the planets to merge and the moon to condense from a silicon-vapor "atmosphere" that existed briefly afterward. For the first several hundred million years of its existence, a heavy bombardment of meteors continuously pelted the new planet with lashing violence.

Both the lava-like temperatures of the Earth's forming surface and the energy released by the barrage of incoming meteors during this heavy bombardment phase would surely have created conditions inhospitable to life.[10] The energy alone produced by this constant rain of gigantic comets and asteroids prior to about 4.4 billion years ago would have kept the Earth's surface regions at temperatures sufficient to melt all surface rock, and keep it in a molten state. There would have been no chance for water to form as a liquid on the surface.

The new planet began to change rapidly soon after its initial coalescence. About 4.56 billion years ago the Earth began to segregate into different layers. The innermost region, a core composed largely of iron and nickel, became surrounded by a lower-density region called the mantle. A thin, rapidly hardening crust of still lesser-density rock formed over the mantle, while a thick roiling atmosphere of steam and carbon dioxide filled the skies. In spite of being waterless on its surface, great volumes of water would have been locked up in the interior of the Earth and would have been present in the atmosphere as steam. As lighter elements bubbled upward and heavier ones sank, water and other volatile compounds were expelled from within the Earth and added to the atmosphere.[11]

The early solar system was a place with new planets and a lot of junk that had not been included in planet formation, all orbiting the sun. But not all those orbits were the stable, low-eccentricity ellipses that the current planets show today. Many of them were highly skewed, and many more crossed between the orbiting planets and the sun. All solar system real estate was thus subjected to a cosmic barrage, and no more so than between 4.2 and 3.8 GA. Some of these objects—the comets in particular—may have contributed to the planetary budget of water, but this is a subject of rather intense debate. We simply don't know how much water was delivered by cosmic impacts to the early Earth. The recent discovery that the trace amounts of water present in samples returned from the moon match those of the bulk on Earth argues that most of our hydrosphere and atmosphere was dissolved in the global magma ocean formed in the aftermath of the giant impact of the Mars-sized protoplanet, Thaea.

But any life then existing surely would have paid a price. NASA scientists have completed mathematical models of such impact events. The collision of a 500 km diameter body with the Earth results in a cataclysm almost unimaginable. Huge regions of the Earth's rocky surface would have been vaporized, creating a cloud of superheated "rock-gas," or vapors several thousand degrees in temperature. It is this vapor, in the atmosphere, which causes the entire ocean to evaporate into steam, boiling away to leave a scum of molten salt on the seafloor. Cooling by radiation into space would take place, but a new ocean would not rain out for at least several thousand years after the event. Such large, Texas-sized asteroids or comets could evaporate a ten-thousand-foot-deep ocean, sterilizing the surface of the Earth in the process.[12]

About 3.8 billion years ago, even though the worst barrage of meteor impacts would have passed, there still would have been a much higher frequency of these violent collisions than in more recent times. The length of the day was also different, being less than ten hours long, because the Earth's spin was faster then. The sun would have appeared to be much dimmer, perhaps a red orb of little heat, for it not only was burning with far less energy than today, but it had to shine through a poisonous, riled atmosphere composed of billowing carbon dioxide, hydrogen sulfide, steam, and methane—and no atmospheric or oceanic oxygen was present. The sky itself would probably have

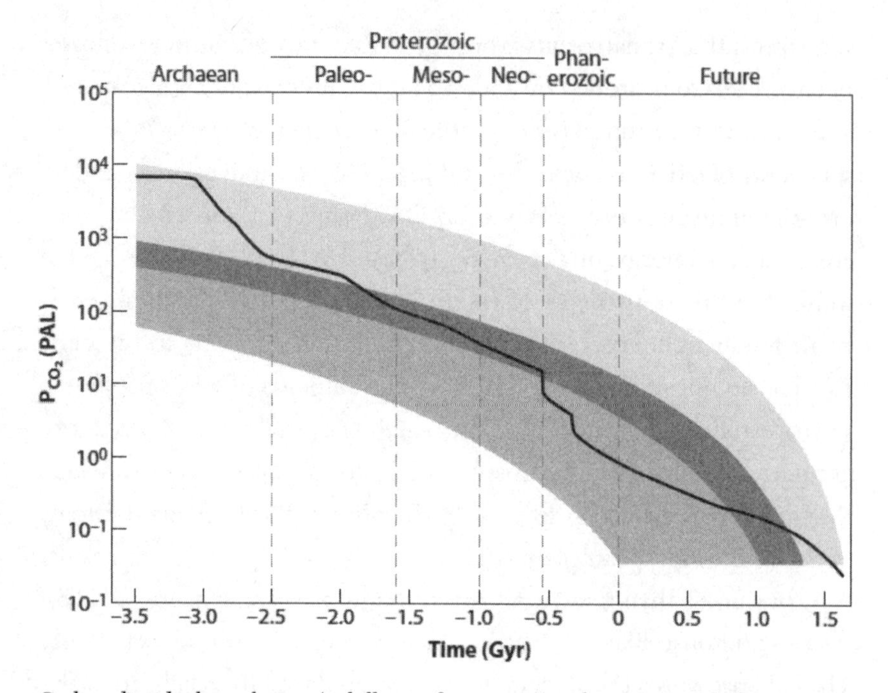

Carbon dioxide through time (in billions of years ago), with estimates of values into the future. Zero represents the present.

been orange to brick red in color, and the seas, which surely covered virtually all of the Earth's surface, would have been a muddy brown in color. But it was real estate with gas, liquid water, and a rocky crust with myriad minerals, rocks, and environments—including those now thought to be necessary for the two-part process of evolving life: producing the many "parts" and then bringing them all together on a factory floor.

NECESSARY LIFE SUPPORT SYSTEMS AND THEIR HISTORY

One of the most critical prerequisites for the origin of life on Earth was to have had atmospheric gases "reducing" enough to permit the formation of prebiotic molecules, the building blocks of Earth life. The chemical processes known as oxidation-reduction can be remembered as "oil-rig." This speaks to whether a compound is giving up electrons (OIL: oxidation is loss) or getting electrons (RIG: reduction is gain). Electrons are like money that can be swapped for energy: in oxidation, an electron loss pays for gain in energy. In reduction, the gain of an electron is money in the bank—and this money is in the form of energy. For example, oil and coal are "reduced." That is, they have a lot of energy in the bank that can be freed when they are oxidized as we burn these fuels. In other words, we oxidize them, which produces energy.

The composition of the Earth's atmosphere early in its history is a controversial and heavily researched topic. While the amount of nitrogen may have been similar to that of today, there are abundant and diverse lines of evidence indicating that there was little or no oxygen available. Carbon dioxide, however, would have been present in much higher volumes than today, and this CO_2-rich atmosphere would have created hothouse-like conditions through a super greenhouse effect, with CO_2 pressures ten thousand times higher than today.[13]

Today our atmosphere is made up of 78 percent nitrogen, 21 percent oxygen, and less than 1 percent carbon dioxide and methane—and this composition seems to be relatively new. As is becoming all too apparent, our atmosphere can change its composition relatively rapidly, especially in that deceptively small 1 percent that includes carbon dioxide and methane, two of the so-called greenhouse gases (along with water vapor) that are of importance far out of proportion to their atmospheric abundance.

ELEMENT CYCLES AND GLOBAL TEMPERATURE

Our human body requires an immense number of complicated processes to foster the strange state we call life. Many of these systems involve the movement of the element carbon. In analogous fashion, the movement of carbon, oxygen, and sulfur are key aspects in maintaining environments suitable for life on Earth. Of these, carbon is most important.

Carbon undergoes an active cycling in and out of solid, liquid, and gas phases. The transfer of carbon between the oceans, atmosphere, and life is referred to as the carbon cycle, and it is this movement that has the most critical effect on a changing planetary temperature brought about by varying concentrations of greenhouse gases. What we refer to as the carbon cycle is really composed of two different (but intersecting) cycles—the short-term and long-term carbon cycles.[14] The short-term carbon cycle is dominated by plant life. Carbon dioxide is taken up during photosynthesis, and some of this carbon becomes locked up as living plant tissue—which is a reduced compound, thus rich in energy that can be liberated. When plants die or leaves fall, this carbon is transferred to soil, and can be again transformed into other carbon compounds in the bodies of soil microbes, other plants, or animals—where the reduced carbon compounds are oxidized with a gain of energy to the organism doing the oxidizing.

At the same time, organisms also convert other carbon molecules to a reduced state, where it can be used for energy. As it passes through a food chain of animals, this same carbon, now in reduced state, can be oxidized and then respired out of the animal or microbe as carbon dioxide gas, and thus the cycle can renew. Other times, however, still locked within plant or animal tissue, the energy-rich reduced carbon might be buried without being consumed by other organisms, to become part of a large organic carbon reservoir within the Earth's crust. In so doing, this carbon is no longer part of the short-term carbon cycle.

The second, or long-term, carbon cycle involved very different kinds of transformations. The most important is that the long-term cycle involves the transfer of carbon from the rock record into the ocean or atmosphere and back again. The time scale of this transfer is generally measured in millions of years. The transfer of carbon to and from rocks can cause changes in the Earth's atmosphere larger than those that can be attained by the short-term carbon cycle, because there is more carbon locked up in rocks than in the ocean, the biosphere (the sum total of living organisms), and the atmosphere combined. This may seem surprising, because the amount of living matter alone is huge. But Bob Berner of Yale University has calculated that if every plant on our planet were suddenly burned, with all their carbon molecules then entering the atmosphere, this short-term carbon cycling would increase atmospheric carbon dioxide by about 25 percent. In contrast, long-term changes in the past have accounted in swings both up and down of carbon dioxide of more than 1,000 percent.

A crucial aspect of the Earth's carbon cycle concerns calcium carbonate, or limestone. This common Earth material makes up the skeletons of most skeletonized invertebrates. It is also found in tiny planktonic plants, called coccolithophorids, whose skeletons accumulate to form the sedimentary rock known as chalk. Coccolith skeletons make up a vital part of Earth's habitability, because they help control long-term temperature at stable levels. Because of the plate tectonic process known as subduction, eventually some of this chalk is carried by the plate tectonic conveyer belt to subduction zones, long depressions in the Earth's crust where oceanic crust sinks downward into the Earth's interior at these depressions. Miles down into the Earth, now well below the surface of the sea bottom, sufficient heat and pressure cause the calcareous and siliceous skeletons to change into new minerals, such as silicates, as well as carbon dioxide gas. These minerals and hot carbon dioxide gas then make their way back to the surface of the Earth as upward-rising magma, rich in gas, where the minerals are extruded as lava, and the gas is liberated into the atmosphere.

This, then, is the key process of the carbon cycle. Carbon dioxide is transformed into living tissue, which eventually decays and helps form the skeletons of other kinds of animals and plants, which eventually fuse into lava and gas deep in the Earth, which is then brought back into the surface to renew the cycle. The long-term carbon cycle thus has a huge effect on atmospheric gas compositions, which itself largely controls global temperature. And since processes of sediment burial and erosion as well as chemical weathering are key components determining how much and how fast carbonate and silicate organism skeletons are produced in the sea, ultimately the amount of minerals going down the hungry maw of the subduction zones will dictate how much carbon dioxide and methane is pumped back into the atmosphere through volcanoes. This entire process is therefore both largely controlled by life and ultimately allows life to exist on Earth. More than just dictating atmospheric concentrations, it produced what might be called a planetary thermostat, for there is a feedback aspect to the cycle that regulates the long-term temperature on Earth.

The thermostat works like this. Let's say the amount of carbon dioxide spewing from earthly volcanoes increases, causing more carbon dioxide and methane to enter the atmosphere. Making their way into the upper atmosphere, many of these molecules cause heat energy rising up from the surface of the Earth (after getting there first as sunlight) to be reflected back toward the Earth. This is the greenhouse effect. With more heat energy trapped in the atmosphere, the temperature of the entire planet rises, in the short term causing more liquid water to evaporate in the atmosphere as water vapor, which itself is also a greenhouse gas. This warming, however, has interesting consequences. With warmer temperatures, the rates of chemical weathering increase. This is most important with regard to weathering of silicate minerals. As we have seen, this weathering process eventually leads to the formation of carbonate or other new kinds of silicate minerals, but the weathering process itself strips carbon dioxide out of the atmosphere.

As weathering rates increase, more and more carbon dioxide is pulled out of the atmosphere to form other chemical compounds that have no first-order effect on global temperature. As atmospheric CO_2 levels begin to drop, so too does global temperature by a less effective greenhouse caused by fewer greenhouse gas molecules in the atmosphere. At the same time, weathering rates decrease as it gets colder, and fewer skeletons are precipitated because there are fewer bicarbonate and silica ions to choose from. Eventually, this results in less skeletal material being subducted, and a lower volume of volcanic carbon dioxide. Now the Earth is cooling rapidly. But as it does so, many ecosystems such as coral reefs or surface plankton regions reduce in size, and thus less atmospheric carbon dioxide is called for. In this world, the volcanoes begin to emit more carbon dioxide than can be used by organisms, and the cycle renews.

The crucial weathering rates are not just affected by temperature. The rapid rise of a mountain chain can cause an uptick in silicate mineral erosion, no matter what the temperature. Rising mountains thus cause a more rapid weathering of these minerals and the removal of more atmospheric CO_2. The Earth rapidly cools. Many geologists believe that the rapid uplift of the massive and rugged Himalaya mountain chain caused a sudden drop in atmospheric CO_2 levels, and thus brought on (or at least contributed to) the cooling that eventually produced the Pleistocene ice age that began some 2.5 million years ago.[15]

A third factor affecting chemical erosion rates is the kind and abundance of plant life. "Higher" (multicellular) plants are highly efficient at causing physical erosion of rock material, thus creating more surface area for chemical weathering to act on. A sudden rise in plant abundance—or the evolution of a new kind of plant with deeper roots, such as found in most trees—has the same effect as the short-term rise of a new mountain chain: weathering rates increase, causing global temperature to decrease. The opposite—the removal of plants either through mass extinction or human-caused deforestation—causes rapid atmospheric heating.

Even the movement of continents can affect worldwide weathering rates, and hence global climate. Since weathering proceeds faster in higher temperatures, even a world in the midst of a very cold interval will get even colder if continental drift moves large continents to equatorial from higher latitudes.

Chemical weathering is quite slow in the Arctic and Antarctic, but high at the equator. Moving continents to equatorial regions will have an effect on global temperature. Another effect of continental position comes from the relative positions of the continents. No amount of chemical weathering can change global temperature if the crucial solutes and mineral species to be used to build skeletons cannot make their way to the sea. Moving water does this, but if all the continents coalesce, as they did in the formation of Pangaea some 300 million years ago, huge areas of the supercontinent interior would have been bereft of rainfall and rivers to the sea. While untold tons of bicarbonate, dissolved calcium, and silica ions would have been produced in the center of this giant continent, much of it never made it to the world ocean.

Eventually, with reduced rainfall, weathering rates would have lowered even in the higher temperatures, and the feedback system may not have worked quite as well as it does with separated continents. The far lower length of continental coastlines produced by the continental amalgamation would have severely affected world climate, as so much of formerly maritime-influenced and wetland areas would have been transformed into regions far from the sea and its water. Deserts and Arctic alike show low rates of weathering, and hence help make the world warmer by lower rate of atmospheric carbon dioxide uptake by mineral by-products of weathering.

THE PHANEROZOIC CARBON DIOXIDE AND OXYGEN CURVES

Perhaps the most influential physical factors other than temperature that most importantly influenced life's history on Earth were the changing volumes (manifested as atmospheric gas pressures) of life-giving carbon dioxide (for plants) and oxygen (for animals). The relative amounts of both CO_2 and oxygen in our planet's atmosphere over time have been (and continue to be) determined by a wide range of physical and biological processes, and it comes as a surprise to most people that the level of both have fluctuated significantly until relatively recently in geological time. But why do the levels of these two gases change at all? The major determinants are a series of chemical reactions involving many of the abundant elements on and in the Earth's crust, including carbon, sulfur, and iron. The chemical reactions involve both oxidation and reduction. In each case, free oxygen (O_2) combines with molecules containing carbon, sulfur, or iron, to form new chemical compounds, and in so doing oxygen is removed from the atmosphere and stored in the newly formed compounds. Oxygen is liberated back into the atmosphere by other reactions involving reduction of compounds. This is what happens during photosynthesis in plants, as they liberate free oxygen as a by-product of the reduction of carbon dioxide through a complex series of intermediate reactions.

There have been a number of models specifically derived to deduce past O_2 and CO_2 levels through time, with the set of equations referred to as GEOCARB being the oldest and most elaborate.[16] This model, used for calculating levels of carbon, was devised by Robert Berner of Yale University. In addition to GEOCARB, separate models have been developed by Berner and his students for calculating O_2. Together, the models show the major trends in O_2 and CO_2 through time. This work represents one of the great triumphs of the scientific method. The importance of the rise and fall of oxygen and carbon dioxide over time is really one of the newest and most fundamental of understandings about life's history on Earth.

Some believe that by 4 billion years ago, conditions and materials on Earth were correct for life to form. But the fact that a planet is habitable does not automatically mean that it will ever be inhabited. The formation of life from nonlife, the subject of the next chapter, appears to have been the most complex chemical experiment of all time. While astrobiologists seem to constantly refer to how "easy" it must have been to start life on Earth, a more nuanced look implies anything but.

Almost more than any other aspect, it has become clear that the interplay and concentrations of the various components of the Earth's atmosphere have been dominant determinants of not only what kind of life (or there being any life at all) on our Earth, but the history of that life. The increasing acceptance of the dominant roles of oxygen and carbon dioxide levels in understanding not only large-scale patterns but nuances of life's progression on our planet is in many respects a twenty-first-century innovation in interpreting Earth history. As is the understanding that two other important gases have played dominant roles in the story of life, and in the pages to come: hydrogen sulfide, or H_2S, and methane (CH_4). Their stories are written in rock, life, and death as well.

CHAPTER III

Life, Death, and the Newly Discovered Place In Between

IN 2006, word began to leak out into scientific circles of a most curious set of experiments dealing with life, death, and what appeared to be a strange and unsettling mixture of the two. First germinating as rumors among colleagues, then slowly maturing at successive talks at various scientific conferences, these findings came into full flower in a series of brilliant papers written by an until-then-unremarkable biologist. Mark Roth was not to remain unremarkable for long, especially after the MacArthur Foundation awarded him a so-called genius grant in 2010 for this work. He is a pioneer, entering a far country, one that could tell us a great deal not only about what "life" is, but what "living" is, and if there could be one without the other not only now, but during the long-ago time when life on Earth first came alive.

Roth had discovered that sublethal doses of hydrogen sulfide put mammals into a state that can only be described as suspended animation.[1] While there is a great deal of popular-culture baggage attached to this appellation (mainly from the science fiction world), in fact these two words quite nicely describe what took place in these gassed animals. Animation, or movement, stopped not only in the observable aspects of the study animals—they no longer moved, had a greatly slowed down respiration rate and heart rate—but also at even more fundamental levels. Normal tissue and cellular functions were greatly reduced in rate. And then even something more surprising occurred: the mammals lost their ability to thermoregulate. They stopped being endothermic, or warm-blooded, and reverted to the more primitive chordate state: ectothermy, or cold-bloodedness. But they were neither dead nor truly alive, for in one of the most basic of mammalian characteristics, they were as if dead. But that death was temporary. It was suspended for a finite amount of time, for when the application of the gas ceased, all normal functions returned. Beyond the obvious medical applications, this new understanding says much about what life is—and is not.

Roth's hunch was simple—that there exists a state between life and death that is both unexplored and of potential medical interest—and it also provided clues to why certain organisms survived mass extinctions. Perhaps death is not so final as generally assumed.[2] His hope was to be able to take organisms to this place and then bring them back. In fact there is no English word that accurately captures the essence of this place. Moviemakers call it zombie land or some such, and maybe stiff-necked science will eventually adopt that term. But we doubt it.

Here was one of his critical experiments. He took flatworms, simple animals, but animals nonetheless. Yet compared to any microbe, no animal can be called simple. He lowered the oxygen content that the flatworms were respiring. Like all animals, flatworms need oxygen, and lots of it. So down went the oxygen content in the closed vessel with the confined worms, and gradually they slowed and then ceased motion. No poking or prodding could get any sort of reaction. But Roth did not conclude the experiment there. In fact, he kept dropping the oxygen content of the worm's water, and they came back to life.[3] The flatworms had entered the state of "dormancy" that is neither alive nor dead. Life and death seem to be far more complicated states than most of us currently believe.

LIFE AND DEATH IN THE SIMPLEST ORGANISMS

Mammals are among the most complex of all animals. Even in these experiments, interesting as they are, the test subjects were obviously alive: their hearts still beat, blood continued to flow in veins and arteries, nerves fired, and the ion transport necessary for life continued to function, if at slower rates. But questions remain about the workings of life in much less complicated and smaller organisms, such as bacteria and viruses, especially when they are put into environments without gas, or in very cold environments. These are not theoretical questions, because every day microbes are flung skyward into the highest reaches of the Earth's atmosphere by violent storms, and find themselves so high that the Earth's protective ozone layer—our major defense against ultraviolet radiation coming from space—can no longer screen them. This is the second frontier in the study of life and death: the study of the Earth's highest life.

After spending days or weeks in the upper atmosphere, these members of the most newly discovered ecosystem on Earth, one not so subtly named "high life" by the scientists who now study this tropospheric biota, come back to Earth.[4] But when they are in space, are they alive?

While it has been known since the dawn of the space age that bacterial and fungal spores could be found at some of the highest altitudes achievable by aircraft, there was very little appreciation of just how many different species can be found in this largest of Earth's habitats, a volume of space utterly dwarfing the volume of the second-largest habitat, the top to bottom of the oceans. But work begun in 2010 demonstrated that at any given time there might be thousands of *species* of bacteria, fungi, and untold viral taxa. It was also discovered by a University of Washington team, sniffing air high atop a mountain in the Cascades of Oregon, that Chinese dust storms routinely drop fungi, bacteria, and viruses onto the West Coast of North America.[5]

Yet beyond an intrinsic biological interest that microbes can be found so high in the atmosphere (or that the atmosphere could be a transport system sending us intercontinental, weaponized viruses), there is a new fundamental understanding that is part of the story of this book: atmospheric transport of life may be how the first life on Earth dispersed away from its site of origin. Why slowly float in an ocean, captive of capricious waves and current, when one can jump from continent to continent through the air in less than a day. Later we will return to the implications of high life for the history of life on Earth; here the issue is whether they are constantly alive during their atmospheric, intercontinental travel, or if they are in dormancy. Here, at the fundamentally basic kind of life, we are finding that the categories of life and death are rather incomplete, if not disingenuous concepts.

High life is collected in three ways: from retired U.S. military high-altitude spy planes, from high-altitude balloons, and when great storms lift off Asia and pass over the Pacific Ocean, and sufficiently "dent" the atmosphere so that air "sniffers" on high mountains can catch a whiff of a descended troposphere. In that air is a zoo full of microbial life. When collected from the immense atmospheric altitudes where cells and viruses are now known to commonly occur, the bacteria are dead. But when brought back to Earth and given some time to react to the altitude they presumably evolved at, they come back to life.

Most of us would agree that for mammals, and perhaps all animals, dead is *dead*. But in simpler life, such is not the case. It turns out that there is a vast new place to be explored between our traditional understanding of what is alive and what is not. And this newly discovered region has important implications about the first chapter in the history of life on Earth, telling us whether "dead" chemicals, when correctly combined and energized, could become alive. Life, simple life at least, is not always alive. But now science seeks to find out if there is a place in between. It could be that the first life on Earth came from the place we call death, or from someplace closer to being alive.

DEFINITIONS OF LIFE

The question "What is life?" is the title of several books, the most famous by the early twentieth-century physicist Erwin Schrödinger.[6] This short book was a landmark, not just for what was written, but also because of the scientific discipline of its author. Schrödinger was a physicist, and before and during his life, the study of biology had been scorned by physical scientists as not worthy of study. Schrödinger began to think of organisms as a physicist would, in physical terms: "The arrangement of the atoms in the most vital parts of an organism and the interplay of these arrangements differ in a fundamental way from all those arrangements of atoms which physicists and chemists have hitherto made the object of their experimental and theoretical research." While much of the book dealt with the nature of heredity and mutation (for this book was written twenty years before the discovery of DNA, when the nature of inheritance was still a perplexing mystery), it is late in the book that Schrödinger considered the physics of "living," when he wrote: "Living matter evades the decay to equilibrium," and life "feeds on negative entropy."

Life does this through metabolism, overtly by eating, drinking, breathing, or the exchange of material, which forms the root of the word from its original Greek definition. Is this the key to life? Perhaps, to a biologist, at least. But Schrödinger, the physicist, saw something much more profound: "That the exchange of material should be the essential thing is absurd. Any atom of nitrogen, oxygen, sulfur, etc. is as good as any other of its kind; what could be gained in exchanging them?" What then is that precious "something" that we call life, contained in our food, which keeps us from death? To Schrödinger, that is easily answered. "Every process, event, happening that is going on in nature means an increase of the entropy of the part of the world where it is going on. Thus a living organism continually increases its entropy." This, then, was his secret of life: life was matter that created an increase in entropy, and in this, a new way of comparing living to nonliving was made.

To Schrödinger, then, life is maintained by extracting "order" from the environment, something that he called (with the self-avowed awkward expression) "negative entropy." Life was thus the device by which large numbers of molecules maintain themselves at fairly high levels of order by continually sucking this orderliness from their environment. Schrödinger suggested that organisms not only created order from disorder but order from order.

Is that all that life is—a machine that changes the nature of disorder and order? From the physics point of view, life could be understood as a series of chemical machines, all packed together and somehow integrated, maintaining order by expending energy to do so. For decades this view was the most influential of all concerning the definition of life. But a half century later, others began to question and amend these views. Some were, like Schrödinger, physicists, such as Paul Davies and Freeman Dyson. But others were trained biologists.

Paul Davies, in his book *The Fifth Miracle*,[7] approached the question "What is life?" by using a different question: what does life *do?* It is *actions* that define life, according to his argument. These main actions are as follows:

Life metabolizes. All organisms process chemicals, and in so doing bring energy into their bodies. But of what use is this energy? The processing and liberation of energy by an organism are what we call metabolism, and they are the way that life harvests the negative entropy that is necessary to maintain internal order. Another way of thinking about this is in terms of chemical reactions. If the organism moves from this state of performing chemical reactions on their own (not in the body of the organism) to a state where the reactions stop, the organism has ceased to be alive. Not only does life maintain this unnatural state, but it also seeks out environments where the energy necessary to stay in this state can be found and harvested. Some environments on Earth are more amenable to life's chemistry than others (such as a warm, sunlit ocean surface of a coral reef or a hot spring in Yellowstone National Park), and in such places we find life in abundance.

Life has complexity and organization. There is no really simple life, composed of but a handful of (or even a few million) atoms. All life is composed of a great number of atoms arranged in intricate ways. It is organization of this complexity that is a hallmark of life. Complexity is not a machine. It is a property.

Life reproduces. Davies makes the point that life must make a copy not only of itself but of the mechanism that allows further copying; as Davies puts it, life must include a copy of the replication apparatus too.

Life develops. Once a copy is made, life continues to change; this can be called development. This process is quite un-machinelike. Machines do not grow or change in shape and even in function with that growth.

Life evolves. This is one of the most fundamental properties of life, and one that is integral to its existence. Davies describes this characteristic as the paradox of permanence and change. Genes must replicate, and if they cannot do so with great regularity, the organism will die. Yet, on the other hand, if the replication is perfect, there will be no variability, no way that evolution through natural selection can take place. Evolution is the key to adaptation, and without adaptation there can be no life.

Life is autonomous. This one might be the toughest to define, yet is central to being alive. An organism is autonomous, has self-determination; it can live without constant input from other organisms. But how "autonomy" is derived from the many parts and workings of an organism is still a mystery.

Action and constitution are one and the same thing for the living system. The system consists of the continuous generation (and regeneration: a protein exists for only about two days) of all the processes and components that put it together as an operational unit. In this view, it is the constant reproduction and renewal of the life form that defines life itself.

This last, the temporary life-span of molecules crucial to living, and thus life, has been underappreciated as a major clue in understanding where life may first have formed. The NASA definition of life is simpler, and is from a definition favored by Carl Sagan: life is a chemical system capable of Darwinian evolution.[8] There are three key concepts to this. First, we are dealing with chemicals, and not just energy or even electronic computing systems. Second, not just chemicals, but also chemical *systems* are involved. Thus, there is an interaction among the chemicals, not just chemicals themselves. Finally, it is the chemical systems that *must undergo Darwinian evolution*—meaning that if there are more individuals present in the environment than there is energy available, some will die. Those that survive do so because they carry advantageous heritable traits that they pass on to their descendants, thus lending the offspring greater ability to survive. The Sagan-NASA definition has the advantage of not confusing life with being alive.

What was the "driver" that caused dead chemicals to combine in such a way to be alive? Was the main driver leading to life a system of metabolism, one that only later added the ability to replicate, or the opposite of this? If it's the first case, primitive metabolic systems—necessarily enclosed in some cell-like space—later gained the ability to replicate and incorporate some sort of information-carrying molecule. In the latter, replication molecules (such as RNA or some variant) gained the ability to use energy systems to aid in their replication, and only later became enclosed in cell. So we see a very stark contrast that this metabolism vs. replication problem poses at the chemical-molecular level: Was it proteins first, or nucleic acids first? Is either alive, and at what point does each pass from chemical reaction to chemical reactions powering life? Yet if the essential characteristic of a living cell is homeostasis—the ability to maintain a steady and more or less constant chemical balance in a changing environment—it follows that metabolism had to come first. Eating before breeding seems to be the accepted view at the present time, but as in so much dealing with the origin of life, disquieting questions remain.

ENERGY AND THE DEFINITION OF LIFE

The role of energy in maintaining life can now be added to our definition of life. We have already defined life as metabolizing, replicating, and evolving. But let us not consider life from an energy flow and order-disorder continuum. Just having energy is clearly not sufficient as a basis of life; there must be an interaction with the energy, and that interaction at a very basic level is needed to maintain a state of nonequilibrium order. Without energy, life goes to nonlife, so life must be something whose very definition is coupled with energy acquisition and energy dumping. Life maintains itself by having states that allow it to become progressively more orderly through the input of energy flow. Our kind of life does this by maintaining a relatively small number of combinations of carbon, oxygen, nitrogen, and hydrogen (and some other elements in smaller volumes). Eventually, a degree of complexity and integration is reached, and maintained, that we call life. The inflow of energy must be sufficient to overcome the tendency of the chemistry within the body that we call life to revert back to its equilibrium condition—nonlife.

One of the universally accepted definitions of life is that it metabolizes. For Earth life, the primary sources of energy are from the heat of the Earth or from the sun, itself the energy arising from the sun's thermonuclear fusion reactions. By far the most common way that life taps into solar energy is through photosynthesis. In this process, sunlight provides the energy to convert carbon dioxide and water into complex carbon compounds with many chemical bonds that store energy. By breaking these bonds, energy is released.

Life on Earth uses a variety of biochemical reactions, and they all involve the transfer of electrons. But this system works only if there is what might be called an electrochemical gradient: the steeper the gradient, the more energy that can be realized. This means that some types of metabolism yield far more energy than others, just as some kinds of environments have more energy to harvest than others. The organic (carbon-containing) compounds containing the greatest amount of stored energy are fats and lipids—long chain carbons that have much energy tied up in their chemical bonds.

Metabolism is the sum of all the chemical reactions occurring within an organism. A virus is very small; typical viruses are from 50 to 100 nm in diameter, where nm stands for nanometer, or 10^{-9} meter. They come in two general types: one group is enclosed in a shell of protein, the second by both a protein shell and an additional membrane-like envelope. Within this covering is the most important part of the virus, its genome, made up of a nucleic acid component. In some there is DNA, in others only RNA. The number of genes also varies widely, with some having as few as three genes and others (such as smallpox) having more than 250 individual genes. In fact, there is a huge variety of viruses, and if they were considered alive, they would be classified across a great taxonomic spread. But common wisdom treats them as nonliving. The viruses that contain only RNA show that RNA by itself, in the absence of DNA, is capable of storing information, and serving as a de facto DNA molecule.[9] This finding is strong evidence that there may have been an "RNA world"[10] before DNA and life as we know it originated. And there is an even more striking implication of the presence of RNA viruses.

Viruses are parasites. They are technically termed obligatory intracellular parasites, as they are unable to reproduce without a host cell. In most cases, viruses infiltrate cells of living organisms and hijack the protein-forming organelles and start making more of themselves,

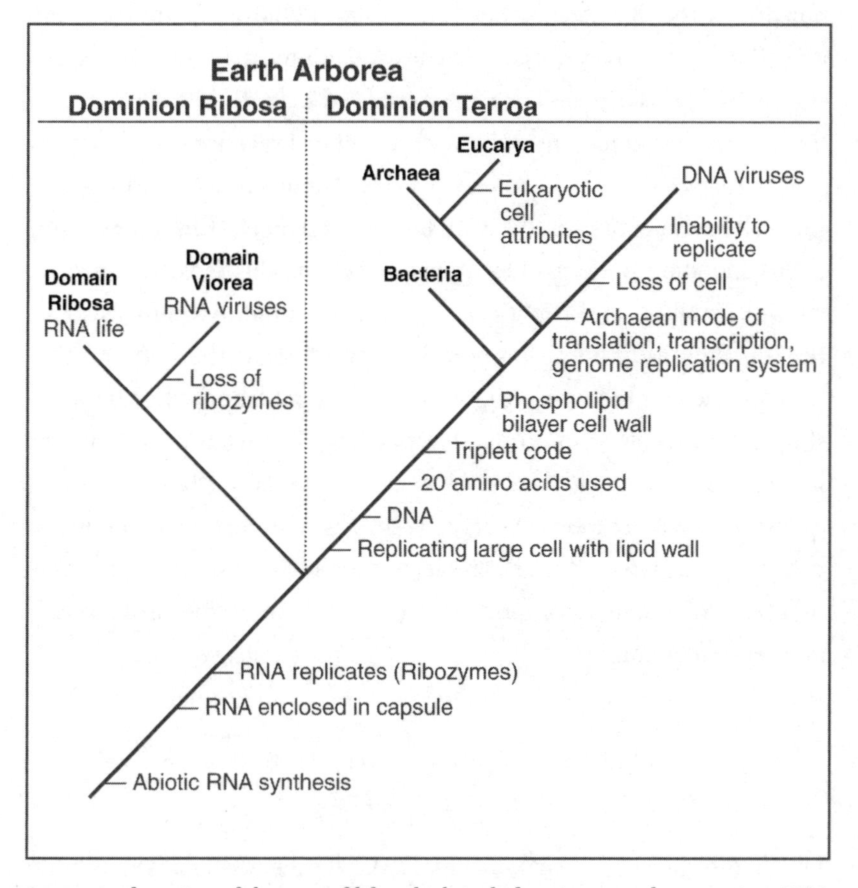

Our revised version of the tree of life, which includes viruses and now-extinct RNA life. This requires a new taxonomic category, one that is higher than domains (which are above kingdoms). RNA life is currently not definable on the accepted tree of life. (From Peter Ward, *Life As We Do Not Know It*, 2006)

turning the invaded cells into virus-producing factories. Viruses have a huge effect on the biology of their hosts.

The greatest argument against including viruses as alive is the fact that they are unable to replicate on their own—and thus seemingly fail this major test of whether or not an object is living. But it must be remembered that viruses are obligate parasites, and parasites tend to undergo substantial morphological and genetic changes in adapting to their hosts.

We can also ask if other parasites are alive. Parasitism, which is essentially a highly evolved form of predation, is generally the result of a long evolutionary history. Parasites are not primitive creatures. But like our viruses, they have stages that do not seem fully alive. *Cryptosporidium* and *Giardia*, both parasites on humans and other mammals, have resting stages that are as dead as any virus outside of its host. Without the hosts, these two organisms (and thousands of other species as well) will not live, perhaps cannot be classified as living. Yet when in their hosts, they show all the hallmarks of life as we know it: they metabolize, they reproduce, and they undergo Darwinian selection. But if we accept that viruses are alive, and this is increasingly accepted, we must radically reassess the tree of life as it is currently accepted.

In studying life on Earth, two questions can be posed: What is the simplest assemblage of atoms that is alive? And what is the simplest life form on Earth, and what does it need to stay alive? To answer these questions we must look at what current Earth life needs to attain and maintain the state of life described above. To do this we must briefly digress into chemistry of the materials that all Earth life uses to attain and then maintain life.

THE NONLIVING BUILDING BLOCKS OF EARTH LIFE

Of all the molecules making up Earth life, perhaps none is more important than water, and water in a single phase: it has to be liquid water, and not ice or water vapor (a gas). Earth life is composed of molecules bathed in liquid, and while the number of molecules that can be found in life is staggeringly large, in fact there are only four main kinds of molecules used by Earth life: lipids, carbohydrates, nucleic acids, and proteins. All of these are either bathed in liquid, in this case water with salt in it, or serve as an outer wall to contain the other molecules and water.

Lipids—what we call fat—are key ingredients in the cell membranes of Earth life. They are water resistant due to an abundance of hydrogen atoms, but they contain few oxygen and nitrogen atoms. Lipids are the major components of the cell boundary or wall that separates the outside environment from the fluid-filled interior of what we call life. These membranes, although delicate, provide control of substances in and out of cell.

Carbohydrates are the second major class of structures that Earth life is made of, and they are what we informally call sugars. By linking together a number of them, we can form a polysaccharide, which means "many sugars." Sugars, be they linked or single, are important building blocks in that they can be combined with themselves or with other organic and inorganic molecules to form larger molecules.

Sugars are also important in forming the next category of building block, nucleic acids. This group contains the stored genetic information of any cell. They are giant molecules that combine sugars to nitrogen-containing compounds called nucleotides, themselves formed from subunits called bases, phosphorus, and more sugars. In this arrangement the bases are crucial, for they become the "letters" in the genetic code.

DNA and RNA are sugars that are among the most important of all molecules of life. DNA, composed of two backbones (the famous double helix described by its discoverers, James Watson and Francis Crick), is the information storage system of life itself. These two spirals are bound together by a series of projections, like steps on ladder, made up of the distinctive DNA bases, or base pairs: adenine, cytosine, guanine, and thymine. The term "base pair" comes from the fact that the bases always join up: cytosine always pairs with guanine, and thymine always joins with adenine. The order of base pairs supplies the language of life: these are the genes that code for all information about a particular life form.

If DNA is the information carrier, a single-stranded variant called RNA is its slave, a molecule that translates information into action—or in life's case, into the actual production of proteins. RNA molecules are similar to DNA in having a helix and bases. But they differ in usually (but not always) having but a single strand, or helix, rather than the double helix of DNA.

Why the enormous complexity of DNA and RNA? The answer lies in the need for information to first build (blueprints) and then maintain the many tasks that staying alive requires. DNA is the blueprint, instruction manual, repair manual, and directions for building copies of itself and all that it codes for. In computer terms, DNA is the software, in that it carries information but cannot itself act on the information. Proteins can be thought of as the computer's hardware, needing the DNA software to provide information of when and where specific chemical changes should occur in time and space, and to produce material necessary for life. RNA has the interesting characteristic of being either hardware or software, and in some cases both at the same time.

Proteins, the last building blocks, perform four functions in Earth life: building other large molecules, repairing other molecules, transporting material about, and securing energy supplies. Proteins also modify both large and small molecules for a variety of purposes and are involved in cell signaling. There are a huge number of different proteins, and we are only now learning how these work and what they do. A new insight is that their topology, or folding pattern, is as important to their function as their chemical makeup.

All proteins used in Earth life are formed from the assembly of the same twenty amino acids. A new twenty-first-century area of research is asking an old problem: are these same twenty used because they are the best building blocks out there—or because they were common where life was first forming and then became permanently "coded" into life? In fact it looks like it is the former; they work the best, at least according to research in 2010.[11] This group is specific to Earth, and perhaps diagnostic of Earth life.

Proteins are constructed in the cell by stringing together the various amino acids in a long, linear chain that folds into its final shape only when all its amino acids have been joined together. Sometimes they fold as they are still being synthesized. Because the assembly of amino acids into a protein is done one at a time in linear and specific order, that protein is often analogized to a written sentence, each amino acid being a word. Within its cell walls, a living cell is packed with molecules, arranged in rods, balls, and sheets, all floating in a salty gel. There are about a thousand nucleic acids and over three thousand different proteins. All of these are going about some sort of chemistry that combined makes up the process we call life. Many chemical processes can go on simultaneously in this one-room house.

There are about also about ten thousand individual spheres within the cell, known as ribosomes, which are distributed rather evenly throughout. Ribosomes are composed of three distinct types of RNA, and about fifty kinds of proteins. Also present are chromosomes, which are long chains of DNA connected to specific proteins. The DNA in bacteria is usually localized in one part of the cell, but is not separated from the other interior material by a plasma membrane, as is the case in higher forms of life known as eukaryotes, which have an interior nucleus. It can be asked just what in this cell is "alive."

A bacterium is composed of inanimate molecules. A DNA molecule is certainly not alive, in any sense that any rational person would accept. The cell itself is composed of myriad chemical workings, each, taken alone, being but an inanimate reaction of chemistry. Perhaps nothing is alive but the whole of the cell itself. If we are to understand how life first arose, we need to find the minimum cell that can accomplish this with the fewest molecules and reactions.

One of the pressing problems in looking at this simple cell is that when examined in detail, it is in no way simple. Freeman Dyson has explicitly looked at this aspect of modern life, asking, "Why is life

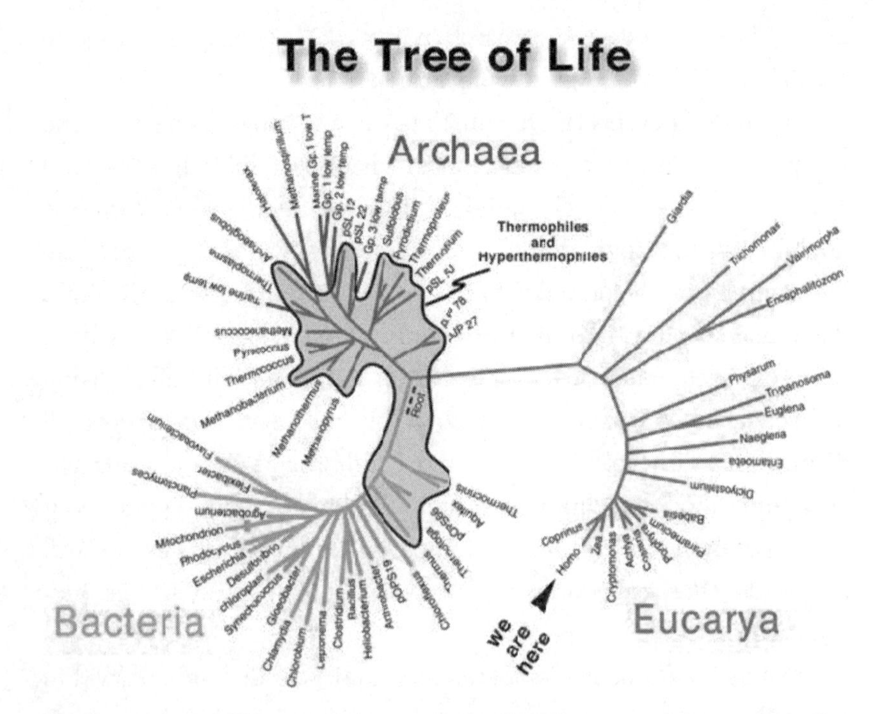

The eventual tree of life as it is now viewed. The shaded areas are those organisms that thrive in high heat. What is missing are the presumably many kinds of organisms and "pre-organisms" that evolved from inorganic chemistry step-by-step to produce the first living cell.

(at least life today) so complicated?"[12] If homeostasis is a necessary attribute of life, and if all known bacteria contain a few thousand molecular species (coded by a few million base pairs in the DNA), it looks as if this might be the minimum-sized genome. Yet all bacteria come to us today at the end of more than 3 (and perhaps more than 4) billion years of evolution. Perhaps the simplest Earth life is among the most complicated of life forms in the cosmos.[13]

CHAPTER IV

Forming Life: 4.2(?)–3.5 GA

On July 28, 1976, a robotic claw extended from a huge one-ton machine that only days before had completed the long, silent flight from the Earth to Mars and then had successfully landed there. The claw scooped Martian soil into the *Viking* spacecraft. This sample collection was the first time such an engineering achievement had ever been accomplished off the Earth. With this sediment now contained in its complicated interior, NASA's *Viking* performed four basic experiments, all designed to look for chemical evidence of life or its processes. That was the entire reason for *Viking* coming to Mars: to search for life.

The initial experiments[1] raised hopes that Mars indeed harbored extant life in its soil, for it was soon found that the soil contained more oxygen than was expected, and furthermore that chemical activity of the soil at least *hinted* at a microbial presence in the Martian regolith. These first-blush experiments created such a wave of optimism in the *Viking* scientific team that one of the mission's chief scientists, Dr. Carl Sagan, was optimistic enough to tell the *New York Times* that he thought that life on Mars, even large forms of life, was not out of the question. By large life, he meant *really* large, for in the same interview he went on to posit the existence of Martian polar bears!

But the onboard spectrograph, after carefully analyzing the Martian soil, could find no evidence of organic chemicals in the soil. Mars, as viewed from this first *Viking* lander, not only seemed dead, but inimical to life, leading to speculation that any life that might be there would soon be killed by the toxic chemicals in the soil. Sagan, ever the optimist, could now only hope that the second *Viking* lander, on that same day already orbiting Mars, would yield telltale evidence for life.

On September 3, 1976, the second lander safely parachuted onto the Martian surface at a place named Utopia Planitia. Like the first, this huge machine functioned perfectly.[2] And also like the first, no evidence of life was found in any of the separate and crucial life-detection experiments. *Viking* had been conceived as a multi-investigative program. But while its study of the chemistry and geology of the soil and atmosphere was important, its primary mission, and most of the instrumentation crammed into the crowded spacecraft, as noted above, was dedicated to the search for extraterrestrial life.

The *Viking* results suggested that Mars was sterile,[3] and NASA began to lose interest in Mars, because NASA was and is driven by the search for life beyond Earth. NASA's lack of interest began to benefit another branch of science, one that also was bent on studying alien worlds, and perhaps alien life: the oceanographers.

In the immediate post-*Viking* years, huge new sums went into the technology necessary for deep-ocean exploration, and soon another kind of spacecraft made its own successful descent onto an alien surface. In this case, however, life was found, but of a kind that was totally unexpected. First in the Atlantic Ocean, and then in rapid succession in the deep sea off the Galápagos Islands followed by dives in the Gulf of California, the small yellow submarine *Alvin* photographed and sampled a kind of life using a radically different source of energy than sunlight.

This discovery of deep-sea "vent" faunas would radically change our understanding of where and how life on Earth came into being, if in fact it originated on Earth at all, for there is a possibility that life formed elsewhere and then was transported to Earth. If life on Earth formed soon after our Earth coalesced into a large and ultimately habitable planet, it suggests that life is not all that hard to make. But how old really is the oldest Earth life—and where was this first life formed?

Usually when historians try to find the "first" of anything, they look into records of ever-older time units, and so it has been with the Earth historians. Their problem has been the paucity of rocks of sufficient age, and the near impossibility of a bacterium-like early cell to actually fossilize.

For more than two decades it has been axiomatic that the oldest sign of life on Earth came from a frozen corner of Greenland, at a place named Isua.[4] No fossils were found. Instead, small minerals called apatite were reported to contain microscopic amounts of two different isotopes of carbon that showed a ratio quite similar to one that is characteristic of life today. The Isua, Greenland, rocks were well dated at about 3.7 billion years in age, and later, new dating suggested that they were even older, about 3.85 billion years, in fact, and this is the date that has long been codified into textbooks.

The date of 3.7 to 3.8 billion years old made a lot of sense to those looking for the oldest life on Earth. As we saw earlier, asteroids bombarded the Earth along with every body in the then-young solar system and other junk left over after planet formation from about 4.2 to 3.8 billion years ago. We mentioned earlier that life, while it may have formed (or have been even older that this), would have been wiped out by the process of "impact frustration."[5] Thus the age of the Isua rocks was perfect; the heavy bombardment would have been just over, and life could start. Unfortunately for this tidy package, new instruments developed in the twenty-first century discovered that the small bits of carbon in the Isua, Greenland, samples were not formed by life at all.[6]

The next-oldest life was 3.5 billion years in age, and in this case, the claim was based on fossils, not just chemical signals. Filamentous forms in an agate-like rock dated to be around 3.5 billion years in age[7] were discovered by American paleontologist William Schopf. The fossils came from a previously obscure and ancient assemblage of rocks located in one of the least habitable places on current Earth, a highly deformed rock assemblage called the Apex Chert in Western Australia. The exact geographic position of these fossils, in the dust-dry enormity of Western Australia, was the "North Pole," a whimsical name given some years earlier because the locality is, in fact, one of the hottest places anywhere on Earth, and about as removed both geographically and especially climatically from the Arctic as a place could be.

Schopf's discovery galvanized science, for it showed that life on this planet began very early in Earth history indeed. For almost twenty years these ancient Australian fossils were accepted as the planet's oldest fossil life. Then these too were cast into doubt by Oxford's

Martin Brasier, who claimed that the so-called oldest fossils on Earth were only tiny crystal traces, not life relicts at all.[8]

What came next was a scientific donnybrook. Scientists on both sides unleashed attacks and counterattacks, most polite (but some less so). Back and forth it went for some years, with Schopf gradually losing ground, not only from attacks from the Oxford crew about the interpretation of the small traces in the Apex Cherts, but also soon after about the age of Apex Chert itself.

Around 2005, Roger Buick of the University of Washington claimed that even if the tiny objects in the Apex Chert are fossils at all, the rocks themselves are far younger than Bill Schopf has claimed, more than a billion years younger, in fact, which would still make them old (any fossil with billions attached to its birthday qualifies for old-age discounts), but nowhere near the oldest life on Earth. With these one-two punches, the Apex fossils were knocked out of the ring.

So matters rested until the summer of 2012, when the same Martin Brasier coauthored a paper[9] demonstrating the presence of life that is at least 3.4 million years old—which, according to the authors, makes it the oldest fossil life ever discovered. What makes the discovery even more important is the identity of the fossils themselves, all microscopic, of the size and shape of a specific kind of bacteria living on Earth today. The oldest life on Earth lived in the sea, appeared to need sulfur to live, and quickly died if exposed to even a small number of oxygen molecules. While this life is still what we might call a carbon-based life form, it brings the element sulfur front and center in our assessment of how life came about.[10]

The fossils described in the Brasier paper appear to be related to minute bacteria still living on our planet—bacteria that need the element sulfur to live, and that die quickly if exposed to the thinnest whiff of oxygen. If this discovery holds, it will confirm that life on our planet began in a place utterly alien to most of the Earth today, and depended on sulfur, not oxygen.

Earth life is usually associated with the forests, seas, lakes, and skies of our present-day Earth—with the creatures living in clear air, clean blue water, on grass-covered hills. Yet the tiny fossils found by Brasier came from an environment of temperatures far higher than those of today, with air composed of the toxic gases methane, carbon dioxide, ammonia, and not a little of the poisonous gas hydrogen sulfide.[11] It lived on a planet certainly without continents, or virtually any land of any consequence at all, beyond strings of ephemeral, volcanic isles. In this setting, life began (or arrived, a major possibility to be explored in the pages to come) and then thrived for billions of years. The majority view is that we are all descendants of this Hades on Earth cradle, bearing the scars and genes of a sulfur-rich origin of life.

Soon after this description of Earth's earliest life originating in an anoxic, sulfur-rich environment appeared, NASA's *Curiosity* rover[12] landed on the surface of Mars. Soon after his own discovery about earliest life on Earth, Martin Brasier was asked if the sulfur microbes whose fossils he had just found could have lived on Mars, or might still live there. His answer, after a brief rumination, was yes.[13]

If the 3.4-billion-year-old life turns out to be the oldest on Earth, it puts into question many of the currently favored "crèches" where life might have first begun on Earth. Our planet at that time was already ancient in its own right—for our Earth coalesced, as we have seen, 4.567 billion years ago. If this was indeed the first life, it suggests to some that life must have been relatively easy to form in the first place.

But how easy? And in what order? Let us look at what it took to get life on Earth. In order for life to come into being there had to be four steps:

1. The synthesis and accumulation of small organic molecules, such as amino acids and molecules called nucleotides. The accumulation of chemicals called phosphates (one of the common ingredients in plant fertilizer) would have been an important requirement, since these are the backbone of DNA and RNA.
2. The joining of these small molecules into larger molecules such as proteins and nucleic acids.

3. The aggregation of the proteins and nucleic acids into droplets, which take on chemical characteristics different from their surrounding environment: the formation of cells.
4. The ability to replicate the larger complex molecules and establish heredity.

While some of the steps leading to the synthesis of RNA—and the even more difficult feat of producing DNA—can be duplicated in the laboratory, others could not. There is no problem in creating amino acids—life's most basic building block—in test tubes, as shown by the Miller-Urey experiment of the 1950s. But it has turned out that making amino acids in the lab is trivial compared to the far more difficult proposition of creating DNA artificially. The problem is that complex molecules such as DNA (or RNA) cannot simply be assembled in a glass jar by combining various chemicals. Such organic molecules also tend to break down when heated, which suggests that their first formation must have taken place in an environment with cold to moderate rather than hot temperatures. Life on Earth has the nucleic acids RNA and DNA. Once RNA has been synthesized, the path toward life is open—because RNA can eventually produce DNA. But how the first RNA came into existence—under what conditions and in what environments—became the central problem facing those trying to work out the where and how of life's origin. There is no shortage of hypothesized sites where life may have begun.

Steps in the Origin of Life

Formation of Earth	Stable hydrosphere	Prebiotic chemistry	Pre-RNA world	RNA world	First DNA/ protein life	Last universal common ancestor
4.5	4.4	4.2–4.0	~4.0	~3.8	~3.6	3.6–present

DARWIN'S POND

The first, most famous, and longest-accepted model for life's first appearance on Earth was proposed by Charles Darwin, who in a letter to a friend suggested that life began in some sort of "shallow, sun-warmed pond." To this day, this type of environment, be it of fresh-water or perhaps in a tide pool at the edge of the sea, still remains a viable candidate in some circles and in textbooks. Other scientists early in the twentieth century, such as J. Haldane and A. Oparin, agreed with Darwin and expanded on this idea.[14] They independently hypothesized that the early Earth had a "reducing" atmosphere (one that produces chemical reactions the opposite of oxidation; in such an environment iron would never rust). The atmosphere at that time may have been filled with methane and ammonia, forming an ideal "primordial soup," from which the first life appeared in some shallow body of water.

Until the 1950s and into the 1960s, it was thus believed that the early Earth's atmosphere, thought to have consisted of methane and ammonia, would have allowed commonplace inorganic synthesis of the organic building blocks called amino acids by the simple addition of water and energy.[15] All that was needed was a convenient place to accumulate all the various chemicals. Seemingly the best place to do this was in a shallow, fetid pond, or a wave-washed tide pool at the edge of a shallow, warm sea. And, the idea goes, in such a place some kind of primordial soup filled with organic molecules sat around just waiting for Dr. Frankenstein.

Many scientists now looking at environments on the early Earth doubt this scenario. The organic compounds necessary to form life are complex and easily fall apart in heated solutions. Furthermore, an enormous amount of energy would be required to keep this soup out of equilibrium, which is necessary. What Darwin could not appreciate in his time was that the mechanisms leading to accretion of the Earth (and other terrestrial planets) produced a world that early in its history was harsh and poisonous, a place very far removed from the idyllic tide pool or pond envisioned in the nineteenth and early twentieth century.

Yet the *Alvin* dives to the deep, oceanic volcanic rifts described earlier in this chapter offered a new possibility in the early 1980s, championed by John Baross, now at the University of Washington: life on Earth began in the newly discovered deep-sea vents.[16] Soon new molecular techniques used to classify the vent microbes added confirmatory information to this idea. DNA tells us either that life spent its first eons in hot water, in fact *very* hot water, or that after forming in a cool place it was somehow scalded nearly to death in some ancient, highly energetic process.

Most of the microbes from the vents were eventually found to belong to the domain Archaea. The archaeans belong to the most ancient lineage of organisms known on Earth, and the oldest are thermophilic, or heat loving, to the point that they thrive in near-boiling water, something not found in ponds. This discovery suggested that the microbes from the vents were of a great antiquity.[17]

During the period of heavy impact, between 4.4 and 3.8 billion years ago, the time of the heavy bombardment described in the last chapter, each successive impact event (caused by comets as large as 500 km in diameter) would have partially or even completely vaporized the oceans. Huge regions of the Earth's rocky surface also vaporized, creating a cloud of superheated rock-gas, or vapors, several thousand degrees in temperature. It is this vapor, in the atmosphere, which caused the entire ocean to evaporate into steam, and thus could sterilize the surface, killing all nascent life. Cooling by radiation into space would follow, but a new ocean would not rain out for at least several thousand years after the event, and it is difficult to conceive that life would survive anywhere on the planet's surface.

Large-body impact on the Earth had never previously entered into ideas about life's origin. But now we know that during the period when life must have been first coming into existence on Earth, the only places that would have been insulated from the titanic energies of the heavy bombardment impacts would have been either in the deep oceans and/or in the Earth's crust itself. Perhaps only depth in the sea or in rock provided the bomb shelters necessary for earliest life's survival.

Even around 4 billion years ago there was little land area. Volcanism and the eruption of lava from the interior of the Earth were far more common and energetic than they are today. Thus the deep-sea ridges and vent systems being explored by the few small submarines of the mid-1970s were, in that long-ago time, much longer and more active than they are today. All of this translates to a very energy-rich volcanic world, with huge amounts of deep-earth chemicals and compounds spewing forth into the oceanic environment. Seawater chemistry would have been enormously different than it is than now. The ocean was what we would call reducing (as opposed to the present-day oxidizing oceans), since there was no free oxygen dissolved in the seawater. The temperatures of the oceans would have been scalding.

There may have been a hundred to a thousand times as much carbon dioxide in the atmosphere as there is today. There was also a constant bath of lethal levels of ultraviolet radiation on the surface. You need land to have a pond, and when life first formed on Earth, there may not have been any land at all. Perhaps there was only a hot, toxic ocean from pole to pole.

MINERAL SURFACES IN HYDROTHERMAL VENTS

The hydrothermal vents and their biota of extremophilic microbes, including abundant and heat-loving archaeans, are still a favored site for the origin of life, and unlike the oceans and atmosphere of the early Earth, it is an environment that is indeed strongly reducing. The vents emit chemicals that are appropriate for the evolution of life, such as hydrogen sulfide, methane, and ammonia amid lots of hot water. The vent chemistry would be largely decoupled from the atmosphere, and thus the evolution of life could have taken place independent of the atmosphere. This removed the problem that the Earth's atmosphere at the time was not chemically correct to form life. But the so-called vent origin had its own problems. How could RNA, that highly unstable molecule, have formed in the vents, with their high temperatures and pressures?[18]

Early life may have formed on the surface of iron sulfide minerals, at least according to respected early-life theorist Günter Wächtershäuser. He named this idea the "iron-sulfur world theory."[19] The hypothesis is that the earliest life, which Wächtershäuser termed the "pioneer organism," was assembled within the high-pressure, high-temperature confines of an underwater hydrothermal vent, one created by undersea volcanism causing hot, mineral-rich fluids to bubble upward through the rock-lined vents along the many thousands of miles of these deep sea fissures. Life would have begun in temperatures that at the surface would boil water (100°C). Under pressure, however, water does not boil as it does on the surface, and the water coming out of the vents was a veritable chemistry set of elements and compounds. But for any sort of organic buildup to occur, the fluids coming from the vents had to have sufficient volumes of carbon monoxide, carbon dioxide, and hydrogen sulfide dissolved in them to provide the carbon and sulfur needed for the construction of amino acids and, eventually, nucleic acids, proteins, and lipids.

Eventually there were buildups of minerals containing iron, sulfur, and nickel as the hot, mineral-rich fluids jetted out of the volcanically heated vents. This allowed the formation of small regions that could capture carbon-carrying molecules, and then chemically change them to first free up carbon atoms and then to link these newly isolated carbon atoms together into ever more complex, carbon-rich molecules. When the poisonous gas hydrogen sulfide came in contact with iron atoms found in various minerals in the same region, the mineral pyrite (fool's gold) was produced. This reaction produces energy-containing molecules, thus uniting two important aspects of life—the correct elements to produce life and sources of energy to fuel the necessary chemistry. But the energy produced by reactions with pyrite is very small and not enough on its own to fuel any sort of primitive life form. Wächtershäuser realized that it would take the reaction of a second gas, carbon monoxide, to serve as a fuel. This energy was the all-important driver for all that followed: the slow accumulation of Lego-like molecules, piecing slowly together to form a final product completely different from the sum of the various chemical parts.

The idea that mineral surfaces could act as templates for life's formation is not new. The faces of these flat minerals such as clay and crystals of silicate minerals or pyrite could have been microscopic regions where early organic molecules could have accumulated. As envisioned by geologist A. G. Cairns-Smith some decades ago, the earliest life would have had several characteristics: it could evolve, it was "low -tech," with few genes (sites on the DNA molecule that code for the formation of specific proteins) and little specialization, and it was made of geochemicals, arising from condensation reaction on solid surfaces, either from pyrite or iron sulfide membranes. Yet many students of early life remain skeptical of this scenario, particularly as the organic takeover lacks a process involving natural selection through which it might evolve.

Carbon monoxide and hydrogen sulfide are animal killers, with the former having taken untold numbers of human lives in poisonings both intentional and not. Yet if this idea is correct, the combination of two killer gases and fool's gold was the pathway to life. This view was stated by Nick Lane: "The Last Common Ancestor of life . . . was not a free living cell, but a rocky labyrinth of mineral cells, lined with catalytic walls composed of iron, sulfur and nickel and energized by a natural proton gradients. The first life was (thus) porous rock that generated molecules and energy, right up to the formation of proteins and DNA itself."[20] William Martin and Michael Russell published a variant on this idea in 2003 and again in 2007.[21] They took the idea of hydrothermal vent origin a step further, arguing that such an environment could provide not only all raw materials and energy necessary, but one of the key aspects of life: a cell. Their view is that life began in highly organized minerals called iron monosulfide. The place where life would have formed would have been between the devil (too hot) and the deep blue sea (too cold)—in this case, somewhere geographically between a sulfide-rich (and hot!) fluid jetting out of the hydrothermal (volcanically produced) vents or seeps and an ancient seawater that would have been full of iron. But this is more than theoretical. There are indeed three-dimensional frameworks near fossil vents and seeps observable today—and these could have been the precursors to cell walls. The "prebiotic synthesis" of organic molecules would have occurred on the inner surfaces of the microscopic compartments in the minerals forming near the vents or seeps. The chemistry of the ensuing "RNA-World" would have taken place within these mineralized cell walls.

By the turn of our new century, many clues were present, and many potential places for life's first origin had been mooted. The oldest surviving life on Earth was certainly heat loving, of a type found still in the hydrothermal vents. All chemicals and energy necessary for life were found in the vents, although it did not necessarily evolve there. And finally, the vents offered a refuge from the rigors of the early Earth's surface, most important as a bomb shelter from the murderous asteroid barrage of the Earth's first billion years. But there was one great obstacle to uniform acceptance of this theory: RNA, and to a lesser extent DNA, are highly unstable at high temperatures such as those found in hydrothermal vents. Once RNA was created, the leap from RNA to DNA would have been more straightforward. RNA serves as a template for DNA. But getting from small molecules to something as complex as RNA, which even in its simplest forms is composed of many atoms in very precise positions, is still a mystery. Yet while a mystery, it is not an impossibility by any means, and rapid progress in the artificial formation in what are essentially test tubes shows us the overall pathway if not yet every detail.

Biologist Carl Woese theorized[22] yet another possible origin of life pathway, with life starting even before the Earth was fully formed and differentiated into the core, mantle, and crust components that we see today. Thus, in these early times, there would have been large amounts of metallic iron present on the surface of the Earth in contact with steam and some liquid water, amid an atmosphere filled with carbon dioxide and hydrogen. It is the latter that is so interesting, since hydrogen is a potent driver of chemical reactions. But because of hydrogen's light weight, it is easily lost to space on small-mass planets like the Earth, Mars, and Venus (the gas giants are so massive that they can hang on to their hydrogen). At this time the Earth was being barraged by space debris large and small, causing the planet to be encircled by a haze of dust particles and water vapor. High clouds of water vapor would form, and these tiny droplets would have served as protocells—tiny cell-like objects. With sunlight serving as an energy source, and the dust thrown up from the surface carrying organic molecules among the many other molecules and elements blasted into the sky by the asteroid bombardment, there would have been plenty of raw materials to make life from. With lots of hydrogen present as well, the first primitive organisms to evolve could have produced methane after using carbon dioxide as a carbon source. Microbes using this pathway today—hydrogen for energy and CO_2 for carbon—are called methanogens. As the Earth cooled, oceans formed, and life would have fallen from the sky to populate the oceans.

IMPACT CRATERS IN DESERTS

One of the newest suggestions comes from the University of Florida's Steve Benner[23] and a coauthor of this book, Joe Kirschvink. As mentioned earlier, the hardest step of all is making RNA. This is because RNA is a very fragile molecule, large and complicated, and thus very easily destroyed. Water attacks and breaks up the nucleic acid polymers (strings of smaller molecules) that make up RNA. In fact, it appears that there are many steps required in making RNA, and each step would require different conditions, or a different chemical environment. Biochemist Antonio Lazcano has described this problem as follows: "The RNA-world model confronts several serious challenges, including the lack of plausible primitive abiotic mechanisms to account for the formation and accumulation of ribose."[24] A possible way out is the hypothesis that ribose can be made under current temperatures from common desert minerals.

Benner noted that the major problem was not making carbohydrates (including ribose), the problem was *preventing* the reactions that created them from continuing madly, producing a sticky brown coal tar that gummed everything up. By looking at the synthesis pattern carefully, and after staring at a table of ionic radii, he realized that the pathway to coal tar could be blocked specifically by reaction with calcium (Ca^{+2}) and borate (BO_3^{-3}) ions. These calcium-borate minerals (e.g., colemanite, ulexite) are often used in soap, and they form by evaporation of salty brines in dry, hot environments. One additional step, a subtle rearrangement catalyzed with oxidized molybdenum, is all that is needed to produce biologically active ribose.

Benner also looked to extant life for clues. He analyzed the stability of various bacteria and found that the most ancient lineage may have formed at 65°C. This is hotter than any "warm little pond," but much cooler than a hydrothermal vent, which typically has temperatures in the many hundreds of degrees. In fact, there are very few places on the surface of the Earth now, or even 3.7 billion years ago, that would have such temperatures—except for deserts.

Desert-like conditions, where the overall environment is alkaline and has calcium borate in abundance, is the only environment where the formation of ribose from borate minerals might be favored. Clay minerals of various kinds are also common in such settings, and increasingly it looks as if templates formed from clay would help bring about the synthesis of the complex organic compounds necessary for life.

To form the borate minerals needed to stabilize RNA, there has to be a liquid system that repeatedly decants and distills the liquids in a series of interconnected steps. Kirschvink, in collaboration with MIT professor Dr. Ben Weiss, has hypothesized a natural setting that could lead to the formation of RNA from borate in the rough fashion that Steve Benner has suggested. A good example is in California, where boron leached from the igneous rocks in the Sierra Nevada mountain ranges passes through a chain of transient lakes, including Mono Lake, Owens Lake, China Lake, Searles Lake, Panamint Lake, and finally into the bottom of Death Valley. Massive borate deposits form in the last few of these reservoirs. The most obvious candidate for such a system, at least on the early Earth, and especially between 4.2 and 3.8 billion years ago, the time during which life may have first formed, would be a series of impact craters linked in a desert setting, with communicating water systems among craters of higher to lower elevation. In this way the same series of distillations and decanting could be accomplished. But such a site would have been unlikely on Earth

4 billion years ago, when all of this early chemistry was taking place. Earth was also strongly reducing then, precluding the presence of the oxidized molybdenum for the final rearrangement of ribose synthesis.

All of the earliest Earth rocks appear to have been produced in a water setting. In fact, there is no good evidence of extensive, subaerial continents on Earth until less than 3 billion years ago—on a planet 4.6 billion years in age—and the oldest detrital zircons suggest oceans going back at least to 4.4 GA. Our best evidence is that the Earth, at the time when life would have first formed, had nearly global oceans, with at most strings of islands. But Earth was not the only inner terrestrial planet. Venus is the same approximate size as our Earth, but is so close to the sun that it is highly unlikely that life could have ever formed there. Yet we know that there is another possibility, one beloved by science fiction: Mars.

There has been great progress in understanding the ancient geological history of Mars during our new century. Mars never had planet-covering oceans; we are quite sure because the older rocks are still there, exposed at the surface. But the immense amount of new data from the various Mars rovers has told us that the so-called Red Planet had large lakes, maybe small seas, and possibly an ancient ocean in the north polar basin. There is also evidence that Mars had larger oxidation-reduction gradients than Earth, which are the important means used by life to gain energy. The deep mantle of Mars is so reducing that methane, H_2, and the other gases needed for prebiotic syntheses of the carbon-rich chemicals needed for life should have been present, thus providing needed raw materials. There are some, coauthor Joe Kirschvink among the truest believers, who support the radical notion that life not only formed on Mars more than 4 billion years ago, but that it came to Earth on meteorites—and it is us. The question is if early Mars life could get to Earth at all.

PANSPERMIA AND THE CASE FOR MARS

Today, the surface of Earth is divided roughly into the larger ocean basins, which cover about 75 percent of the surface, and continental masses, which stand up above the mean sea level. We know from the simple age dating of the continents and a variety of other geochemical proxies that the continents have been slowly growing through time. New granitic basement rocks are added along the margins of the continents at subduction zones, where moist, sediment-laden rocks are carried down several hundred kilometers and are melted partially to form granites. Thus, as we go back deeper into geological time there is a good expectation that we would have less land area versus ocean area.

But there are even more constraints. We know from geophysical models that immediately after the moon-forming giant impact event at 4.5 billion years ago that the entire Earth was molten. A gigantic magma ocean existed, the result of the intense heat of the collision as well as the segregation of nickel-iron metal down into the Earth's core. The first half billion years or more after this event was a time of intense heat flow coupled with the gradual solidification of the surface crust in the uppermost layers of Earth's lithosphere. This increased heat flow limits the elevation that any landmass can reach above mean sea level. A continent stands high above the seafloor simply because it is underlain by less dense material that causes it to "float" upward. If the heat flow is high, the root underneath the continent will melt. That prevents high mountain ranges from forming.

Finally, geochemists suspect that the volume of Earth's oceans may be slowly decreasing with time. After the giant Earth-forming event, it is likely that a lot of the water vapor present in the system condensed out as steam on the surface of the young Earth, and has been gradually worked back into the mantle through the process of plate tectonics. This reworking is certainly seen in the chemical fingerprint of the 4.4-billion-year-old zircons mentioned earlier. Estimates on the size of this initial ocean vary from a minimum of about equal to what we have today to three or four times more than presently exist. Given all of these constraints, it is extraordinarily unlikely that anything other than the tippy-top peak of some volcano ever stuck itself above sea level before about 3.5 billion years ago.

A water world is not a very good place to form ribose. It is also a terrible place to form large molecules like proteins and nucleic acids, which release a little bit of water each time they add a new subunit. For these reasons, Earth was probably not a very good place anywhere for the origin of life until about 3.5 billion years ago. And even then it was unlikely to have had a series of lakes like those in Death Valley capable of enriching calcium borate minerals to the levels needed to stabilize ribose and other carbohydrates that early life absolutely needed until much later. It certainly did not have large chemical characteristics producing enough energy to have fueled sloppy, early metabolism.

Extensive experiments conducted during the past decade have unequivocally showed that meteorites can go from the surface of Mars to the surface of Earth without being heat sterilized—and thus they could carry life from Mars to Earth.[25] Over 1 billion tons of Martian rock has made this transition to Earth over the last 4.5 billion years. It is therefore important for the origin of life to consider the possibility that it arose first on Mars and was carried here by meteorites.

Mars is only about half the diameter of Earth, and about 10 percent of our mass. As a smaller planet, it has a smaller gravitational field. It is therefore easier for something like a meteorite or a molecule of gas to escape completely. For this reason when a small asteroid impacts into the Martian surface (traveling at 15 to 20 km/second) it can eject a lot of surface material into orbit around the sun, and the Mars rocks thrown off their planet would not suffer sufficient heating or "shock" to sterilize them. On Earth, the stronger gravity means that a lot more energy is required to launch material into deep space, making it very probable that material launched in this fashion will be melted. There is no record of unsterilized materials ever having been launched from Earth by natural processes.

Life, if it ever evolved on Mars, would thus escape easily. On the other hand, the stronger gravitational field of Earth means that it is much better than Mars at keeping its hydrosphere and atmosphere intact over geological time. The atmospheric pressure on Mars is so low that liquid water will simply boil away at room temperature. Data from the most recent Mars rover, the 2012-landed *Curiosity*, make it clear that there were bubbling streams merrily percolating down alluvial flans toward a large lake or perhaps an ocean at Gale crater, where *Curiosity* landed. A world with volcanic rocks, replete with bubbling streams and oceans and an active hydrological cycle, ought to have had life. Or it certainly *could* have had life. We argue that it was possibly the place where life, the life now of Earth, in fact first evolved.

If we go further back to the Hadean record on Earth it is clear that oceans did exist as far back as 4.4 billion years ago. A Martian setting for life's first formation, using the borate pathway hypothesized by Benner, but then passing through linked craters in a desert setting, is this new possibility advocated by Kirschvink and Weiss[26] earlier in this century. A number of experiments now confirm that complex organic molecules, and even the resting stages of microbes, could be transported from Mars to Earth through a process known as interplanetary panspermia—where a large impact on the Martian surface, say 3.6 billion years ago, hurls a great number of Mars meteorites onto the Earth—and in so doing seeded our planet with Mars life.

There is one more bit of evidence to support a Martian origin, based on new research by David Deamer of the University of California at Santa Cruz.[27] One of the great problems in arriving at an RNA strand long enough to do anything is getting it to link to other of the component pieces of RNA to form a "polymer," a long strand of RNA made up of many of the subunits, called RNA nucleotides. Deamer showed that freezing a dilute solution of single nucleotides forces many together along the edges of ice crystals. There was no ice on Earth back then. But Mars would have had plenty of polar ice, especially early in its history when the sun was dimmer, just as it does now.

FORMING LIFE—A 2014 SUMMARY

Advancing our understanding of how life first formed from nonlife on early Earth to some extent has depended on how close are we to producing life in a test tube. Even five years ago the answer would have been not very close at all. But thanks to a group at Harvard, led by biochemist and 2012 Nobel laureate Jack Szostak, we are closer than is perceived by most of the public.[28] Szostak and his colleagues have for nearly two decades been experimenting with the chemistry of RNA. The earliest information molecule was either RNA or something much like it that later evolved into RNA as we know it. And it is in the study of RNA that the Szostak group has made great strides in this century.

The trick has been trying to get nucleotides in solution to link one to another into short lengths of RNA. Getting them to link into a chain is easier than getting them to reproduce, once formed. Yet they will do this if around thirty of the nucleotides are linked, because with such a length and longer, the RNA molecule attains an entirely new property: it becomes a chemical known as a catalyst, which is a molecule that helps speed up chemical reactions. In this case the reaction to be sped up is nothing less than the reproduction of the RNA molecule into two identical copies.

Getting RNA strands at least thirty nucleotides long somewhere on (or in) the early Earth perhaps required clay to serve as a template. The clay mineral montmorillonite seems the most favorable. According to this hypothesis, single nucleotides, floating in liquid, bumped into the clay. They became weakly bonded to the clay and held in place. On some parts of the clay mineral, chains of thirty nucleotides or more were produced. As they were only weakly bonded, they were easily detached, and if there were some sort of concentration of these strands that then became engulfed in a small bubble of lipid-rich liquid, much like a soap bubble, there would have been the makings of a first protocell.

The two major components necessary for life are a cell that can reproduce itself and some sort of molecule that can carry information, as well as performing chemical catalysis (changing conditions so that a chemical reaction that would not otherwise occur does take place because of the action of the catalyst being present). If enough new components of RNA can be brought into the cell, the catalyzing action of the RNA will make more RNA as appropriate new chemicals are brought into the cell itself. The old idea was that cells and the small information-carrying molecules formed separately somewhere and then later merged. Now it seems that they evolved in tandem.

Many biologists have argued that the first life was just that: a "naked" RNA molecule, floating around in a soup of nucleotides and reproducing itself, over and over. But a more favored view is that cells and RNA evolved as a single unit—double-walled cells of fat with small RNA nucleotides within them grew by obtaining more fat and more nucleotides, which could have passed through gaps in the fat of the cell wall, whereas the larger linked nucleotides of the interior would be too large to pass out of the walls. The material available on the early Earth necessary to make the protocells were chemicals that would have combined to form fatty (lipid) molecules, which themselves would readily link together to form sheets and then spheres.

The newly discovered Lost City mid-ocean ridge vents in the north central Atlantic Ocean, discovered by University of Washington oceanographers, are composed of lime-rich rocks and hence are whiter in color than the more common black smokers of the Pacific Ocean. These sites are considered prime possible places where life was first assembled on Earth. (Image from University of Washington, with permission)

Because of its chemical properties, accumulations of sufficient fatty molecules will easily form hollow spheres when agitated, just as water will form tiny drops on its surface for a short period. As these spheres form, they will be filled with the molecules that can form RNA if these molecules (the nucleotides) are present in the liquid. Again, this is where concentration is crucial, and why the analogy of a "prebiotic soup" is used so incessantly. There would have to be a great number of nucleotides caught up in a suddenly forming protocell sphere if there were to be any chance for RNA to form within. Unless, of course, some property of the new protocell either actively or passively moves nucleotides that are outside of the cell through its walls into the interior.

The cell wall would not only be "feeding" on nucleotides. It would also be accumulating more of the fatty molecules, and in so doing elongating into a sausage shape. Eventually it would split and two spheres would be present, with each now carrying around half of the RNA—and a lot more than just RNA, of course. To function for any length of time, the cell would then have to obtain energy, and that requires chemical machinery—made of proteins. So the interior would have to have a lot of chemicals within it, functioning in some orderly fashion so that needed chemicals can be brought in, unneeded chemicals tossed out, and there would have to be plenty of spare parts (molecules of various kinds) readily available.

This is the stage where evolution begins. Some of the cells might reproduce faster based on the nature of the molecules within the new cell. Natural selection thus kicks in, and the engine of life as we know it has been turned on, cells that are autonomous, metabolize, reproduce, and evolve. The rest, as the great Francis Crick so famously said, long ago, was history.

THE DARWINIAN THRESHOLD

Early Earth-life cells might have been like modular homes, with each part installed as a separate component in a different place and then transported to a single place. The transport system could have been through water or air. The latter case is receiving strong support from new work, begun in 2010, looking at the amount of life and life material found in the upper atmosphere.

The earliest life might have been composed of cells with very porous cell walls, allowing the swapping of whole genomes, a process known as horizontal gene transfer. But there came a time when the cell systems went from ephemeral to permanent. This is the point that biologist Carl Woese called the "Darwinian threshold." It is the point where species, in something approaching the modern sense, can be recognized, and when natural selection—evolution in other words—takes over. Natural selection favored more functionally complex, integrated cells than simpler precursors, and they flourished at the expense of the simpler modular varieties.

Modern Earth life was born when the radical changing of genes stopped. Some who study the evolution of the first life, such as Carl Woese, believe that arriving at this grade of organization is the most important event in all of evolutionary history. Yet those first cells were surely not alone, for there were probably ecosystems packed with all manners of complex chemical assemblages that had at least some life aspects. We can think of a giant zoo of the living, the near living, and the evolving toward living. What would that zoo contain? Lots of nucleic acid creatures of many kinds, things no longer existing and having no name because of this. We can imagine complicated chemical amalgamations that have been roughly defined as RNA-protein organisms, RNA-DNA organisms, DNA-RNA-protein creatures, RNA viruses, DNA viruses, lipid protocells, protein protocells. And all these huge menageries of the living and near living would have existed in one thriving, messy, competing ecosystem—*the time of life's greatest diversity on Earth*, perhaps 3.9 to 4.0 GA (billion years ago), but with our new view being that it was later rather than sooner. Natural selection whittled what might have been a thousand really different kinds of life down to one.

Nobel laureate Christian de Duve stated that once the ingredients were in place with the right amount of energy present in the early Earth stove, life would have emerged from nonlife very quickly. Perhaps in minutes.

CHAPTER V

From Origin to Oxygenation: 3.5–2.0 GA

ONE of the least visited (and populated) corners of the Earth is the northern half of Western Australia. Covering a land area close to that of the western United States from the Rocky Mountains to the Pacific Coast, this gigantic region is arid, mainly red in color—and contains some of the most important sites for understanding the history of life on Earth. The most important of these are the sites where the earliest known life on Earth (to date) has been found. In a desolate region known as the Pilbara, ancient hills rich in oxidized iron lend a burnt umber canvas to the remains of life's first chapters—on our planet, at least. The red hills of the Pilbara are created by massive amounts of iron ore—and because of this, the region is the site of massive open-pit mining of its ancient iron-bearing strata, most of it being shipped to China as fast as it can be loaded onto the endless succession of freighters.

Yet there is more than iron ore to be found in the Pilbara's ancient hills. In the treeless landscape are rocky outcrops that have long been thought to contain the Earth's oldest fossils—including the Apex shale described in a previous chapter, as well as the newest entrant in the "oldest life on Earth" derby, the Strelley Pool site, less than twenty miles from the Apex Chert locality in the Pilbara.

The Apex Chert and Strelley Pool sites do not trumpet that they bear fossils (or not, in the case of the Apex Chert). Yet in the surrounding countryside there is unmistakable evidence of early life, for the landscape is rich in stromatolites, the layered, humped deposits created by shallow water and intertidal bacterial slicks that proclaim the presence of life—and indeed were the most common kind of life on Earth from some time after its origin until about a half billion years ago. It is ironic, and entirely coincidental, that Western Australia, at the end of a long estuary known as Shark Bay, is also the place where one of the last oceanic remnants of a far more ancient world, one without any atmospheric or dissolved oxygen at all, still lives today.

The presence of both the oldest fossil life known and the coincidental existence of the best examples of what that oldest life may have looked like have left an indelible impression of Western Australia as the world's most important "museum" of early Earth life. From the first time life arose, until the first snowball Earth episode essentially ended the Archean era, the fossil record of this long time interval, more than 1 billion years, is known mainly from the presence of stromatolites, as well as rare, exceptional fossils found within agate-like rocks called chert. The stromatolites present in the two regions that yielded the most information about the nature of the oldest life on Earth—the North Pole region of Western Australia, and an area in South Africa called the Barberton Greenstone Belt, located near the famous Kruger National Park in South Africa—both show the presence of very ancient forms of stromatolites.

Through most of the 1900s, we all thought that these structures were formed as a by-product of algal mats, which can induce carbonate precipitation as a result of photosynthesis. But over the past two decades, many Earth scientists have concluded that some (but not all!) of the finely laminated structures can also form from direct chemical precipitation from salty brines. To distinguish those that did come from the processes of life, it is necessary to study the modern representatives, which, in fact, are few indeed.

The best place to observe living stromatolites today is in the World Heritage site, again in Western Australia, called Shark Bay, mentioned above. There, large, sometimes meter-wide mounds of interbedded sediment (mainly sand and mud) are found atop and below communities of photosynthetic bacteria. If one of these stromatolites is sectioned in half with a rock saw, the two halves show finely layered intervals—layers that show some very characteristic undulations. Stromatolites are generally round at the top, but the cuts show a wonderful diversity of shapes and structures.

The stromatolites of Shark Bay have been long lauded as one of the best ways to understand the Archean. Once again, we see

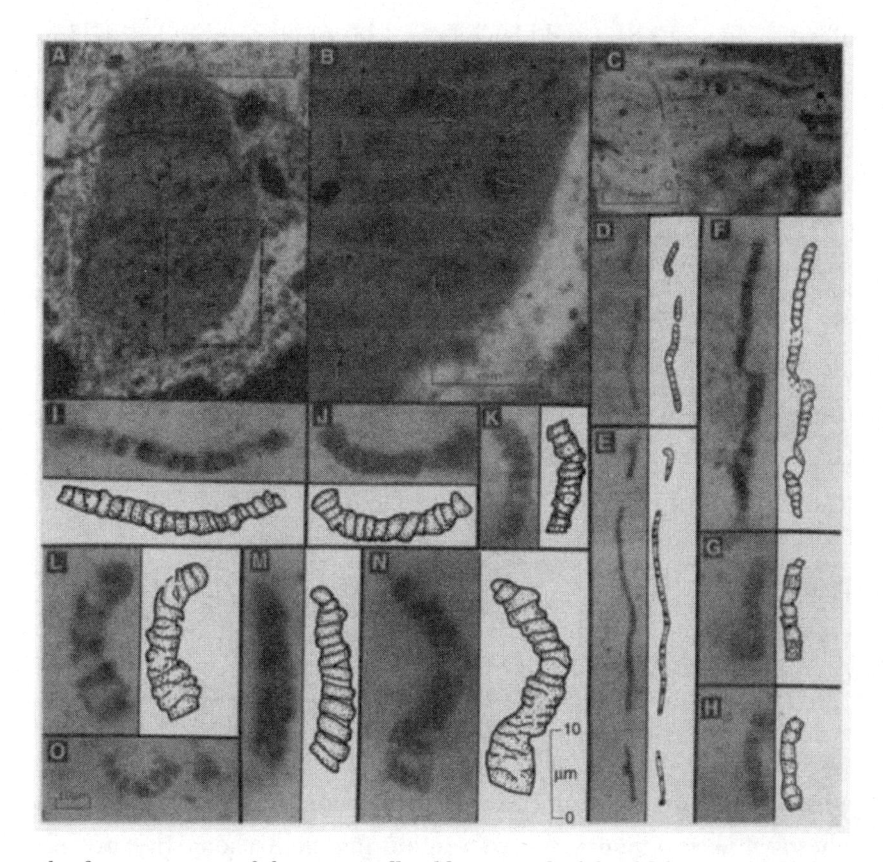

The famous images of the supposedly oldest record of fossil life, as published by Bill Schopf of UCLA in the 1980s and 1990s. These fossils were then dated at over 3.5 billion years in age. Subsequently both that age date (now thought to be a billion years younger) and even their identification as fossils is under attack.

uniformitarianism at work in this: the structure, chemistry, and biology of these structures living in the broiling corner of Australia are unquestioned as windows into the deep past, and their presence there is of inestimable value in interpreting the fossil stromatolites. But there are things about Shark Bay not mentioned or described in the endless television specials and written and photographic treatments of this site that are decidedly *not* a model for Archean oceans. Principal among these is the identity of other organisms inhabiting the most important of stromatolite-bearing regions in Shark Bay (the bay is huge, and covered more than 2.2 million square acres). They are also reminders of what life on Earth would have been like for at least the first billion years of its existence.

ARCHEAN LIFE AND THE PATH TO OXYGEN

Major changes affected the Earth and the history of life around 2.5 billion years ago—changes so consequential that it marks a new era on the hierarchical geological time scale. The oldest era is the Hadean, which began with the Earth's formation (4.567 billion years ago) and ended with the appearance of the first rock record, at about 4.2 billion years ago. The Archean era succeeded it, a violent time in Earth history, which began with the start of the heavy bombardment and ended ~2.5 million years ago, with the succeeding era named the Proterozoic. The changeover from the Archean to the Proterozoic coincided roughly with the rise in oxygen—and that oxygen was created by photosynthetic life.

Photosynthesis is the process that life uses to turn inert carbon dioxide into living cell matter (and thus changes inorganic carbon to what we call organic carbon). There is evidence that some kind of photosynthetic organism was present during the Archean, the time 4.2 to 2.5 billion years ago when life first evolved. It also seems clear that the evolution of photosynthesis postdated the oldest life. The first life probably used hydrogen in a compound in which the hydrogen is chemically merged with a sulfur atom, producing the very important (to the history of life) compound called hydrogen sulfide for its energy needs.[1] Hydrogen is energy rich, which is why human technology is trying to harness it for everything from cars to power plants. We also do know that the Archean organisms appear to have used the major life-required elements still used by life today: carbon, sulfur, oxygen, hydrogen, and nitrogen.

We have some information on the nature of the oceans and atmosphere at 3.5 billion years ago. Carbon dioxide was probably at much higher concentration than we see today. There was probably a great deal of water vapor in the atmosphere, as well as the gas methane, the kind of atmosphere that holds in heat, thus warming the planet at a time when the sun was far less energetic. Without those Archean greenhouse gases—water vapor, methane, and carbon dioxide—there would have been no liquid water on the planet. Greenhouse gases create a warming mechanism that the planet otherwise would not have, an atmosphere that could trap heat. But it was also an atmosphere without oxygen.

Much of our understanding of life during this long Archean chapter comes from studying modern-day environments that appear to be useful analogues. Low-oxygen environments are relatively rare in our oceans today, but they're much more common in smaller lakes. In fact, many modern-day lakes are essentially stratified, with a thin layer of oxygen (absorbed out of air), underlain by water that has no oxygen at all. The study of the microbial communities living in these types of environments has provided insights into what life must have been like in the deep past. One of the important groups of organisms that are necessary for carbon cycling in the modern-day lakes, and most likely in the ancient waters of the Archean ocean, is associated with the chemical methane. As noted earlier, methane gas helps trap heat reflected off the Earth by sunlight from escaping into space.[2] Some bacteria can break down methane and use it for food. Much of the Earth's early life used methane in this way, telling us that soon after life evolved on Earth, it quickly diversified in the ways that it acquired energy, just as the evolution of the car evolved along lines in which cars get energy—first from steam, then diesel fuel, then gasoline (both diesel and gasoline are carbon compounds that contain energy, just as methane is), and soon hydrogen fuel. While our civilization comes last to this fuel source, life came to it first.

Much of the evidence telling us about the history of early life on Earth comes from the sedimentary rock record. For instance, one of the characteristics of Archean sedimentary deposits is the frequent appearance of brightly colored red layers within some Archean-aged sedimentary rocks. These are called banded iron formations, or BIFs, and these interesting sedimentary rocks have not been produced on the Earth's surface in any sort of significant accumulation for the past 1.85 billion years, except during one or two snowball Earth intervals at the end of Precambrian time, which we will describe in far greater detail in the next chapter.

There is a long-standing puzzle associated with these BIF sediments—in order to be distributed so widely and gently, the iron has to have been dissolved in water—and that means it should have been in the greenish reducing form called *ferrous iron*. On the other hand, for it to have been precipitated out means that it was rusted into the red *ferric* form, which is not soluble in water at all: it simply falls out of water as particles, rather than dissolving in water, as a cube of sugar would. The problem is oxygen: ferrous iron reacts instantly with free molecular O_2 to form the red ferric state. Any iron or iron mineral that is bright red in color tells us that the iron has undergone this chemical change, which we commonly call rust, and that almost always requires molecular O_2. How could oxygen levels in the ocean waters be low enough to allow the iron to stay in the green soluble form, and yet then be available to make it rust? This was a long and perplexing scientific mystery.

Over fifty years ago, one of the important figures in Precambrian paleobiology, Preston Cloud of the University of California at Santa Barbara, hypothesized that the oxygen needed to turn dissolved ferrous iron into the rusted, particular ferric iron in the oceans came from a group of primitive photosynthetic microbes known as the blue-green algae, which are now called the cyanobacteria.[3] This is the only organism on Earth that ever learned how to perform the life-giving process of oxygenic photosynthesis, which is literally the ability to cleave a water molecule and liberate its oxygen atom. Some of their descendants were enslaved by other organisms, and now serve us all as the green light-gathering organelles in plants and other algae. Every plant on Earth now has tiny "capsules" that evolved from those first cyanobacteria, but are now "endosymbiosis" slaves doing the bidding of the multicellular plant. Preston Cloud envisioned a floating "oxygen oasis" of these first tiny photosynthesizers, the cyanobacteria, each excreting tiny amounts of oxygen, and over hundreds of millions of years radically changing the nature of not only life on Earth, but the chemistry of our planet's oceans, atmosphere, and even rock cover. With each tiny trace of new oxygen liberated into the ancient Archean sea, tiny flakes of rust would then settle to the ocean bottom, slowly but inexorably accumulating to make the banded iron formations.

Molecular oxygen is one of the most toxic compounds around. Anyone who takes antioxidants along with their vitamin supplements knows that they fight cancer—and cancer is usually caused by oxygen wrecking delicate cell chemistry at the wrong time, in the wrong place, and changing it into a new zombie-like killer cell as a result. Antioxidants are not just an advertising slogan. Oxygen is a cell wrecker, cell changer, and often cell killer because of its chemical ferociousness. So how could the organisms producing this poison survive as soon as the oxygen molecules were liberated?

This now leads to a classic "chicken and the egg" problem: any early life form that evolved a system to release O_2 without having protective antioxidant enzymes would have killed itself, so the systems to control oxygen would have had to evolve first. But all of the oxygen in our atmosphere is produced by oxygenic photosynthesis, so there should not have been any oxygen before this to drive the evolution of the protective enzymes! Thus, there must have been some nonbiological source that produced trace amounts of molecular oxygen and then exposed primitive cells to it in an environment where they could gradually evolve enzyme systems to protect them from this poison, in a way analogous to how we protect ourselves from killer diseases by exposing ourselves to tiny amounts of the disease when we are very young, letting our body gradually build up defenses.

But where did this early oxygen "vaccine" come from if not from photosynthesis? It is very difficult to produce oxygen in nonbiological ways, but one way that works is through photochemical reactions involving ultraviolet radiation, the same UV that causes sunburns on unprotected skin. UV radiation hitting water and CO_2 molecules in the atmosphere will generate trace levels of O_2 and other chemicals. Today, solar UV radiation is mostly blocked by a layer of ozone high in the atmosphere, far above the layers with water vapor (that freezes out). But early in Earth history there was no oxygen and thus no ozone, and hence no UV screen. Thus very strong ultraviolent radiation from the sun blasted the Earth, creating a tiny number of oxygen molecules. Unfortunately, reactions similar to those that generate the oxygen quickly snuff it out, making it unlikely to survive long enough to generate a biological effect, particularly as this is all happening in a UV radiation bath that is very good at scrambling DNA and sterilizing anything it hits. What is needed is a mechanism that allows the oxygen to be separated from the other products (hydrogen and CO in particular) before it is snuffed out.

Two processes are known that can do this. First, if water gets high up in the atmosphere, a significant fraction of the hydrogen atoms liberated from the UV light will be traveling faster than Earth's escape velocity and can be lost to space. That will leave a small trickle of oxygen, ozone, and hydrogen peroxide diffusing down from above (which are too heavy to escape), but it is really only a trickle. Reducing gases produced by biological activity and from volcanic eruption will easily squelch these oxic compounds long before they could reach the biosphere. The second process occurs on the Earth's surface—but on the surface of a glacier! In Antarctica today the "ozone hole" permits a wider spectrum of UV radiation to reach the surface, where it can blast water molecules apart, eventually generating H_2 gas and H_2O_2 (hydrogen peroxide). That peroxide gets locked in the ice, separated from the H_2 gas. Working with a graduate student at Caltech, Danny Liang, we calculated that up to 0.1 percent of the ice during a Precambrian glaciation could be made of H_2O_2, which, when the glacier melts, would be converted into O_2 and water.[4] Although this is not enough to breathe, it is enough to cause life with its powerful tool kit we call evolution to react. As noted below, we think the first oxygen-releasing cyanobacterium evolved during one of these Precambrian glacial intervals, and it certainly must have had evolved protection from oxygen.

In a 2008 paper, one of the most experienced researchers about life and the early Earth, Roger Buick of the University of Washington, broke down the alternatives to the "when" of oxygenation as follows: First, oxygenic photosynthesis (such as is common today in all green plants) evolved hundreds of millions of years before the atmosphere became significantly oxygenated, because it took eons to oxidize the continued production of reduced volcanic gases, hydrothermal fluids, and crustal minerals. Second, it arose at ~2.4 billion years ago in what we refer to in these pages as the great oxidation event, causing immediate environmental change. Third, oxygen production from photosynthesis or any other means began very early in Earth's history, before the start of the geological record, leading to an Archean (greater than 2.5GA) atmosphere that was highly oxygenated. To choose between these alternatives, let's look at the record as we now know it, because this is so critical to a good understanding of the history of life, and indeed there is a great deal that is "new" about this history in terms of our knowledge.

GEOLOGICAL CONSTRAINTS ON THE GREAT OXYGENATION EVENT

Despite the widespread agreement that the evolution of the cyanobacteria was the most profound biological event on this planet (even more so than the evolution of the eukaryotic cell, and then multicellular life), there is a surprising disagreement about exactly when this seminal biological innovation happened. Over fifty years ago, geologists realized that some of the oldest stream-deposited sedimentary rocks on Earth contained rounded grains of the common mineral called pyrite (fool's gold) as well as another mineral containing tiny amounts of uranium (the mineral named uraninite). These minerals are extremely unstable in the presence of oxygen (like iron, they quickly rust), and are never found in our open, oxygenated oceans and land areas unless completely cut off from contact with our normal, oxygenated atmosphere. That led to the initial concept that the atmosphere contained very little oxygen until some time near the end of Archean time, perhaps as late as 2.5 billion years ago or even later. Most of the geological community agrees that even at these dates, oxygen concentrations in the atmosphere were so low that both pyrite and uraninite grains could exist on land and in the sea without rusting, and in fact in rocks created as late as 2.5 billion years ago we find both pyrite and uraninite in abundance, telling us that at those dates the amount of oxygen in air and sea would have been nil. Yet by 2.4 billion years ago both kinds of minerals disappear from rocks created underwater or on land. Does this mean that cyanobacteria thus evolved only after 2.5 to 2.4 billion years ago? This has spurred a profound debate of great importance to understanding the history of life.

How to solve this quite important question was to take years of research. The disagreement centered on whether the cyanobacteria evolved around 2.5 billion years ago, or perhaps 1 billion years earlier, nearer 3.4 billion years ago, and hence around almost as soon as life on Earth appeared in the first place. In the late 1990s the then-novel use of chemical fossils, also known as biomarkers, seemingly solved the problem: Australian geologists found what they concluded to be clear biomarker evidence that there *had* to be something creating oxygen in shallow oceans during the Late Archean (before 2.5 GA) time interval. They reported trace levels of biomarkers in Archean rocks that—at least in the modern biosphere—require molecular oxygen in the biosynthetic pathways; a class of organic molecules called sterols are a prime example.

This discovery was singular enough that we will paraphrase the abstract from the paper itself: Molecular fossils (biomarkers) from 2,700-million-year-old sedimentary strata found from cores taken from an ancient part of the deep and old Australian sedimentary rock record indicate that when these ancient strata were actually deposited, they were in an environment shared by photosynthesizing bacteria called cyanobacteria, putting far back in time the oldest-known occurrence of these tiny, oxygen-producing microbial plants. But even more surprisingly, a second kind of biomarker called steranes found in the sampled strata provided persuasive evidence that not only the prokaryotic life forms were present, but that eukaryotes were there too—a group whose first fossils come from strata as much as a billion years younger than the age of the rock cores of this study.

This paper, printed in the prestigious journal *Science*, hurled a revolutionary new finding at the scientific world for two reasons—the presence not only of photosynthesis producing oxygen at a very early date, but also the even more surprising discovery of one of the three great groups, or domains, of life, the Eukarya (the other being Bacteria and Archaea, both microbial and dominantly single celled) in the old rocks too. All this evidence came from cores extracted from deep in the Earth. The take-home point: both photosynthetic bacteria and eukaryotes existed far earlier than previously thought, all the way back to 2,700 *million* years ago. This electrifying paper in one fell swoop rewrote scientific history, and the history of life as well.

But science is about doubting and questioning. Let us jump almost ten years, to 2008, and look at another paper on this subject, with one of the coauthors being the same J. Brocks who was senior author of the 1999 *Science* paper mentioned above. Here are the salient two sentences here: "The oldest fossil evidence for eukaryotes and cyanobacteria therefore reverts to 1.78–1.68 billion years ago and around 2.15 billion years ago, respectively. Our results eliminate the evidence for oxygenic photosynthesis at about 2.7 billion years ago and exclude previous biomarker evidence for the long delay (circa 300 million years) between the appearance of oxygen producing cyanobacteria and the rise in atmospheric oxygen 2.45–2.32 billion years ago."

Quite a difference! So what happened between 1999 and 2008 causing this abrupt scientific volte-face?

The original biomarker studies from the late 1990s were criticized on several fronts, including the fact that many ancient biochemical pathways that do not use oxygen are known to have been "updated" after the great oxygenation event to incorporate enzymes that do. However, the real problem with the biomarker studies was the methodology used to get the samples, not the analyses of what was in the samples. The investigators were finding the precious biomarkers, all right. But when, exactly, did the biomarkers get into the cores? Rocks are not the impermeable, hard, and durable objects we usually take them for, but actually exist often in environments where chemical changes—and later contamination—occur. In the late 1990s there was not yet sufficient appreciation of the intense need for testing for—and eliminating—the chance of younger contamination in these ancient samples, particularly when the putative biomarkers are present in concentrations less than that of the surrounding air.

Thus it was to the horror of the mainstream biomarker community that one of their rising stars—Jochen Brocks of the Australian National University—suddenly changed his tune in 2005 (ultimately leading to the 2008 article we have cited from above), arguing that his own thesis work documenting the presence of Archean biomarkers was confounded by contamination! That, in turn, led one of the major geobiology funding agencies (the Agouron Institute) to support a critical repeat of the biomarker scientific drilling projects, with new means of testing for contamination. The result (as of this writing in mid-2014) is that no biomarkers were found. In fact, at a meeting late in 2013 the source of the contamination was revealed to be a stainless steel saw blade that had been made "stainless" (by the manufacturer) by high-pressure impregnation with petroleum products! As of this writing, the biomarker community has not developed intellectually rigorous tests to prove that the organic biomarkers in Archean rocks—any of them—date back from the time the sediments accumulated.

Another big picture in the great debate about the origin of molecular oxygen in Earth's atmosphere was framed using a new kind of Earth history tool: comparing the concentrations of sulfur isotopes. We have already seen (and will see again, in the sections on mass extinction) that comparing the compositions of the isotopes of carbon is useful for studying life, and was even used to try to decide when the first life appeared on Earth, since living cells favor specific isotopes of the same kinds of atoms (such as carbon or oxygen, and as we show here, sulfur) over the others of that same element. In normal chemical reactions, light isotopes move through reaction series slightly faster than heavy isotopes because the lighter elements have slightly weaker chemical bonds that can be made and broken faster, producing higher reaction rates, and because of this, plants prefer the lightest isotopes of carbon and oxygen over their slightly heavier, more massive sister isotopes. James Farquhar, Mark Thiemens, and colleagues at the University of California in San Diego came up with a new method in 2000 to use the relative numbers of sulfur isotopes found in rocks of known age to tell us when particular kinds of life might have arisen.

Farquhar and Thiemens analyzed the pattern of sulfur isotopes in sedimentary rocks from Archean to Paleozoic time, finding large variations in sulfur isotopes prior to about 2.4 billion years ago. But in rocks younger than these the fluctuations disappear, and the best interpretation is that this change was caused by a lack of ultraviolet radiation hitting molecules of SO_2 in Earth's atmosphere. This could have happened only through the formation—the first formation at that—of the ozone layer that exists to this day. If there is no oxygen, there is no ozone screen, and we now have evidence that there was no ozone layer before about 2.4 billion years ago. After this, many other sedimentary indicators start to suggest the presence of atmospheric oxygen.

So there was no oxygen before 2.4 billion years ago, at least not enough oxygen to create an ozone layer. But were there any cyanobacteria anywhere at all? Probably not. When it became clear that the major scientific drilling program in South Africa (funded by the Agouron Institute mentioned above) had missed the great oxygenation event, they allowed the team to drill two more holes through slightly younger sediments in South Africa, which certainly did cross this event. This is the time interval between ~2.4 and 2.2 GA, the earliest part of what is called the Paleoproterozoic. And they found something rather *peculiar*. As noted above, the minerals pyrite and uraninite, and the sulfur isotopes, are very strong indicators of the lack of oxygen. At the other end of this spectrum is the element manganese, which is usually an equally powerful indicator for the *presence* of free molecular oxygen. The new data show copious levels of sedimentary manganese oxide, but in the same rock that has the other indicators of the lack of oxygen!

But it was more complex. Our junior colleague at Caltech, Woodward Fischer, working with graduate student Jena Johnson and Caltech alumnus Sam Webb (in charge of one of the microanalytical beam lines at the Stanford linear accelerator), decided to look further.[5] It turns out that the same sediments that have this slug of sedimentary manganese *also* have silt-sized grains of detrital pyrite and uraninite, and the isotopic sulfur signature that demands *no* free oxygen (well, less than 1 ppm). This was completely unexpected, but it gets worse. Working with another Young Turk colleague at Caltech, Mike Lamb—an expert in the geophysics of mineral transport during sedimentation—they extended this *no oxygen* constraint to the entire depositional system. The silt at the edge of the delta where we sampled it had to have originally been eroded from a continent somewhere, then transported through a river system, through meandering streams, coastal estuaries, near-shore sedimentary environments, and out to the distal toe of the delta. None of these environments could have had even 1 ppm of free molecular oxygen[6] (and so were obviously not affected by glacial meltwater, which might have had a little bit). Oxygenic cyanobacteria have well-known nutrient requirements—principally iron and phosphorous[7]—that would have been provided in many places along this depositional pathway. They produce copious quantities of oxygen—bubbles—when they grow. If any of these "islands of oxygenic photosynthesis" had actually existed, then *where were they?* The worst place for them to grow would be far out at sea,

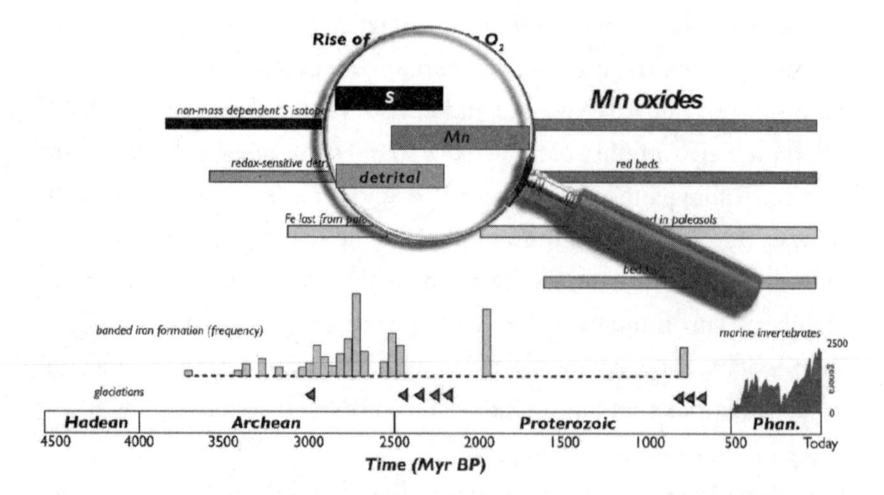

Overlap interval of contradictory geochemical signals for the rise of oxygen. Silt-sized rounded grains of pyrite and uraninite, which are quickly destroyed by the faintest whiff of oxygen, are associated with the first pulses of sedimentary manganese, which normally requires molecular oxygen. This overlap interval (inside the magnifying glass) may be the hint of a manganese-precipitating photosynthetic bacterium, which would be an important evolutionary stepping-stone on the path to oxygenic photosynthesis. (Diagram courtesy of Woodward Fischer, Caltech)

away from these nutrient sources. That was the vision of Preston Cloud mentioned above, but it frankly does not make sense in this context. The survival of those sedimentary indicators of anoxia is totally incompatible with the presence of oxygen—and cyanobacteria—anywhere in the environments those grains traveled through.

So what could the resolution of this paradox be? We think that the oxygen-emitting system of cyanobacteria had not evolved at this time (2.4 billion years ago), but that many of the evolutionary steps needed to get there had already been taken. It turns out that the actual biochemical complex in oxygen-releasing photosynthesis that collects the energy to split water, releasing oxygen, relies on a cluster of four *manganese* atoms, with a calcium atom thrown in for good luck. When this protein is made from scratch in living plants, the manganese atoms are sucked into the complex, one at a time, with the aid of photons that oxidize them. We suggested that these unique bursts of manganese in the sediments (not timid whiffs) might be the product of an evolutionary ancestor of the cyanobacteria that fed on reduced manganese dissolved in the water, using it as a source of the electrons needed to do photosynthesis.[8] Many primitive photosynthetic bacteria are known to do this with H_2S, organic carbon, and ferrous iron, but none have yet been found that can use manganese. Photosynthesis of this sort would leave copious amounts of a waste product—manganese oxide—behind in the sediment, but would not release the molecular O_2 that would destroy the sedimentary pyrite or uraninite, or create an ozone screen to change sulfur chemistry. This overlap interval where sedimentary manganese exists with rounded, detrital grains of the minerals pyrite and uraninite are present happens in one—and only one—brief interval of geological time, between ~2.4 and ~2.35 billion years ago.[9] If that is indeed the time that this protein evolved, all of the other indirect suggestions for earlier oxygenic photosynthesis must be wrong. This is a new and controversial interpretation we are posing here. But we are confident it is the correct one.

In our model, this manganese-oxidizing microbe, then new to the world through some random new mutation in all probability, dominated the ecosystem for a few million years until it managed to deplete the surface waters of soluble manganese. Through some biochemical rearrangement, this tiny new kind of microbe became capable of grabbing electrons directly from water molecules, releasing copious quantities of O_2 in the process. That would have been the first true cyanobacterium. Because water is basically everywhere, its growth would no longer be limited by the supply of electron donors in the environment. Only trace levels of iron and phosphate are needed for it to grow. But during this interval of time, there are clear records of glacial deposits, and those deposits contain plenty of iron, phosphate, and other nutrients for these new cyanobacteria to grow on. In fact, this glacially fertilized growth would be capable of destroying the planetary greenhouse in less than a million years by removing two important gases—CO_2 and methane—too rapidly for the system to recover.[10] The result of the sudden destruction of the greenhouse would be a global glaciation, termed a "snowball Earth" event.

We apologize for the complex chemistry necessary in the preceding section. But to get this story right requires complexity. As we see now, the world was unalterably changed from this point onward.

A SNOWBALL FROM HELL

In all of Earth history, we have rarely seen ocean stratification (where the ocean has a thin upper layer that is oxygenated, but a much thicker layer underneath that is not) when the Earth's high latitude poles are glaciated. Cold water sinks at the poles, driving circulation. On top of that, the glaciers themselves are very good at grinding up continental rocks into powder and throwing them back into the oceans, where the tiny particles of rusted iron and phosphorus are two of the same key ingredients of fertilizer we use on our lawns and gardens today. Satellite images of melting icebergs show a plume of photosynthetic activity in their wake, confirming the powerful effect on oceanic productivity that a little ground-up rock can have. And a great debate is raging even today about the effect of an illegal iron-dumping experiment in the Pacific Northwest in 2012, commissioned by Haida Gwaii (formerly the Queen Charlotte Islands), which was followed only two years later by a massive increase in salmon.

During Archean and Early Proterozoic time there were several major glacial intervals before the great oxidation event, including three minor episodes from ~2.9 to 2.7 billion years ago, and several more between 2.45 and 2.35 GA. A simple calculation suggests that the amount of iron and phosphate dumped into the oceans during any of those glacial advances would have been more than enough for cyanobacteria—if they had evolved by then—to completely overwhelm the anoxic surface environment, and flip the planetary atmosphere and surface ocean into a stable oxygen-rich situation like today; it would have taken less than 1 million years to do so.[11] The fact that it did not happen then is another strong line of reasoning that oxygenic photosynthesis had not yet evolved.

The *youngest*, and firmest, constraint on the presence of copious oxygen in the atmosphere comes from the presence of a vast deposit of the mineral manganese, known as the Kalahari manganese field in South Africa, dated at 2.22 GA, in the same basin where the Agouron drilling project sampled. This deposit is enormous, a blanket fifty meters thick that covers nearly five hundred square kilometers, deposited on a continental shelf. There is no trace of detrital pyrites, uraninite, or weird sulfur isotopes. It could only have formed in an oxygen-rich atmosphere, and thus this gives us the oldest date at which we are sure that the world of cyanobacteria, the ozone shield, and oxygen in both the sea and air existed.

Between this deposit and the underlying manganese overlap interval is another peculiar beast—a glaciation so severe that it marched into the tropics[12] and most likely froze the entire ocean surface, producing the first of the snowball Earth episodes.[13]

This first snowball Earth episode, actually named by coauthor Kirschvink, may have lasted nearly 100 million years.[14] So what is a snowball Earth? In fact they were first discovered in younger rocks.

We now know that glacial deposits were produced between 717 and 635 million years ago, and can now be found on virtually all the continents. Two geologists working in the first half of the twentieth century, Brian Harland of the UK and Douglas Mawson of Australia, recognized early on that there was a great infra-Cambrian ice age that seems to have had an unusually large, global extent. Although they recognized clear features of unambiguous glacial origin—like drop stones, tillites, and glacially striated pavements at the bottom of the units—there were several features of these deposits that were puzzling. Many of the clasts were composed of shallow-water limestone, much as if the glaciers had marched out over the carbonate platforms like those in the Bahamas (which today form only in the tropics), ripping up pieces and carrying them away. They were also associated with an unusual occurrence of banded ironstones, similar to those that had disappeared from Earth nearly a billion years earlier, and the glacial sediments were usually covered by layers of limestone (again, a "fingerprint" of low-latitude formation). In a 1964 review article published in *Scientific American*, Harland argued that the glaciers must have reached the equator because some of the deposits would have been at low latitudes no matter where Earth's rotation axis had wandered. Harland also specifically rejected the idea that the oceans might have frozen over, as it would invoke the "ice catastrophe" from which climate modelers assured him the planet could never have escaped.

Measuring the latitude of continents in the past is a specialty of a branch of geophysics called paleomagnetism, which studies the fossil record of Earth's magnetic field. Earth's field is vertical at the poles, but horizontal at the equator. Hence, measuring the angle of the magnetic field at the time a rock formed with respect to the (horizontal) bedding planes provides an estimate of the latitude at the time the rocks formed. Unfortunately, it is necessary to actually prove that the magnetism one measures is as ancient as the rock, and was not acquired during recent weathering or some metamorphic event. (To be meaningful, we must study things that really and truly date to the time that the rocks formed. This is the flaw with the Precambrian biomarker studies noted earlier.)

The possibility of testing this low-latitude glaciation hypothesis attracted many early attempts at paleomagnetic analysis. However, in

Example of a striated cobble from the first "snowball Earth" event in earth history, the Makganyene glaciation of South Africa. This rock has several sets of parallel striations, in different orientations, carved on all surfaces. Patterns like these are known to form only on cobbles that are dragged along the basement rock at the bottom of actively moving glaciers. The sets of differently aligned groves form each time the rock acquires a different orientation at the bottom of the ice. Most such stones are ground down into glacial dust; this one was lucky enough to survive.

1966 a new paradigm for the geological sciences was proposed—that of plate tectonics. If the continents could move relative to each other, it was then possible that all of the infra-Cambrian glacial sediments actually formed at the poles, and plate tectonics could have moved them down to their present position in low latitudes. The idea of low-latitude Precambrian glaciation basically dropped off the geophysical radar screen. It just seemed too far-fetched to the scientists studying the early Earth.

That was the situation until 1987, when detailed analysis of new samples directly from glacial rocks in Australia proved the low-latitude magnetic direction had been there before the sediments turned from mud to rock. This was the first bulletproof result demonstrating an equatorial position for a sea level, widespread glaciation. And if Earth was frozen on the equator, it must have been even colder toward the poles. With this impetus, a change of scientific view took place. Once there was acceptance that perhaps it was possible to have world-covering ice in the deep past, the available information from fossil distributions, rock types, and even the paleomagnetic data made more sense, but it kept putting the major continental masses on the equator. The commonly accepted model of glaciers creeping along the continents (and never covering the oceans) from high latitude till they reached the equator simply did not agree with the data.

As the various possibilities of how the world had produced glacial deposits at the equator were reexamined, it became clear to at least some of the scientists studying this time that the Earth actually *had* frozen over. Once that great leap of faith was made, the rest fell in line. Floating pack ice would seal off the ocean surface, curtailing photosynthesis, stifling gas exchange with the entombed ocean beneath it, and causing the sea bottoms to go anoxic. Hydrothermal vents at the seafloor would then gradually build up concentrations of iron and manganese in solution, which would supply the metals needed for deposition of the banded iron stones mentioned above. Without access to sunlight, photosynthesis would be restricted to a few hydrothermal areas that would manage to break through the ice, as is done today in Antarctica and Iceland. Photosynthetic life could survive there. In a short seven-paragraph chapter in a 1,400-page book published by a project at UCLA in 1992 (four years after it was written), coauthor Joe Kirschvink marshaled this data for the first time, and gave it a new name: snowball Earth. At the same time, he took an additional step, by hypothesizing that the aftermath of one or more of the snowball Earth episodes of the Proterozoic era might have produced environmental conditions that would have resulted in rapid evolution—what we now accept as the evolutionary drive for the radiation of animal phyla.

So what was wrong with the climate models, all of which gave solutions suggesting that once in this kind of global glaciation, the Earth would never escape from global ice? The problem was that they had not incorporated the increase of carbon dioxide over geological time that would gradually increase the greenhouse effect. Climate scientists, particularly James Walker and Jim Kasting, had noted ten years earlier that CO_2 could eventually cause an escape from the ice catastrophe because of a pressure broadening of its infrared absorption spectrum. However, their suggestion was only one paragraph in a long paper and it had never been included in global climate models, simply because no one ever suspected that this has actually happened!

In the two decades that have followed publication of this idea, numerous geologists, geochemists, and climate scientists have conducted intense debates on, and tests of, this hypothesis, expanding the concept and clarifying predictions of models. Paul Hoffman and colleagues at Harvard, for example, contributed an enormous amount of stable isotope data showing that the elevated carbon dioxide concentrations in the atmosphere most likely wound up being converted into the limestone and carbonates that smother the glacial deposits. Geochronologists using high-resolution uranium-lead dates were able to show that both of the major low-latitude glacial intervals in the Neoproterozoic ended synchronously, a clear prediction of the model.

Here, again, we see a major refutation to the principle of uniformitarianism. A snowball Earth would inevitably cause a severe decline in marine organic production because the sea ice would block out sunlight. A succession of snowball glaciations and their ultragreenhouse terminations must have imposed a severe environmental filter on the evolution of life. The pre-Ediacaran fossil record provides few clues, but the diversity of microfossils in the sea known as acritarchs (planktonic organisms of small size, but definitely eukaryotes) waxed and waned dramatically. Many living organisms are known to respond to environmental stress by wholesale reorganization of their genomes. The developmental and evolutionary significance of such genomic changes are hot topics of research in molecular biology. The fact that diverse Ediacaran fossils first appear in the immediate aftermath of the snowball glaciations supports the hypothesis of an ecological "trigger" for their abrupt appearance. However, molecular sequence comparisons of extant organisms imply that the major metazoan clades evolved prior to some or all of the snowball events, but such "molecular clocks" assume uniform rates of genetic change. If the climatic shocks associated with snowball events caused greatly accelerated rates of gene substitution in most ancestral metazoan lineages, then the molecular and fossil evidence may be reconciled.

A frozen ocean, however, is a bad place for surface-dwelling organisms, and thus it was that the great oxygenation event could not have started, ironically, until the signal event that would allow it to melt away. During this snowball Earth, the cyanobacteria survived, probably in local hot springs. Earth was lucky that it was close enough to the sun and had enough volcanic activity releasing greenhouse gases to let it eventually escape the snowball state or else we might be frozen still, and not get liquid oceans until some time in the future when our ever-heating sun finally melts through the ice. If it had been slightly farther from the sun, CO_2 could have frozen at the poles as dry ice, robbing Earth of the snowball escape and making it more like Mars. Surface life might have died completely.

The Earth with its new oxygen atmosphere was a bizarre place, at least in terms of what was happening, or not happening, to life. It is clear that aerobic respiration, our biochemistry that allows us to breathe oxygen, could only have evolved *after* oxygen was present. There had to have been a time gap between the presence of oxygen and the first organisms capable of breathing it. In fact, evolution would have immensely favored any organism that could use oxygen, since no other molecule lets the chemical reactions we call life take place faster, with more precision, and liberate as much energy as those where oxygen is used.

The time gap between the evolution of oxygen release and the presence of organisms in the biosphere that could breathe it is identifiable in the geological record. The cyanobacteria that suddenly found themselves in a world no longer covered in ice would quickly have invaded the new and warm surface waters of every ocean, and because the amount of land area more than 2.2 billion years ago was vastly less than now, and the planetary ocean would have had millions of years to load up on raw nutrients from hydrothermal vents, they would have multiplied to numbers almost incomprehensible, rapidly increasing the amount of oxygen. They would have been floating in the marine ecosystem on shallow subsurface horizons where light could reach, and even on what little land area was present. While these organisms would be madly excreting this molecular oxygen, they would also be rapidly depleting the carbon dioxide that had built up in the air during the snowball Earth event they caused, producing a wealth of hydrocarbons in the ocean environment. For every molecule of O_2 released by photosynthesis, one atom of carbon is incorporated into the stuff of life. Today, light hydrocarbons of that sort are eaten by oxygen-breathing organisms and converted back into carbon dioxide. But if organisms had not yet evolved the ability to breathe oxygen, the question arose as to where all of this floating organic material would have gone. There would have been so much of it that there would have been major changes to the surface of the Earth's chemistry and its oceans and air.

Oil and oxygen when mixed together in the air form an explosive cocktail; a single spark of lightning would cause a reaction to go without stopping. But oil dispersed in water, as little particulates, can only be degraded by the action of microbes. Without efficient recycling, Earth should have experienced a huge imbalance in the carbon cycle. In particular, a large amount of oil should've been produced, and an equal amount of oxygen should've been pumped into the atmosphere. At this time, we do have evidence for a massive oxidation event at 2.1 GA; it formed one of the world's largest deposits of pure hematite (Fe_2O_3) iron ore—the Sishen mine in South Africa.[15] Earth's atmosphere must have been supercharged with oxygen at that time, to levels not encountered since, and probably impossible to reach without some deviant biosphere driving it there. If planets orbiting other stars went through the same process, the hyperbaric oxygen in their atmospheres would be waving a spectral flag proclaiming, "We are here, and we solved the photosynthesis problem!"

In fact, the record of carbon isotopes for the period of time between 2.2 and 2.0 billion years ago is so wildly out of balance that geochemists have given it its own name, with the jaw-breaking name

"Lomagundi-Jatuli excursion," and it is the biggest and longest such event yet found in the entire history of our planet. Most of the carbon being emitted from volcanoes was being sequestered as organic material, releasing oxygen to the air; today this ratio is only about 20 percent. This is the evidence of an Earth with oxygen but without organisms capable of breathing it: wild swings in the carbon cycle caused by cyanobacteria excreted lots of carbon compounds as waste, but with no organisms using these chemicals as food. In fact, the remnants of this sludge appear to exist in the Russian province of Karelia, as a weird rock type called shungite. Today, most of these oil-like compounds would have been quickly biodegraded by microorganisms that breathe oxygen, like the fate of most of the Deepwater Horizon spill. This is direct evidence that the environment choked on hydrocarbons, rather than recycling them directly. As a result oxygen kept rising in content until it was so abundant that it produced an atmosphere supersaturated in oxygen, existing at pressures much higher than today. Had there been any forests, the first spark from lightning would have caused a global forest fire of heat and scope beyond anything that has ever occurred on Earth in the time of forests.

This weird episode in the history of life ended abruptly when evolution produced the first organisms that could *breathe* oxygen efficiently. Special copper-based enzymes evolved to do this, but copper deposits themselves require oxygen-rich environments to form. An entirely new kind of intracellular body came into existence, and it exists still, the organelle called the mitochondria, the major source of energy for eukaryotic cells, which are cells that are larger than their prokaryotic (bacterial) ancestors, as well as being cells that contain walled-off interior "rooms" in the giant (compared to all that came before) cells. The mitochondria has its own little piece of DNA, left over from when it was once a free-living bacterium, a microbe that learned to breathe oxygen efficiently. As a result, it has been enslaved for the past 2 billion years.

It is intriguing to note that the best estimate for the age of the last common ancestor of all eukaryotes is about 1.9 billion years, and that may mark the time that eukaryotes finally evolved to restore balance to the global carbon cycle. It would seem that the biosphere required over 200 million years of evolution to come up with an adequate response to the presence of the intrinsically poisonous oxygen.

CHAPTER VI

The Long Road to Animals: 2.0–1.0 GA

THE time between the great oxygenation event (culminating at ~2.3 GA) and the first appearance of common multicellular life has been called the boring billion. The reason is that (supposedly) virtually nothing happened in terms of major biological change. It is as if the history of life took a snooze. A billion years is a long time for almost nothing to happen. But like so much else, the boring billion has recently been shown to be not so boring. New discoveries are showing us that life was not resting at all. But at the same time, in spite of repeated suggestions to the contrary, there are no animals a billion years in age. Instead, this long interval begins with the first significant oxygen in the atmosphere, and by 2 billion years ago a major revolution in life had occurred—the common occurrence of eukaryotic life, our kind of life, large cells with a nucleus. And while the greatest diversity of these new creatures during this period of time were protozoa, familiar to us as the still-living amoeba, paramecia, euglena, and their cohorts, there appeared some strange larger fossils as well, including one of the most bizarre fossils ever recovered.

The various experts agree that there was probably not enough oxygen between 2.2 and 1.0 billion years ago to support animal life.[1] (This is a good time to quickly summarize the difference between animals, (metazoans) and protozoans. All three are eukaryotes—organisms with large cells that contain a nucleus as well as other smaller organelles, such as mitochondria. Animals and "metazoans" are the same thing. All are composed of more than a single cell during their lives except at fertilization. Protozoans can seem animal-like, in that many are capable of movement and relatively complex behavior. But all of them are composed of only a single cell. Nevertheless, they are far larger and far more complex than bacteria.) Yet if that was agreed on, the reason for this was not. Life was capable of oxygenic photosynthesis, but there should have been far more life than all evidence suggests. Animals need a good 10 percent of the atmosphere to have oxygen (we are at 21 percent today) and the "photosynthesizers" were not doing their job. The answer, when it finally came, was once again the element that runs through the history of these pages in an ever-repeating pattern: sulfur, usually in the guise of its most toxic and at the same time life-giving form—hydrogen sulfide, molecule of life and death. In a 2009 paper published in the *Proceedings of the National Academy of Sciences*,[2] Harvard paleobiologist Andy Knoll and his colleagues showed that oxygen levels *should* have been higher during the boring billion, but were not. Something was holding them back. The long interval devoid of any kind of real intermediary between the single-celled organisms of the 2.3-billion-year-old great oxidation event and the appearance of larger, multicellular creatures of far, far later in time was real.

There were no life forms that we might call complex in this long interval of time (although we hope it is clear from preceding chapters that even the simplest life forms on Earth are unbelievably complex when viewed at the molecular and chemical scale!). And the reason was an overabundance of single-celled sulfur-using bacteria that were competing with the oxygen-releasing forms. Thus it was that two very different life forms competed for resources coveted by all life—space

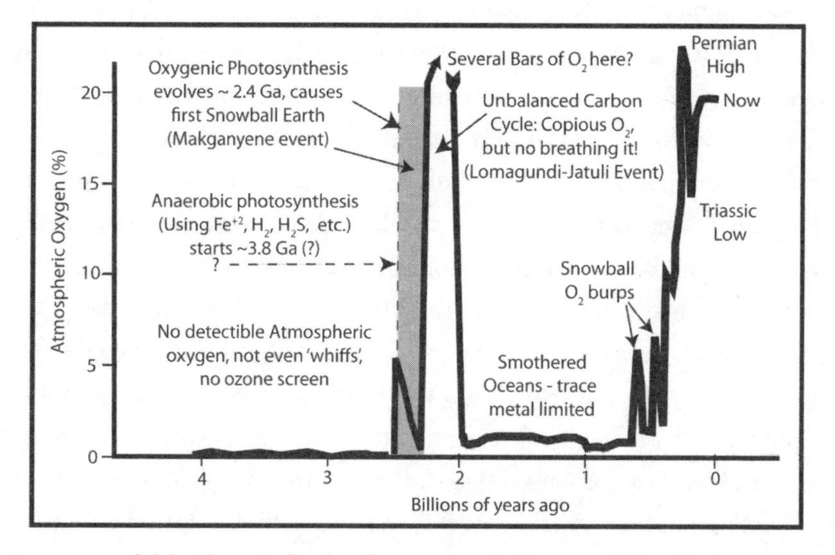

Our new model for the rise of atmospheric oxygen and some of the related events.

and nutrients. The sulfur-requiring microbes, called green and purple sulfur bacteria, are still alive today, but only in the most toxic of places—shallow-water lakes and some seaways that have no oxygen but are shallow enough so that sunlight can penetrate to the levels of the bacteria, allowing photosynthesis. But the problem is that this kind of photosynthesis does not split water apart, and thus does not produce oxygen as a by-product.

Fundamentally, it seems that life was *lazy*. Splitting water is actually a difficult task, which generates all sorts of nasty, toxic compounds. Using H_2S for photosynthesis instead of H_2O results in less toxic sulfur compounds, and even many strains of cyanobacteria—if given the choice—will shut down their oxygen-generating machinery and use H_2S rather than water.

For most of the boring billion the oceans were stratified, with a thin top layer of oxygenated, clean surface water where single-celled green algae took sunshine and used that energy for cell growth, all the while releasing oxygen. But beneath them, perhaps only ten or twenty feet below, was a totally different layer of seawater, and this layer would have extended all the way to the deepest ocean bottom. It would have been purple in color in its uppermost, shallowest regions, stained this color by the untold numbers of the purple sulfur microbes. The water they lived in would have been a fatal poison for most ocean life of our world, as it was filled with toxic hydrogen sulfide all brewed in a near-boiling miasma of liquid brimstone. Even in death they would have helped rob the world of oxygen (unconsciously, of course, although some microbial specialists actually seem convinced that the microbes have always been some kind of sneaky smart). After death, their tiny bodies would have sunk to the bottom, or even stayed in place in water salty or sediment filled enough, and in rotting would have taken even more of the precious few oxygen molecules being produced by the thin layer of oxygen-producing microbes in the surface layer above them. Precious oxygen molecules, destined for the atmosphere and clear oceans, were used up instead in the rotting of a purple demon.

While rare on Earth now, this same stratified system still exists in a few places. One of the most famous is in the Micronesian island of

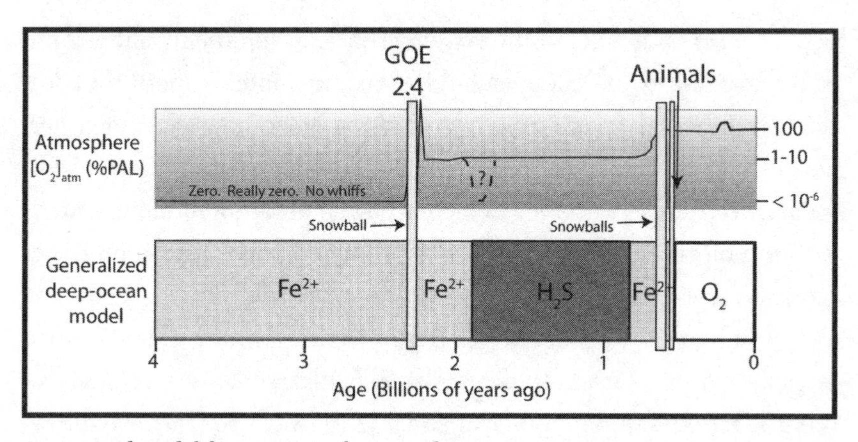

Our revised model for oxygen in the air and sea.

Palau, in the famous "jellyfish" lakes. Here, large freshwater lakes are filled to bursting with enormous, abundant jellyfish, swimming gracefully through aquamarine and well-oxygenated water. Yet some tens of feet below this crystal lens of clean, oxygen-life-filled water rests a second and deeper stratum, which is dark, and to us creatures of light and oxygen, vile to the extreme. It has little or no oxygen, but is saturated with hydrogen sulfide. And it is dark purple in color, stained by untold numbers of the same purple sulfur bacteria that kept the world unsafe and unavailable for anything needing abundant oxygen for what was to them probably not boring at all.

The purple sulfur bacteria and their world needs were finally sent to dank, poisonous back rooms of our world. But they were always there, always ready to take back the world they lost when oxygen finally broke through to higher levels, some 600 million years ago. They can be thought of as the evil empire. And in the Devonian, Permian, Triassic, Jurassic, and middle Cretaceous, this empire struck back, as we will see in subsequent chapters.

Eventually, the balance of sulfur photosynthesizers to oxygen producers changed in oxygen's favor, possibly triggered by a gradual increase in the area of subaerially exposed continents. Iron eroding from the continents and washing down into the ocean would react rapidly with the sulfur, precipitating it into a heavy, sinking solid mass of pyrite, keeping it out of the system. This loss would have starved the sulfur bacteria of the one element they could not do without. In addition, continental weathering and erosion generates clay minerals, which bind strongly to organic molecules and bury them in the sediments. If an atom of organic carbon is buried before something can eat it, the molecule of oxygen that was produced when it formed hangs around in the environment, raising the oxygen level and destroying H_2S. Prompted by two snowball Earth events, each of which seemed to jack up oxygen levels by the postsnowball algal bloom, the environment reached a tipping point of some kind. After the last event, 635 million years ago, the first traces of big animals appeared. It did not take so long to evolve them after all, once hell on Earth was banished.

THE BIZARRE, FIRST MULTICELLULAR CREATURES

Most life during the now not-so-boring billion was composed of the long-running champions of life on Earth, the longest-running show of all—stromatolites. Microbes still held sway, just as they had from their first appearance on Earth. But appearing about 2.2 billion years ago a strange new kind of life form appeared. It looks like a thin black spiral, but is certainly not microscopic. Its name is *Grypania*, and its appearance demonstrates that life had made an important advance: the ability to live as "colonies" of cells, held together and bound by membranes. These were the first multicellular organisms.

Grypania has long been known. But in 2010 a strange series of fossils from Gabon, Africa, changed our view of things.[3] While *Grypania* might be a colony of prokaryotes (in this case probably bacteria), the new fossils, still unnamed, look too large and too complicated. Whatever they were, we know what they were *not*. They certainly were not the first animals.

The first true animals are much younger than *Grypania* and its ilk. Animals are less than a billion years old, and while the exact age of the first animal keeps getting put back into older-aged rocks, based on ever more sophisticated means of detecting their presence, there is still no known fossil evidence of animals much older than that last snowball event. But in a way this is an argument over rather small chunks of time, compared to the vast interval that life has been on this planet. There are, of course, many types of *multicellular* organisms, including a considerable diversity of prokaryotic forms, and there is no doubt that the evolutionary invention of life with more than a single cell goes back to more than 2 billion years ago. But in most cases these multicellular prokaryotes are composed of only two cell types, and none would be mistaken for an animal.

Cellular slime molds are multicellular, as are some cyanobacteria and one group of magnetotactic bacteria. In a way, however, these are evolutionary dead ends (unless, of course, you are a slime mold; however, this group ultimately gave rise to little else but slime molds). They have existed on Earth for more than several billion years, and are highly conservative in an evolutionary sense. More complex are the multicellular plants that appeared more than a billion years ago, species probably looking very much like the green and red algae found on any seashore, from the intertidal zone down to the levels that light can penetrate. Animals, however, are younger yet.

The size of organisms seems to show some relationship to the appearance of oxygen in the atmosphere. Oxygen has allowed larger size than times prior to oxygen, and biological adaptations increasing the rate and/or volume of oxygen acquisition have often lead to gigantism.[4] The best example of this will be described in a later chapter, showing how the gigantism in dinosaurs was caused by a new kind of highly efficient lung and respiratory system design.

Fossils of true animals first appear in *abundance* about 600 million years ago. About this time the rock record shows the first evidence of "trace fossils," the trackways or feeding records of ancient animals preserved not as body fossils in sediment, but as activity fossils—the record of ancient behavior. By that time oxygen levels were approaching (but not yet reaching) modern levels. Not only free oxygen but ozone levels had also reached relatively high concentrations, and thus much of the hard ultraviolet and other radiation reaching the Earth's surface in earlier times was muted.

Geobiologist Andy Knoll of Harvard University on Neoproterozoic rocks exposed in East Greenland on an anomalously sunny day. (Copyright Andy Knoll, used with permission)

THE CURIOUS ORGANISMS KNOWN AS ACRITARCHS

In any discussion of Precambrian life, acritarchs actually make up a fair bit of the conversation. They appeared early on Earth: some of the oldest seem to appear around 3.2 billion years ago, and they then continue all the way into the time of animals. Yet the fact that they are a "garbage can" taxon, meaning that any number of not only different species but even different kingdoms and domains of organisms get placed in this catchall name, is just one more indication of how poorly we know the history of life before fossils became common in the time of animals and higher plants.

While being among the oldest-known multicellular fossils, first appearing at the almost unfathomable time period of 2 billion years ago, they remained relatively rare. But halfway through the Proterozoic era, or about a billion years ago, they started to increase in diversity, size, abundance, and morphological complexity in shape. The increase in complexity was generally marked by an increase in the number of spines extending from their small, spherical bodies. From 1 billion to 850 million years ago they remained common, and then the Cryogenian period began, with the enormous global changes that gave this time interval its name, the onset of the Greek word "cryo," all right: a great freeze. The result of the Proterozoic snowball Earth episodes was that there must have been a great mass extinction in the oceans, and perhaps on land as well. Their populations crashed during the snowball Earth episodes—when all or very nearly all of the Earth's surface was covered by ice or snow—but they proliferated in the Cambrian explosion and reached their highest diversity in the Paleozoic.

It is a given that any young aspiring paleontologist will gravitate to dinosaurs above all fossils. Since professional paleontologists always start as fossil-mad kids, in fact the supposedly less exciting fossil groups attract far less attention, even among the eventual professionals. Few indeed are the young scientists wanting to study tiny microfossils. And yet some of the most important of all scientific questions can be answered with them. So it is with larger questions about the history of life, as the acritarchs and other microfossils of a billion years ago are rich in question-answering information, and only recently have provided whole new insights into what was, in fact, a hugely important time period in the history of life, beginning at a billion years ago.

From 2 to 1 billion years ago, the microfossils of Earth were simple and long ranging through the rock record. They must have been formed by both prokaryotic and small (compared to later) eukaryotes of the single-celled variety, such as the still-living protozoans. But about a billion years ago a strange thing happened. The formerly unornamented microfossils began to acquire ornamentation.

The increased spinosity of acritarchs, starting about 1 billion years ago but then continuing through the Cambrian period, could have several causes. First, spines on a small sphere would increase surface area to volume relationships, and thus slow the rate of settling of these tiny spheres in the ocean. Many planktonic species extant today use this method of staying high in the water column rather than sinking

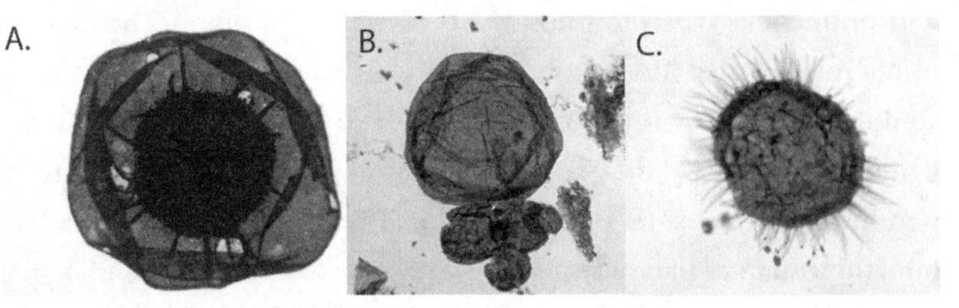

Changes in the morphology of acritarchs, enigmatic microfossils that were composed of a number of distinct kinds of small, marine, floating organisms. Notice the change from Proterozoic (A), which are smooth, to the very spiky forms of the late Neoproterozoic (B) and Cambrian (C).

onto a deep bottom and sure burial under the constant snowfall of sediments that typifies most ocean bottoms. But a second use of spines is defense against predators. Perhaps the oceans of a billion years ago began to harbor an ever-greater rogues' gallery of carnivores (or for acritarchs, these might technically be considered herbivores). In any event, eaten is eaten, no matter what one calls the eater. Yet the new work of Knoll and his group now show that spiny microfossils became ever more diverse and abundant soon after the end of the last snowball Earth, some 635 million years ago, but then utterly disappeared about 560 million years ago, a time when the evolution of animals was well under way. In the next chapter we will see how an understanding of what we might call the Ediacaran revolution is importantly fleshed out by the record of spiny microfossils, as well as by their disappearance. We will return to the story of these spiny microfossils in the next chapter.

THE END OF THE BORING BILLION

Here is a view of a shallow sea bottom, some 1 billion years ago: Kelp-like plants and green algae wave in the currents, as do shimmering mats of rainbow-hued microbial life, multicolored sheaths of the softest chiffon covering all of the sunlit portions of the bottom.[5] Stromatolites peak out from the bottom sheaths, large to small domes and hummocks punching upward out of the microbial sheaths. The water is thick with life, single celled to multicellular. There is nary an animal anywhere on the planet. But a genetic and atmospheric clock is ticking down toward catastrophe and icy crèche.

In the oceans a revolution was brewing a billion years ago, while on land there may already have been a vast biomass of life: the ever-resourceful microbes, invading first ponds and swamps, but ultimately covering wetlands, bogs, and anywhere that was exposed to sun, had a least a modicum of water, and might get windblown dust with enough phosphates and nitrates to allow these tiny, single-celled plantlike microbes to grow their land-covering tarps of green snot. Life, colonizing the land in exuberance. And in so doing ultimately nearly extinguishing itself from the Earth.

CHAPTER VII

The Cryogenian and the Evolution of Animals: 850–635 MA

THE Australian city of Adelaide is a well-kept secret. Isolated on this island continent from the rest of the world, and even isolated from the rest of Australia, this coastal city has evolved its own culture artistically and scientifically. The latter has been importantly influenced by an enormous paleontological discovery, made right after the end of World War II—the discovery in the arid hills inland from Adelaide of the first acknowledged larger animal fossils, the Ediacarans. Adelaide pays homage to this fossil record in many ways, including the naming of buildings and institutes for two of the scientific giants who brought clarity to the time period of a billion to 600 million years ago: Douglas Mawson, a hardy Australian who survived harrowing Antarctic expeditions and the killing fields of World War I France, a man who also went on to discover proof of late Precambrian glaciations in Australia, a concept highly doubted at the time, and Reg Sprigg, who discovered the fossils,[1] as we will recount below, followed, like Mawson, by another geology professor at the University of Adelaide, where coauthor Ward now lives and works—Martin Glaessner.[2] But new generations of workers have kept this tradition of the study of the origin of animals alive, and one of the most important is Jim Gehling of the South Australian Museum, which sits next door to the University of Adelaide. Gehling has overseen a new exhibit of Ediacaran fossils in a newly refurbished, large, and modern room of the museum, and there, unlike so many new museums where actual fossils are kept away from the public, substituting plaster casts or other reproductions instead, the Ediacaran exhibit that Jim Gehling[3] oversaw has real fossils, real Ediacarans on display. The surprise is how large and complex they are. But another surprise is how they are interpreted. Until recently the party line has long been that these were sedentary, strange, and mainly flat creatures, like stuffed pillows sitting on the seafloor (and some as large as a large, if flat, pillow). But overhead, on television screens, the animated reconstructions are anything but sedentary. Some even swim; others move robustly. Herein lies the controversy. This view is new. But is it correct?

The time interval of this chapter is the long period beginning about a billion years ago and ending with the start of the Cambrian period around 540 million years ago (MA). In that interval, far more than great changes in life's history took place. Just as it had in the period of around 2.5 to 2.4 billion years ago, at around 717 million years ago the Earth cooled. It cooled so much that—as it had near the end of the Archean era—the oceans began to freeze, starting at high latitudes, but continuing toward lower and lower latitudes until the entire ocean from pole to equator was ice-covered. Once again the Earth had become a snowball. The first time that singular event caused a great revolution in the history of life by leading to an oxygen-rich atmosphere. This second time, the Proterozoic snowball Earth also produced momentous if very different effects. This time the snowballs led to animals—but not without danger to all life on Earth. Once again life was in the balance. The overriding question is whether the snowball Earth episodes of this time interval were the key reasons for the sudden rise of animals, a case we will make.

LIFE AND SNOWBALL EARTH EVENTS

As we saw in an earlier chapter, the first snowball Earth episode (beginning at about 2.35 billion years ago) seems to have been caused by life: the explosive rise of cyanobacteria caused a reduction in the greenhouse effect of the atmosphere's methane and carbon dioxide content. The start of this second and final series of snowball Earth events of Earth's long history to date occurs within the Cryogenian time period described in chapter 1. Due to recent work on calibrating the Cryogenian, we know now that there were most likely two major events beginning 717 million years ago and ending 635 million years ago. The start of this second and final series of snowball Earth events is essentially in the middle of what is now formally defined as the Cryogenian period in the geological time scale (it begins prior to a pair of sharp isotope shifts slightly older than 800 million years ago, which are the result of a true polar wander oscillation).

Both of the differing snowball Earth episodes (each made up of ocean freezing and then thawing events) caused a severe decline in marine organic production, because the sea ice would block out sunlight. Thus, the amount of life on Earth, as measured by its overall mass (known as biomass), shrunk to tiny values compared to both before and after the events themselves. The succession of snowball glaciations and their ultragreenhouse terminations during both the periods from 2.35–2.22 billion years ago and from 717–635 million years ago must have imposed a severe environmental filter on the evolution of life. The fossil record provides few clues, but the acritarchs first described in the last chapter (planktonic organisms of small size) waxed and waned in both diversity and abundance.

Many living organisms are known to respond to environmental stress by wholesale reorganization of their genomes, and any snowball Earth event would have been stressful, to say the least. The

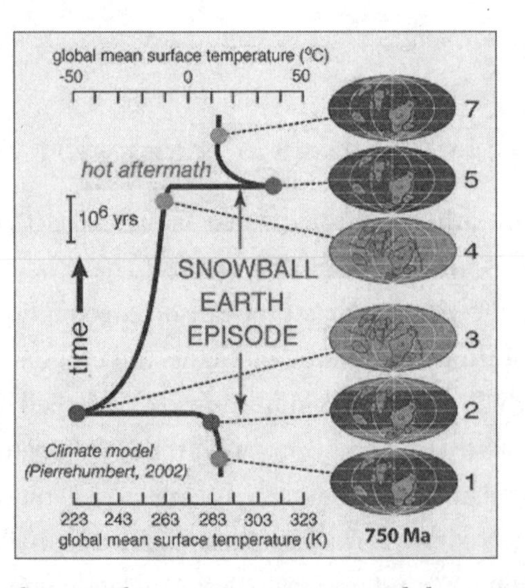

Diagram showing the rate of temperature increase and decrease over time in the Snowball Earth episodes.

developmental and evolutionary significance of such genomic changes are hot topics of research in molecular biology. The fact that diverse fossils of more complicated organisms than were there before the onset appear in the immediate aftermath of the snowball glaciations supports the notion that the snowball events created some sort of an ecological trigger for vast changes in the complexity of life and its diversity.

One of the most profound of all questions relating to the snowball Earth events relates to their cause. Earlier we noted that the first snowball Earth episode might have been triggered by life itself: the invention of oxygenic photosynthesis, which would have caused a rapid depletion of greenhouse gases. But there may have been a quite different reason for the onset of the second episodes, occurring well more than a billion years after the first. The second snowballs may have been triggered by the movement and tectonic activity of continents of the time.[4]

The so-called Neoproterozoic snowball events, the most recent of the two grand snowball Earth episodes, occurred around 40 million years after the great continental amalgamation—called the supercontinent Rodinia (an amalgamation of every continent into one continuous landmass)—began to disintegrate. Supercontinents tend to have arid climates because most of their land area is far from the ocean. Conversely, when continents and especially supercontinents separate, maritime climates displace formerly arid regions, creating the potential for increased chemical weathering. Chemical weathering of silicate rock minerals causes a rapid reduction of carbon dioxide levels in the atmosphere. As CO_2 drops, so too does temperature. This second time it may not have been life so much as inorganic chemical reactions. Interestingly enough, the onset of the second snowball event (called the Sturtian after exposures in Australia) coincides rather precisely with the eruption of a massive volcanic province in Canada, at 716.5 million years ago.[5] Although some CO_2 is emitted from eruption of these large igneous provinces, when they erupt on land the drawdown of gases far exceeds the volcanic input, bringing the system closer to a planet getting so white that most sunlight is reflected back into space. And that produces ever more cold.

But perhaps this is not the whole story. If it could be shown that some new kind of plant life suddenly and radically increased in numbers across the globe, once again the possibility arises that the sudden reduction of carbon dioxide by photosynthesis, rather than chemical weathering, was involved. In fact this may have been the case. Some of the newest of all understandings about the history of life is that land plants, still only single celled but nevertheless potentially extending over vast areas of land, appeared around 750 million years ago. This would have done the trick.

A SNOWBALL MASS EXTINCTION—AND SNOWBALL-PRODUCED STIMULUS FOR THE ORIGIN OF SO MANY KINDS OF ANIMALS?

What would have happened to the life on Earth of about 750 to somewhat more than 600 million years ago by the change from a world of ocean and land to one of snow, ice, and bare rock? A simple thought experiment suggests that both the abundance and the diversity of life on Earth found just before these Proterozoic-era snowball Earth events would have diminished. The life then was largely of the single-celled variety, although by this time multicellular plants such as the common kelp and alga (green and red) that adorn so many seashores of our world would have been present too. But much of life was composed either of single-celled protozoa, all eukaryotes, or vast sheets of bacterial slicks and growths both as near-shore stromatolites and other masses of cyanobacteria, and also huge biomasses of single-celled, photosynthetic microbes in the seas. On land, we speculate that single-celled, perhaps even more complex assemblages of photosynthetic organisms, including great sheets of microbes, would have inhabited freshwater, and would perhaps appear liberally on damper land surfaces. Soils as we know them would not have yet existed, but certainly the chemical weathering of rock surfaces, incorporating the dead and rotting bodies of what plants there were would have added organics to the clays and sand of the surface of the land. And then onto both the surface of the sea and that of the land came ice for the former, and for a while ice and certainly cold to the latter.

The extinction potential in terms of biomass is easy to imagine and fathom. Kilometer-thick ice covering the sea surface would have greatly reduced sunlight. While there is microbial life in ice, and in fact some sun does filter through sea ice, surely the biomass of plant life would have plummeted. The loss of sunlight was one part, but perhaps as significant would have been the loss of important nutrients, the all-important iron, nitrates, and phosphates of our world. As the land surface cooled and in many parts came to be covered in snow and ice, chemical weathering slowed, as did the vigor and abundance of land "plants" of whatever kind there were (this is, of course, hundreds of millions of years before true, complex land plants with stems and leaves). But the land would have produced far less fertilizer getting to the sea. Ocean productivity plummeted, and as it did so, surely mass extinction not only of individuals but also of whole species followed.

Yet from this scenario comes a model that perhaps answers the question of why there are so many kinds of animals. Although the entire ocean surface would have frozen with pack ice, in fact the world then had far more volcanic activity than it does now. There would have been many hot springs, geysers, and especially active volcanoes blasting heat into oceans, and in so doing producing small warm bodies of open water, free of ice. Surrounded by icebergs and finally frozen sea, these small "aquaria" would have been isolated, and being scattered around the world, subject to many kinds of different environmental conditions. Evolution works best on small, isolated populations. Thousands of these small marine and even freshwater refuges would have been evolutionary incubators, using the principle of "genetic bottlenecks" (where tiny populations, when isolated, can quickly evolve because of their small number of genes). In this way, protozoa, those small single-celled eukaryotes, may have evolved into many different kinds of *metazoans*—animals. With the release of the snowball conditions, caused by the eventual buildup of greenhouse gases from all of those active volcanoes, there would have been rapid melting of the ice, as well as a rapid release of these thousands of new evolutionary experiments.

Earth came out of its last snowball 635 million years ago, a place very different from the planet we know today. But forces—both evolutionary and physical—were under way that would make our Late Proterozoic Earth much more Earthlike, in the sense that we know it. The oceans were teeming with life: most were single celled, but largely composed of the complex Protozoa, such as amoeba, paramecia, and the enigmatic half plant half animals such as multicellular Volvox and single-celled Euglena. The shores and sea bottoms were festooned with various kinds of kelp, more formerly the large, multicellular red and green algae so common on Earth, *still* so common on Earth. The stage was set for the evolution of the first animals. Around 635 million years ago that process began. We think. The newly named Ediacaran period began at the end of the last snowball and ended with the appearance of creatures that were unquestionably animals. It also is the last formal time interval immediately before the start of the Paleozoic Era. The time is named after its most important denizens, then the most complex organisms to have ever evolved. We call them Ediacarans.[6]

These iconic fossils of this latest Precambrian time—the last part of the Proterozoic era—reveal a wide variety of peculiar body types unlike anything alive today. Once known only from the Ediacaran Hills of South Australia, there are now numerous places on Earth where

these enigmatic fossils are known to be found. But the best remains the low hills north of Adelaide.

The Ediacaran Hills are part of the largest mountain range in the southern part of Australia, the Flinders Ranges. Like so much of Australia away from the more verdant coast, much of the Flinders Ranges is composed of sand, rocky outcrops, and scattered vegetation adapted to a semiarid environment. Here and there larger trees dot the landscape, including sugar gum, cypress pine, and black oak. Year-round water holes are scarce, but when found, a rich assemblage of the iconic Australian fauna is abundant; red and western gray kangaroos have flourished in the area since the eradication of the carnivorous dingoes, their most dangerous predator at one time. Even the once-endangered yellow-footed rock wallabies can now be seen with regularity. But it is not the kangaroos and the other smaller marsupials that make this place special; it is the ancient fossil fauna.

Along with the Burgess Shale of Canada, Solnhofen Limestone of Germany, and the Hell Creek Formation of North America, the Ediacaran Hills is arguably one of the four most famous fossil sites in

Late Precambrian imprint of a segmented worm-like animal, called *Spriggina*, an Ediacarian fossil from South Australia. This is thought to be a primitive annelid and possibly an ancestor of the trilobites.

the world. Ranging between 560 and 540 million years in age, these hills contain the record of what most paleontologists agree are the first-known body fossils of animals.

The discovery was made when geologist Reginald Sprigg was examining old mines in the Ediacaran Hills region of South Australia. Sprigg was a government geologist for the state of South Australia. He was walking through a desolate area countryside of eroded hills as part of his state's reassessment of the mineral resources. His job was to decide whether this particular area should have been a focus for new mining activity. However, Sprigg had been an ardent amateur fossil collector during his student days and was able to recognize that the strange markings he encountered by chance within the slabs of coarse sandstones scattered across the rolling Ediacaran Hills had to have been produced by some life. But what kind?

Sprigg was confronted with what looked like the casts and impressions of jellyfish. But he knew jellyfish are rarely if ever fossilized, with "rarely" a euphemism at best. The strata that Sprigg was looking through were extremely old, and in fact he correctly surmised that the strange fossils he collected had to have been among the oldest direct records of animal life in the world: which was his statement when he first announced the discovery a year after first finding them. Sprigg noted that the fossils appeared to represent animals of varied affinities.[7]

Soon after this first announcement, Sprigg collected more bizarre fossils, this time accompanied by Professor Douglas Mawson of the University of Adelaide and his students. In 1949, Sprigg released a full account of the discovery from a very much larger collection at the same locality, as well as the first detailed description of these curious fossils.[8] They all came from the Pound Quartzite, a geological formation that had never had a satisfactory age determination. If Cambrian, they would be nothing of great interest. But if Proterozoic, the strange fossils would indeed be the oldest-known animal remains ever found on Earth. Subsequent work indeed showed that they were older than the classical Cambrian fossils (the trilobites) that were then used to define the Cambrian (a definition that has since been revised).

When examined in detail these fossils are indeed different from any known living animal, and according to some scientists in the late twentieth century, in fact came from animals with body plans no longer living, with no known descendants, a view first espoused by the great and sadly late Dolf Seilacher.[9] But it was their nature as fossils that was perhaps the oddest aspect of their mystery. First of all, organisms without hard parts rarely produce fossils. When they do, it is generally only in very fine-grained rocks, such as mudstones or shale, sedimentary rocks that have been deposited on the bottoms of quiet, stagnant bodies of water. But Sprigg's clearly skeletonless creatures were preserved in sandstone, rather than in a finer-grained kind of rock.

To determine whether Sprigg's fossils indeed came from the closest-living match, jellyfish, sea anemone, and soft colonies of anemone-like creatures called sea pens, experiments and tests were conducted to see whether such fossilization could happen at all. One such test was conducted by Martin Glaessner, Australian geologist and author of *The Dawn of Animal Life: A Biohistorical Study*.[10] He describes a series of experiments using newly captured, very large jellyfish, placed on thin beds of sand. He noted that the jellyfish indeed left impressions within the sand. But there's still the problem of the sand itself. Sprigg's fossils should never have been preserved.

Sand grains are deposited in places with relatively high energy. Sandstones today are found near shore localities, in rivers, in sand dunes, all places where moving water can carry these fairly heavy grains. In such environments the finer mud and clay particles are never deposited; they are just too light to settle and not be picked up again by currents, waves, or wind and carried to some other locality. Yet the Ediacaran fossils are both large and numerous, and are found in such sandstone settings.

To further test this dilemma, in the summer of 1987 coauthor Peter Ward invited students enrolled in an advanced paleontology class at the University of Washington's Friday Harbor Marine Labs on San Juan Island in Washington State to attempt to re-create the conditions that led to the formation of the Ediacaran fossils. Several kinds of

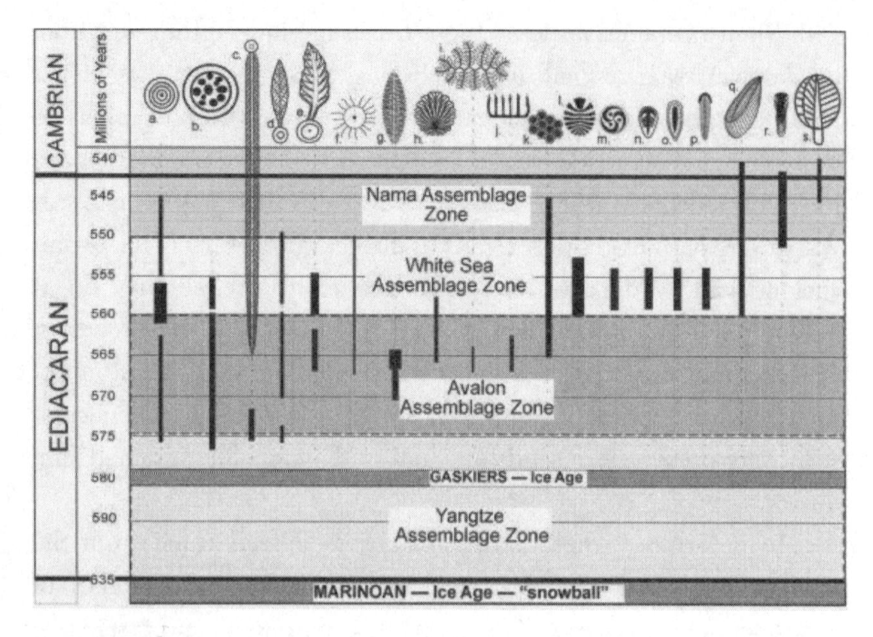

Species in Ediacaran zones.

experiments were conducted. The rich inland sea around the San Juan Islands contains a large diversity and abundance of cnidarians, the phylum seemingly most similar to the apparent Ediacaran body plan. To mimic a 600-million-year-old shallow water bottom, large buckets filled with sand of various screen sizes were then covered with seawater. These experiments were similar to those earlier conducted by Martin Glaessner, but in this case the bodies used were larger, and body types other than jellyfish were studied as well.

Bodies of newly dead sea pens, anemones, and some of the world's largest jellyfish were placed on the sand. More sand was then put over the top to the bodies, and then the experiments were left for a time, and after some days the top layer of sand was removed. In fact, none of these experiments left any sort of mark in the sand; the cnidarians would rot away, leaving nothing.

Finally, one student had a quite different idea. A square piece of very fine mesh nylon, from a nylon stocking, was placed over the top of the sandstone, and then a very large jellyfish was gently placed on top of the nylon. More fine sand was then added over the top of it all, with the entire jellyfish. Sea pen and anemone sandwiches were then covered with seawater. After several weeks, when the top layer of sand and the nylon sheet were removed (the soft parts of the animal put there had already rotted away), it was discovered that just underneath the nylon stocking there was a beautiful impression of the animal that was put there, including extremely detailed morphology that matched the underside of the animals used in these experiments.

Perhaps these experiments mean nothing. But what if the world of that time were covered with something of similar thickness and material properties to nylon stockings, properties that allowed sand grains that would otherwise be picked up by the slightest current to be held in place. We can envision a world where the shallow marine environments became covered with a thin sheet or multiple sheets of microbial life. Although fragile and easily destroyed by storms, these sheets would stabilize sediments and also leave soft-part impressions in the sand below when animals would die, fall onto the bottom, and then be covered by more sand, which would allow for the sanitation of new beds of sand.

We no longer have such marine environments today, ones that can preserve the outlines and impressions of tissue-rich but skeleton-free organisms. The evolution of mobile animals, which both tore and ate the resource-rich microbial sheets, would destroy these. Just as the stromatolites all disappeared with the evolution of animal herbivores, so too would many of the microbial mats and sheets of perhaps all of the world's shallow-water environments have been eaten out of existence.

THE WORLDWIDE EDIACARAN FAUNA

Today, the "Ediacaran biota" is known from about thirty localities on six continents, and its fauna is classified into seventy different species, all restricted in age to the latest Neoproterozoic[11] (although there might be a few of the species that do survive into the earliest Cambrian). The Ediacaran organisms seem to have evolved toward their full diversity in an evolutionary event called the Avalon diversification of

575 million years ago, which would have been as much as 50 million years following the cessation of the last of the Proterozoic snowballs.

From that time they seem to have thrived, whole communities of them in fact. Then, at about 550 to 540 million years ago, when the first evidence of animal locomotion appears in the fossil record of this age as trace fossils (activity fossils of animals, including locomotion and feeding marks preserved in sediment), the Ediacarans rather suddenly disappeared. A large, diverse group of organisms disappeared just as the first animals rapidly appeared on Earth, in an event known as the Cambrian explosion.[12] This disappearance is really the first major mass extinction marked in the fossil record (although certainly not the first mass extinction). While first thought to have been isolated on the Australian continent, it is now clear that the Ediacarans had a worldwide range.

There has been no end of suggestions about how energy flowed through the Ediacaran's ecological communities.[13] In modern ecosystems, photosynthetic plants make up the base of the food chain, and these are then grazed upon by several levels of consumers, which in turn are the prey of several levels of predators. The biomass of each of these steps is only about 10 percent of the "trophic" level below it. The Ediacarans, to some, showed a very different type of community structure. No jaws have ever been found, and no indication of predation at all, yet the most common assignment of most of the Ediacarans is to the phylum Cnidaria—which are all predators! There have been suggestions that the Ediacarans might have contained microscopic symbiotic algae (dinoflagellates) in large numbers, just as modern corals do. But no proof of this exists. Because of the seeming lack of predators, one of the most memorable descriptions of this long-ago time was that it was the garden of Ediacara, the last time larger life lived in a predator-free world. By 540 million years ago this garden was gone, its serpents being a wide diversity of crawling, swimming, predaceous (and herbivorous) animals.

Why did it take so long for these first mobile animals to evolve on Earth? External environmental factors such as low atmospheric oxygen may have been at fault, or very high temperatures of air and sea. What we do know is that in the time between about 635 and 550 million years ago, a whole new category of organisms had evolved, ones with internal water-filled spaces that could act as an internal or hydrostatic skeleton, as well as creatures with muscles, nerves, specialized sensory cells, germ cells, connective tissue cells, and the ability to secrete precipitated skeletal hard parts. Animals or not, the Ediacarans were the first on Earth to evolve skeletons, albeit nonmineralized. Skeletons allow for the attachment of muscles, and muscles allow locomotion. Locomotion then creates other needs that continue to drive the evolution of ever more complexity. Once moving, an animal needs sensory information to find food and mates, as well as to avoid predators. Sensory information needs a brain to process it. All of these developments were intertwined, and were triumphs of the eukaryotic metazoan revolution, which is really what happened near the end of the Proterozoic.

We can now hypothesize about the appearance of what we might call the "stem metazoan," the single ancestor of all the complex organisms now on Earth. It would have been small, composed of relatively few cells. Internally there would be no cell walls. It would have an epithelium sealed highly against the external environment, but here would be internal cavities filled with collagen, giving stiffness to the organism. It would also have a "genetic toolbox" *allowing* it to increase in size and complexity. Large, ecologically specialized, sexually reproducing, multicellular eukaryotes: these were the organisms producing life's greatest adaptive radiation, resulting in the crawling, writhing, swimming, walking, and sessile animal biodiversity marking today's Earth. Numerically dominant among today's animal kingdom are animals, like us, with bilateral symmetry. In the Early Cambrian, however, these were few in number but aligned to take over the Earth.

PALEOECOLOGY OF THE LARGER EDIACARANS

Generally, science resolves interesting problems easily. But the nature of the Ediacarans seems to have resisted a great deal of vigorous effort. They remain mysterious. However, new work in the past few years has begun to chip away at the biggest mysteries, and some of the most important have utilized a field of paleontology that has fallen into a bit of disrespect over the past few decades—a field known as paleoecology. While a brilliant driver of paleontological research from the 1960s on, it failed to yield major new generalizations and was cogently dismissed by Stephen Jay Gould in one of his State of Paleontology addresses published in the last (twentieth) century. But this old-fashioned kind of sleuthing was used in this century by Mary Droser of the University of California, Riverside, and Jim Gehling of the South Australian Museum to arrive at perhaps the best understanding yet about the larger Ediacarans and their world.

The crux of the Gehling and Droser's (and vice versa) work is that we need to look at the Ediacarans in the context of their association with what must have been pervasive microbial mats lining the sea bottom. The profusion of microbial mats would have been the dominating control on the ecology and especially the sedimentology of these communities. Because there were few or no burrowing organisms compared to the sea bottoms of today where burrowing is pervasive, the ecology of these communities would have been nothing like we know.

Four kinds of animal lifestyles that associate with the microbial mats could have been present: mat encrusters, forms that sat upon the mats and perhaps secreted digestive enzymes sufficient to dissolve the mats upon which they rested for food; mat scratchers, forms actually and actively grazing on the mats; mat stickers, partly emerged in the mats and growing upward as the mat level changed (because the mats would have grown upward toward the sun, as stromatolites do); and undermat miners, tunneling beneath the mat. Several of these strategies also seem to have persisted into the earliest Cambrian, but by then the world was rapidly changing because of the profusion of burrowing and larger organisms as well as active carnivores and herbivores with skeletal or hard jaw apparatuses.

This very strange and weird world of organisms can also be understood only in the context of how they were preserved. One of the interesting generalizations made by specialists who study Ediacaran organisms is that the fossils are analogous to the plaster "death masks" of previous centuries, used by dead and dying royalty and nobility of European and other civilizations. Soon after death, the face of some (then) famous person would have an impression made. The fossils we see of Ediacarans may be the same thing. Not an actual fossil from the animal, but a reproduction of the top and bottom surfaces of the creature. Making a death mask required rapid hardening of whatever material the mask was being made of, and so it is thought that the Ediacaran fossils were made of materials that hardened quickly on top of their dead bodies.

THE SPINY MICROFOSSILS OF THE EDIACARAN WORLD

In the last chapter we mentioned the work of Andy Knoll and his group at Harvard, concerning their study not of the bigger fossils of Ediacaran age, but of the microfossils. For a billion years, single-celled life dominated the world, and what fossils they left were mainly tiny, smooth-walled globes. But as the world came out of the last of the Neoproterozoic snowball Earth episodes, the fossil record becomes filled with spiny, ornamented microfossils. This somewhat short-lived episode in the fossil record may tell us important things about the nature of the overall rise of animals to complexity (these microfossils appear no more than 600 million years ago, and then are gone about 560 million years ago, and were thus survived by another twenty million years by the larger Ediacaran macrofossils). Prior to this point microfossils came exclusively from single-celled organisms, but these "spiny" microfossils may in fact be from multicellular animals. In these cases, we are seeing tiny resting stages, like cysts.

There have been several important studies of these tiny fossils, including by paleontologists/developmental biologists Nick Butterfield and Kevin Peterson,[14] who suggested that the appearance of the heavily ornamented microfossils early in the Ediacaran period was in response to small early animal carnivores, such as the earliest tiny nematodes and roundworms. The spines of the microfossils were thus defensive adaptations, serving to buttress the skeletons of these fossils, interpreted to have been single celled. But the Knoll group suggests that the complexly ornamented microfossils are resting stages of early animals themselves. This suggests both a complex and early evolution for animals well before the larger Ediacaran fossils appeared at all. It also suggests that the early environments of animals were anything *but* like the Eden-like garden of Ediacara supposed by paleontologists in the late 1900s. Needing a resting cyst suggests a challenging environment of varying oxygen, including times when the water column had no oxygen at all, and possibly occasional doses of hydrogen sulfide. This view of life poses a world faced by early animal evolution that was challenging, extreme, and often poisonous.

The spiny microfossils disappear around 560 million years ago, and are then replaced by what was the flowering of the large, classical Ediacaran fossils, which themselves lived as the largest creatures on Earth until overthrown by a different set of animals at the base of the Cambrian period, slightly more than 540 million years ago.

THE SEARCH FOR "BILATERIANS"

If the spiny microfossils are small animal resting stages, rather than large protists (single-celled organisms) of some kind, what kind of animals were they? About the same time that the ornamented microfossils appear in the geological record, it is supposed that another great evolutionary event took place: the first animals with bilateral symmetry, something that greatly improved locomotion. The advent of a bilaterally symmetrical body plan was another great evolutionary milestone. A bilaterally symmetrical animal has a distinct "front" and "back," with internal organs roughly symmetrical on either side of this front to back, tubelike body. It was the kind of ancestor we would expect the diversifying animal phyla to have sprung from. But the age of these enigmatic fossils was long debated.

Genetic work suggests that this ancestor should have been alive well between ~570 million and ~660 million years ago.[15] But the fossil record has been opaque to what must surely have been a tiny (perhaps a millimeter in length), wormlike creature without a skeleton. While this is a case deserving not a little of the scorn heaped on it ever since Darwin, the fossil record should be cut some slack: the chance of a tiny, soft, wormlike creature without hard parts leaving any fossil record of itself is low indeed.

Fossils from China came to the rescue.[16] Rocks of an age considered to be the best guess of when the first bilaterian may have lived were found in China early in the twenty-first century. These rocks were then slowly and laboriously dated with higher precision, so that a very specific time interval, when it is thought that bilaterians must have first appeared, was identified. When this was completed, the search for the theorized fossils began. None of it was easy.

It took three years and the completion of more than ten thousand individual "thin sections" (in which a hunk of rock is sliced to a very narrow width and then polished, so that light can be transmitted through it while on a microscope stage), and just such an animal was found. And it was much smaller than an eighth of an inch: tiny fossils that were as long as a human hair is wide were found, examined, and studied. The age of these tiny wonders, named *Vernanimalcula*, was nearly 600 million years old.

Here again is a missing link no longer missing. Small, unassuming, and true revolutionaries, these early bilaterians paved the way for what was to come. And there was more from these strata. In addition to the bilaterian fossils, the Doushantuo Formation of southwest China yielded both eggs and embryos of earliest animals. It has also given us a new window into the world of 600 million years ago, and how animals changed the very nature of the sedimentological record.

Prior to animals, there was no "bioturbation," the disruption of newly accumulated sedimentary layers by the action of organisms. This is so pervasive today that it is hard to envision a time before it was the rule rather than the exception. Only strange environments today have this preanimal mode of sedimentological preservation, such as the bottom of the Black Sea. There the bottom is firm, and the sediments for the first meter below this surface show both lamination (layering) and a very low water content. Contrast this with any modern oxygenated sea bottom: the few centimeters above the bottom's substrate is filled with organic goo—mucus, feces, pseudofeces, dissolved organic material, etc. Going deeper you find a lack of lamination; all has been burrowed and consumed, over and over. The slow-moving invertebrates are either feeding while moving (sediment in, sediment-rich feces out) or escaping and leaving locomotion burrows. A significant thickness of the bottom sediment has a high water content because of all the business of all the locomotor animals.

As changes go, this one was huge. In the late twentieth century it was dubbed the "agronomic revolution," and it is the main characteristic feature of the Proterozoic and Phanerozoic sea bottoms—and the stratigraphic records they left behind.[17] The new bilaterians were moving, and not just atop the sediment-water interface they were increasingly colonizing. A vertical component to burrowing also began. Our own take on this is that could not have taken place without high levels of oxygen in the sea: oxygenation is difficult at best when burrowing through sediment, and surely would have been impossible at global oxygen levels less than, let's say, 10 percent. The old view is that the newly evolved animals increasingly ate the stromatolites and microbial mats out of existence near the Proterozoic-Cambrian boundary. The new view is that the tiny bilaterians were not just eating the nutrient-rich microbial mats: they were also making the firm substrates required for the mats to change from ubiquitous to virtually nonexistent.

By the latest Proterozoic time, the world was primed for animals. The genetic "toolboxes" necessary for the evolution of larger sizes, skeletons, and the many kinds of tissues necessary for activity were in place. Only one thing was lacking: oxygen. After the last snowball 635 million years ago, animals were poised, but oxygen levels were too low. Yet by (approximately) 550 million years ago, that had changed: oxygen levels had risen.

Making oxygen levels rise permanently requires increasing the fraction of organic carbon buried in the sediments, rather than buried as limestone. Most organic carbon is sequestered by clays eroding from the continents, so any factors that increase the clay flux—particularly in the tropical oceans where productivity is highest—will notch up the atmospheric oxygen level. One suggestion is that the rise of a terrestrial biosphere of some sort may have increased the production of clays through weathering,[18] which is certainly true after vascular land plants evolved the ability to make deeply penetrating roots. However, shifts in the position of continents relative to the equator also have a big effect, as physiochemical weathering is much higher in the warm tropics than at the cold poles. Near the beginning of Cryogenian time (but before the snowballs, at about 800 million years ago) there was a stepwise change in the carbon cycle that lasted about 15 million years, during which time the fraction of organic carbon being buried dropped precipitously. This Bitter Springs event was first discovered in Central Australia, and has since been found at many other sites around the globe. It presumably caused a transient drop in surface oxygen. The cause of this event was a mystery until a group led by Adam Maloof at Princeton University discovered that the onset and termination of this event coincided with a pair of very rapid ~60° oscillatory shifts in Earth's rotation axis (from an unpronounceable package of rock in Svalbard called the Akademikerbreen Group).[19] This type of shift is termed a "true polar wander event" (discussed at length below), and involves the geologically rapid motion of the entire solid Earth, right down to the liquid metal at the core-mantle boundary. These particular shifts, however, moved a large chunk of the supercontinent of Rodinia off of the equator into the mid-latitudes, and then back again, fluctuating carbon burial and oxygen production in sync. When paleomagnetic and geochemical data from vastly different parts of the globe show the same shifts, at the same time, and in sync, we might learn something about how the planet works. In this case it is oxygen. We now think that there might be as many as thirty of these TPW events during the past 3 billion years,[20] many of which coincide with interesting events like the Cambrian explosion.

CHAPTER VIII

The Cambrian Explosion: 600–500 MA

PHOTOGRAPHS of Charles Darwin when he had reached seventy years old seem to show a man weathered well beyond this chronological age. This seems a man perhaps eighty or older. Yet at seventy, Darwin was in his last years, and perhaps this physical antiquity was a product of stress as well as the diseases he may have contracted in the tropics when, as a young man on HMS *Beagle*, he slowly circumnavigated the globe. Perhaps it was being so vexed by his many critics, as well as his own distress at his inability to understand how organisms inherited traits—genetics was not accepted until the early twentieth century when the work of Gregor Mendel was "rediscovered"—and especially the nature of the Cambrian explosion must have taken their emotional as well as physical toll. Darwin hated the fossil record in general and the Cambrian in particular. A vexation with the Cambrian fossil record followed him to his grave. It was this, and his inability to know how genetics worked, that surely were among his greatest regrets.

Since well before the time of Darwin it was known that fossils of animal life seemed to appear suddenly in the fossil record: the great English geologist Adam Sedgwick, the author and definer of the Cambrian period itself, mapped its base as the strata containing the first trilobites. While we now first think of the various geological periods as time, in fact they came into existence as a succession of strata, with a bottom bed defined by some fossil first appearing, and its top defined either by the extinction of some fossil, or better yet, a different species first appearing. In this case this was the Cambrian System, based on piles of strata in Wales. The Cambrian period is the time during which these Cambrian strata accumulated—no more, no less.

Sedgwick found that over short stratal intervals, sedimentary rocks seemingly bereft of fossils were found to be overlain by rocks with a profusion of highly visible fossils, the most common being trilobites. Trilobites are fossil arthropods, and thus their fossils are the remains of highly evolved and complex animals. This observation was vexing to Darwin (and hugely comforting to his critics), as it seemed to fly in the face of the then newly proposed theory of evolution.[1]

Charles Darwin thus went to his grave cursing the fossil record. His genius was such that he knew he was right, yet till the end of his life, he was bedeviled by critics who pointed out that the "first" life on Earth was of such complexity that it was inconceivable that the evolutionary processes so eloquently argued by Darwin in the many editions of his great work *On the Origin of Species* could have produced such complications as—a trilobite. Yet the great irony is that trilobites did not appear until the Cambrian was at least half over.[2]

One of the iconic fossils, trilobites were arthropods that dominated oceanic habitats relatively early in the history of animals on Earth. But how early? In Darwin's time, trilobites were thought to be the earliest of all animals. Yet they are undoubtedly complicated, with three body sections, complex eyes and limbs—and large size. Some of the earliest could be up to two feet long. This was not what the earliest animals *ought* to have looked like—small and generalized, not large, complicated kinds of animals. We now know that trilobites were not—not even close, in fact, to being—the first animals.[3]

The history of the origin of animals on Earth is one of life's most fascinating chapters, and also one of the most controversial. There is also a great deal of new information that has been gleaned even in the last ten years. There are two distinct lines of evidence giving quite different views on the timing of the first diversification of animal phyla. One of these lines comes from the pattern of appearance of animal fossils in rocks, the second from molecular clock studies on extant animals. They give important clues to one of the greatest of all paleontological mysteries: the rapid diversification of animals.

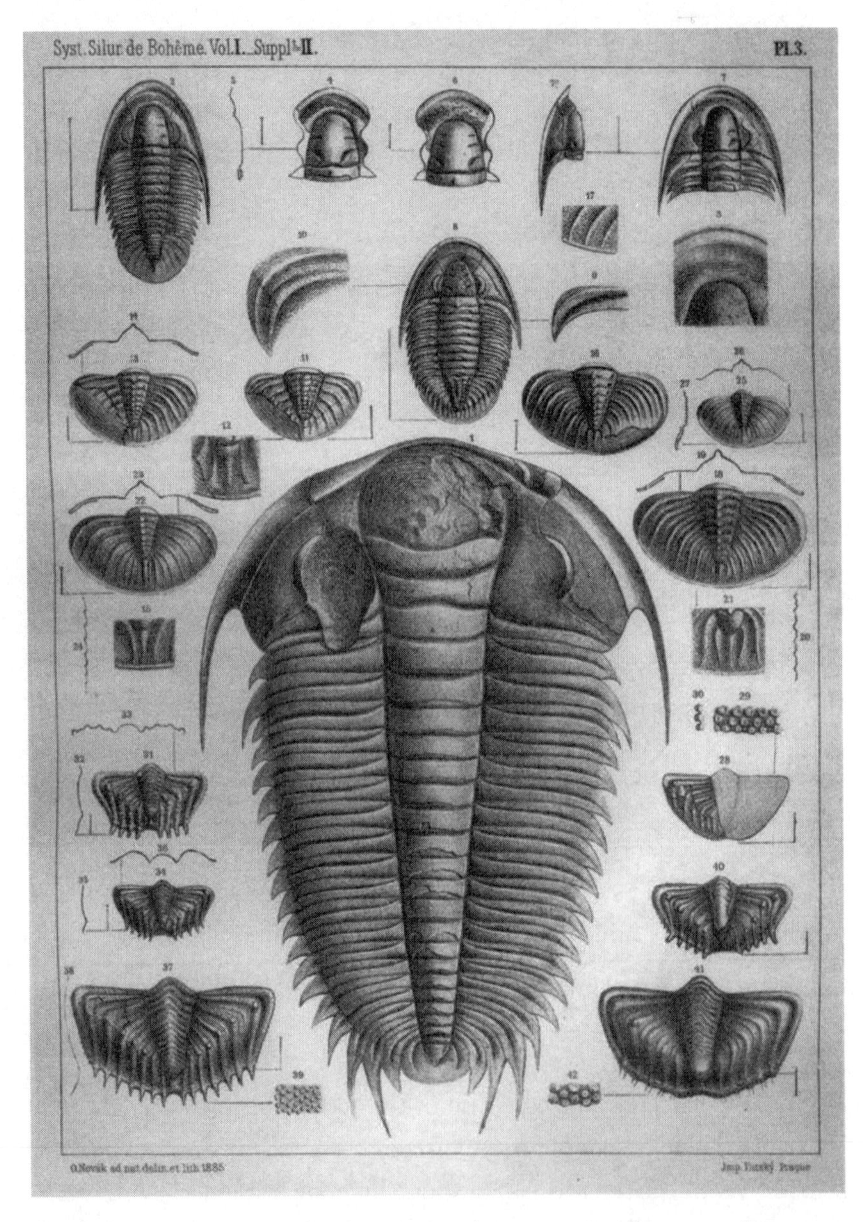

Illustration of trilobites from the nineteenth century. At that time, these were thought to be the oldest fossils on Earth. Trilobites were used to "mark" the start of the Cambrian Period.

The first major line of evidence about the Cambrian explosion comes from fossils. The appearance of animals leaving evidence of themselves in the rock record came in four successive waves. The first began around 575 million years ago and has been called the Avalon explosion, a name coming from the part of eastern Canada where the oldest of this group were found. The second wave is coincident with the almost complete disappearance of the Ediacarans, and is characterized not by actual fossils but with accurate traces of their locomotion. These numerous "trace fossils" could only have been formed by active locomotion of multicellular organisms—animals. These are as old as 560 million years, but most are about 550 million years in age. The sea bottoms would have been alive with actively moving, small wormlike forms.[4]

The third breakthrough was the appearance of skeletons, great numbers of tiny skeletal elements, in strata less than 550 million years in age. They are very small spines and scales of calcium carbonate that would have covered the animals with a coating of these small skeletons, almost like tiles. Finally, the larger fossilized animals appeared, including trilobites, the clam-like brachiopods, spiny echinoderms, and many kinds of snail-like mollusks, all in strata younger than 530 million years in age. In Darwin's day, none of the earlier three were known, and the Cambrian was marked by the first appearance of trilobites in sedimentary strata. The reasons for this sequence might be deceptively simple: oxygen levels, which rose to their highest levels of the world up until then.

Today we know that this succession of animal life originations appeared comparatively rapidly in the fossil record, and new dating techniques now puts the time of the first complex *fossils* (the small skeletal fossils, which are 20 to 10 million years younger than the first trace fossils) at slightly older than 540 million years ago, with the first trilobites appearing in the record some 20 million years after that.

The appearance of animals in the fossil record recorded a significant event, which has been called the Cambrian explosion. To paleontologists the Cambrian explosion marked the first appearance

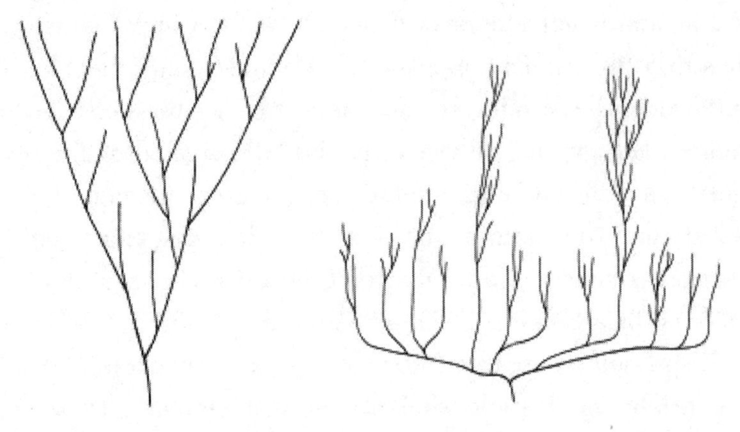

Left, the cone shape represents the traditional model for increasing disparity. Right, the inverted cone represents diversification and decimation.

of most major animal phyla large enough to leave remains in the rock record. To molecular geneticists, the Cambrian marked the first evolution of animals. The controversy raged through the 1990s, to be solved in the early years of this century when new molecular studies[5] using more sophisticated analyses essentially confirmed the younger date for the origin of animals that had been championed by paleontologists. There is now agreement that animal life on Earth did not predate 635 million years ago,[6] and might be closer to 550 million years in age.

The Cambrian period is now dated from 542 to about 495 million years ago (although the latter date, for the base of the Ordovician, might be slightly older). However, the vast majority of animal phyla first appeared in a small portion of this interval, between 530 and 520 MA. All specialists agree that this is the third or fourth most important event in the entire history of life, superseded in importance only by the first appearance of life on Earth, the adaptation to molecular oxygen, and the origin of the eukaryotic cell.[7]

According to our best new information, the oxygen level soon after the start of the Cambrian explosion was about 13 percent (compared to 21 percent today),[8] but then fluctuated. During this time carbon dioxide levels were far higher than they are in the world today—hundreds of times higher, in fact, and such high levels would have produced an intense greenhouse effect, sufficiently high to overcome the fact that the sun at this time was ~5 percent less intense than it is today. Even with the drop in CO_2 levels at the end of this interval, temperatures of this time would have been perhaps the highest of any period in the history of animal life on Earth. Since less oxygen is dissolved in seawater with higher temperatures, the already anoxic conditions of the oceans would have been exacerbated.

The panoply of fossils that have been preserved showing both hard and their soft-parts fossils from the fantastic and newly discovered deposits in the Chengjiang region of China has given us a new window into the origin of the animal phyla on Earth, and the nature of life on the Cambrian planet prior to the most famous of all fossil deposits, the Burgess Shale of British Columbia. The Chengjiang beds are now known to have been deposited between 520 and 515 million years ago, whereas the Burgess Shale is now thought to be no older than 505 million years in age. The approximately 10 million years separating the age of these two deposits thus gives us a new view of how animals diversified.

Because both Chengjiang and the Burgess preserve soft parts as well as skeletonized animals,[9] we have a good picture of what was there, in what relative abundance. Without this added view yielded by the preservation of soft parts, we would never be sure about the relative abundance of various kinds of animals, for perhaps there was a huge abundance of creatures like soft worms and jellyfish, forms that did not have skeletons. Thus our surprise at what appears to be a clear view of the nature of the fauna at both sites. There have now been over fifty thousand fossils collected from the Burgess Shale (and a lesser number of from Chengjiang). In their masterful summary of the Burgess fauna, Derek Briggs, Doug Erwin, and Fred Collier (in their 1994 book *The Fossils of the Burgess Shale*[10]) list a total of 150 species of animals. Almost half are arthropods or arthropod-like. But an even more interesting number relates to the number of individuals. Well over 90 percent of all fossils are from arthropods, followed by sponges and brachiopods. Like the earlier Chengjiang, the Burgess sea bottom was dominated both in kinds and numbers of animals by the arthropods.

Arthropods are among the most complex of all invertebrates, and yet, in these almost earliest of fossil deposits in the time of animals, they are diversified and common. It speaks to a long evolution prior to their first appearance in the record—perhaps seabeds crawling with millimeter-long (or less) arthropods, with many more species swimming or floating in the open sea itself.

One of the great surprises of a visit to the Burgess Shale (which both of the authors of this book have been fortunate enough to do) is the realization that the most common fossils come not from the exotic taxa, the many exquisite, soft-bodied creatures that fill the pages of the many books devoted to the Burgess Shale fauna and flora, but the fact that most of the fossils come from trilobites. They, and the less numerous but highly diverse arthropods of the Burgess dominate the assemblage,[11] in sheer numbers of individuals and species, and in sheer numbers of different kinds of body plans, which is described by a measure called disparity (and compared to diversity, which refers to the number of different kinds of taxa). The arthropods seem to have been the most successful of Cambrian animals. How much of this success was due to their principal body plan characteristic: segmentation?

Segmented animals are the most diverse of all animals on the planet, and most are arthropods. All arthropods, including the highly diverse insects, show repeated body units and body regions based on groupings of individual segments that have specific functions for the animal. The feature uniting the group is the presence of a jointed exoskeleton that encloses the entire body. This exoskeleton even extends into the gut. The exoskeleton cannot grow, so it must be periodically molted and replaced by another slightly larger one. The body has a well-differentiated head, trunk, and posterior regions in varying proportions. Appendages are commonly specialized. On terrestrial arthropods the appendages are usually single (enormous), but the marine forms generally have two branches or parts per appendage, an inner leg branch and an outer gill branch, and are thus termed biramous. The exoskeleton encloses the soft parts like a suit of armor, and that may be its major function: protection. But the consequences of this kind of skeleton are huge: there can be no passive diffusion of oxygen across any part of the body. To obtain oxygen the first arthropods, all marine, had to evolve specialized respiratory structures or gills. Segmented animals are the most diverse of all animals on the planet. Arthropods are not alone in this trait: all annelids are segmented, and some members of generally nonsegmented groups, such as the monoplacophoran mollusks, show at least some segmentation. It appeared early in the history of animals, and indeed in the Cambrian trilobites we see that the most common of these early preserved animal fossils show this trait.

In his 2004 book, *On the Origin of Phyla*,[12] James Valentine also reflects on what is a major evolutionary puzzle: why are there so many, and so many kinds of arthropods in the Cambrian? It is worthwhile to look at what he has written on this subject:

> Although many early arthropods had non-mineralized cuticles, a marvelous diversity of early arthropod body types has come to light, so many and so distinctive as to pose important problems in applying the principles of systematics. These disparate arthropod types are phylogenetically puzzling . . . This evidently sudden burst of evolution of arthropod-like body types is outstanding even among the Cambrian Explosion taxa.

What we call arthropods are composed of what appear to be perhaps many independently evolved groups that have, through convergent evolution, produced body plans of great diversity save for one aspect: all have limbs on each segment that are biramous—each appendage carries a leg of some sort, and a second appendage, a long gill.

Why would basal animal groups opt for segmentation? Perhaps this is the wrong word, for Valentine and others note that the arthropods are not so much segmented—which at least in annelids is composed of largely separated chambers for each segment of the body—but repeated. Valentine proposes that this striking body plan arose in response to locomotor needs, stating, "Cleary, the segmented nature of the arthropod's body is related to the mechanics of body movement, particularly to locomotion, with nerve and blood supplies in support." There is no doubt that this type of body plan is an adaptation aiding locomotion. But a consequence of this kind of body plan is to allow repeated gill segments, each small enough to be held in optimal orientation beneath the segments. In these positions, flows of water can be actively pumped over and through the feather-shaped gills, thereby increasing the availability of oxygen molecules hitting the gills each second, a position suggested by Ward in 2006.[13]

Another animal found in abundance in the oldest of the Cambrian-aged deposits are sponges. Like the cnidarians, sponges show no respiratory structures, nor would we expect any. With a body plan built around a series of sacs (like the cnidarians, but with even less organization: there are no true tissues in a sponge), all sponges show a very high surface area to volume. In fact, sponges are like agglomerations of numerous single-celled organisms, with each cell essentially in contact with seawater. But even with this advantage, sponges show an even more efficient way of gaining oxygen. Their main feeding cell, called a choanocyte, causes large volumes of water to pass through the structure. Some sponge specialists have suggested that a sponge passes as much as ten thousand times its volume in seawater through its body each day. Consequently, sponges are capable of living in extremely low oxygen conditions because they so efficiently move large volumes of water through their body, getting enough oxygen even from water that has little.

The major groups of animals with hard parts in the Cambrian are obviously the huge tribe of arthropods, followed in numerical importance (in most Cambrian marine strata) by brachiopods, and then smaller numbers of echinoderms and mollusks. Brachiopods are a still-living group related to bryozoans that are routinely mistaken for bivalve mollusks. Yet while the shells of bivalves and brachiopods show a superficial similarity, the internal anatomy of the two groups are radically different. The major feature of a brachiopod is a feeding organ known as a lophophore, composed of a large loop with numerous long, thin fingers producing a delicate fanlike shape within the shell. This organ filters seawater for food—and as it is filled with a body fluid, and is very thin, it serves also as an exquisite respiratory organ. For some of us, the brachiopods are a tragic group. Perhaps the most common inhabitants of Paleozoic sea bottoms, they were nearly wiped out by the Permian extinction ~250 million year ago, and never regained dominance.

Cambrian echinoderms make up a weird assemblage of small boxlike animals. Among the earliest echinoderms were peculiar, pinecone-shaped helioplacoids, with some primitive stalked eocrinoids and edrioasteroids found in some deposits as well. More common than echinoderms were mollusks. Most during the Cambrian were small in size, and each of the major classes (gastropods, bivalves, and cephalopods) is found in Cambrian strata. The most common mollusks, however, were monoplacophorans, a minor class today, but common in the Cambrian. They had a limpet-like shell and a snail-like body with a broad, creeping foot. Most interestingly, alone among mollusks of the time they showed a body organization that suggests segmentation. From looking at muscle scars on the fossil shells and comparing anatomy from the still-living forms, we think the Cambrian monoplacophorans had multiple gills. Modern-day gastropods have a single pair of gills, or even just a single gill. But the Cambrian monoplacophorans, which lived a very snail-like existence in all likelihood, found it necessary to have multiple gills. They are celebrated as the ancestral mollusk that would give rise to all the rest: the gastropods, cephalopods, bivalves, chitons, and more minor molluscan classes.

Long thought to have gone extinct at the end of the Permian, the discovery of living monoplacophorans in deep-sea settings in the 1950s led to a much greater understanding of the life of the early mollusks. The living forms confirmed what muscle scars found on the interior of the earliest monoplacophoran fossils asserted—that there was more than a single pair of gills. In fact, multiple pairs of muscles line the entire length of the interior of the shell, leading to the conclusion that these early forms showed an evident segmentation or at least repeat of the gill–blood vessel system. Since it is only the gills (and supporting blood and filtering systems) that show this repeated pattern, it can be surmised that as in arthropods, this repeated pattern was an adaptation for increased respiratory surface area of the gills. A somewhat similar pattern of repetition, extending even to the shell, is found in the chitons, today commonly found on intertidal beaches.

Like the body of an echinoderm, the interior of a brachiopod shell is almost all water. There is very little flesh, and what is there stays in contact with a steady flow of seawater. The brachiopod lophophore creates several currents of seawater that pass into the sides of the shell, move across the lophophore, and are then sent out the front of the shell. This constant stream of new water entering a brachiopod has the same effect as the current passing through a sponge. The small volume of flesh to the great surface area of the lophophore, coupled with the steady flow of water (many times the volume of the interior of the shell), makes the brachiopod consummately adapted for a world of low oxygen.

PHYSICAL AND CHEMICAL EVENTS CAUSING THE CAMBRIAN EXPLOSION

Earlier in this book we noted the advance of entirely new disciplines of science, most notably astrobiology and its allied field, geobiology. But another field, this one a traditional mainstay of the biological sciences, mainly evolutionary development, has undergone a renaissance so important that it can almost be considered a new field as well. Its practitioners now call it evo-devo, and breakthroughs in this field have had a lot to say about the Cambrian explosion in the last decade. One of the greatest of evo-devo practitioners, Sean Carroll, has given us an exquisite tour of this newly revivified area of science in his 2005 book *Endless Forms Most Beautiful*.[14] If there is any single theme in this work, it is that science can now understand far better one of the previously intractable problems in evolutionary biology: the origin of novelty. How evolutionary innovation took place over relatively short periods of time just could not be explained by traditional Darwinian concepts of evolution. The radical breakthroughs—be it the appearance of wings, legs for land, segmentation in arthropods, or even large size, the hallmark of the Cambrian explosion—could not stand up to stories about many and sudden mutations all working in concert to somehow radically change an organism. Evo-devo now seems to have solved this, and in his book, Carroll lists four aspects that combined can explain sudden evolutionary innovation that nicely encapsulates the new way of explaining how radical changes did take place.

The first "secret to innovation," as Carroll puts it, is to "work with what is already present." The concept that "nature works as a tinkerer" is central to this. Innovation does not always need a new set of equipment to build, or even a new set of tools. What is already present is the easiest route. Second and third are two aspects understood by Darwin himself: multifunctionality and redundancy.

Multifunctionality first is using an already present morphology or physiology to take over some second function in addition to that for which it was first evolved. Redundancy, on the other hand, is when some structure is composed of several parts that complete some function. If one of these can be then co-opted for some new kind of job, while the remaining parts are still able to function as before, there is in place a clear path for innovation that is far easier to use than the total de novo formation of some entirely novel morphology from scratch. Cephalopod swimming and respiration are like this. Cephalopods routinely pump huge quantities of water over their gills, and like many invertebrates used separated "tubes" or designated channels for water coming in and water being expelled, to ensure that oxygen-rich water is not rebreathed. But with minor morphological "tinkering" with this excurrent tube, a powerful new means of locomotion came about. Breathing and moving could now take place using the same amount of energy by utilizing the same volume of water for respiration and movement.

The final secret is modularity. Animals built of segments, such as the arthropods, and to a lesser extent we vertebrates, are already composed of modules. The limbs branching off arthropod segments have been amazingly modified into feeding, mating, and locomotion, as well as many other functions. Arthropods are like a Swiss army knife, with each segment bearing limbs evolved to do a very specific function. The same is true in vertebrates with our digits, which have been modified to tasks as varied as walking on land to swimming to flying in the air. Not bad for some primitive fingers and toes! Where does the evo-devo come into play? It turns out that these morphologies are the soft putty for morphological change because they are underlain by systems of genetic "switches," geographically located on the developing embryo in the same positions as the various limbs are found in the arthropod—or vertebrate.

Switches are the key here; they tell various parts of the body when and where to grow. One of the great discoveries is that the exact sequence of different body regions on an arthropod from its head to midregion to abdomen are lined up first on chromosomes in the same geographic pattern, and then on the developing embryo itself. Much of this is done by the crown jewels of the evo-devo kingdom: the Hox genes, and their differently named but equivalents in other taxonomic groups.

The many new discoveries of evo-devo have certainly been brought to bear on the many questions to be solved about that central mystery in the history of life, the Cambrian explosion, and the most important understandings of all: the timing of when and how the various animal phyla and thus separate body plans that we see today originated.

There have long been two schools of thought. The first is that the fossil record gives us a true picture of when the great differentiation of animals actually took place—phyletic divergence somewhere about 550 to perhaps 600 million years ago. But the second line of evidence comes from comparing genes of extant members of the ancient phyla, and using the concept of the "molecular clock," mentioned earlier. At issue is when the most fundamental divisions in the animal kingdom take place—the split between an aggregate of phyla called protostomes and those called deuterostomes. These two groups are separated by fundamental anatomical and developmental differences in embryos.

The protostomes are composed of the arthropods, mollusks, and annelids among others, and they are characterized by embryos that as they develop and grow following fertilization form a mouth out of a central opening in the growing larva called the blastopore. In deuterostomes (echinoderms, us vertebrates, and a number of minor phyla), the mouth and the blastopore remain separate. There is a third group, the very primitive phyla that split off from the main stem of animal evolution prior to the great protostome-deuterostomes split: these include the Cnidaria, sponges, and other jellyfish-like minor phyla.

The first to appear were the simplest forms, the cnidarians and sponges, which appear to be represented, as we have seen, in the Ediacaran assemblages of as much as 570 million years ago, the time interval before the Cambrian period (which began at 542 million years ago). But recognizable protostomes and deuterostomes are not seen until a short interval into the Cambrian period itself.

If the protostomes and deuterostomes split, what was the last animal before that split like? Many lines of evidence indicate that this creature was bilaterally symmetrical and was capable of locomotion. Many who ponder this time and its animals imagine this last common ancestor of both the protostomes and deuterostomes as a small featureless worm, perhaps like the modern-day *Planaria*, or the tiny and extant nematodes. But one of the great new discoveries is that this last member of the as yet undivided stock already had a genetic tool kit allowing it to begin some radical new engineering—and had such a tool kit for at least 50 million years before it was put into use! This worm would have had a mouth at front, anus at the rear, and a long tubelike digestive system in between. It may have had stubby projections sticking out of its side, perhaps for sensory information (touch and chemical sensing?). But the point is that all of this was set up in such a way that rapid transformation could—and did—take place. This is new. All the tools and features necessary for the Cambrian explosion sat around for 50 million years.

As noted above, the base of the Cambrian is dated now at 542 million years ago. The base of the period has been defined as the place in rock where the first identifiable locomotion marks are found in strata—a certain kind of trace fossil showing that animals, moving animals, were present and could make vertical burrows in the mud.

Yet for the next 15 million years, there seems to have been little formation of new body plans at all—or at least that we can find evidence of in the fossil record. The first real indication that a great diversification was taking place comes from the spectacular fossil beds only recently discovered in Chengjiang, China,[15] dated as 520 to 525 million years in age and mentioned above. It is an older version of the Burgess Shale in having common preservation of soft parts.

Both the Chengjiang and Burgess Shale faunas are dominated by arthropods—lots and lots of different kinds of arthropods. They soon became the most diverse animals on Earth—and have stayed that way ever since. There are some estimates that in our modern day, there may be as many as 30 million separate species of beetles alone!

Evo-devo tells us why. Of all the body plans, none can be so easily, quickly, and radically changed as arthropods. The reasons are just those listed above by Carroll: arthropods have modular parts, they have redundant morphologies that can be co-opted for new functions, and they have a series of Hox genes that allow ready transformation of specific regions in the overall body plan of segments throughout.

The old view has been that new animals mean that there must have been new genes coming into existence. There is sound logic in this. Surely a primitive sponge or jellyfish would have fewer genes than the more complex arthropods: it was argued that the common ancestor of all arthropod groups somehow added new genes—new Hox genes, as these are those that are the "switches" that tell the various parts of a body how to form and when. But such is not the case. Carroll and others showed that the last common ancestor of the arthropods did not evolve new genes; it already had them, and that the subsequent and amazing diversification of so many kinds of arthropods was done with existing genes. As Carroll put it: "The evolution of forms is not so much about what genes you have, but about how you used them."

Ten different Hox genes were all that were necessary to utterly change and diversify the arthropods. Their secret was discovered by comparing the distribution of the product of Hox genes—proteins that are specific to a particular Hox gene—and where these proteins can be found on a developing embryo. The old idea that some gene or genes of an arthropod coded for the construction of a leg is false. The Hox genes make proteins. These proteins then become the means of starting and stopping the growth of particular regions of a developing embryo. Some of these proteins are concerned with making specific kinds of appendages. If those Hox gene proteins are somehow moved to different geographic regions on the developing embryo, the product that is produced will move as well. In this way a leg that was formerly in one part of the body might suddenly be found in a totally new place—if, however, the Hox gene protein was somehow moved to the corresponding place on the embryo long before the leg was formed. Innovation came from shifting the geographic places or "zones" on an embryo that a specific Hox gene protein could be found in.

Shifting the Hox gene zones in arthropod embryos resulted in the many different kinds of arthropods that we see. There are thousands, perhaps millions of different kinds of arthropod morphologies—and all of this was evolved using the same tool kit of ten genes. Arthropods are nothing if not body plans with repetitive parts. The specialization of these parts requires that each falls into a separate Hox gene zone.

STEPHEN GOULD VS. SIMON CONWAY MORRIS: THE SHAPE OF DISPARITY

There has been no end of ideas about why there was a Cambrian explosion at all. Sometimes events of the past seem as if they could not have been otherwise. Yet why not a long slow formation of the many animal phyla, instead of the seemingly compressed duration that we do see? And just how diverse were the major animal players in the Cambrian explosion? All of the current animal phyla (variably listed as about thirty-two) first appeared in the Cambrian explosion. Surprisingly, there has not been a single animal phylum added to the world since, even after the devastating Permian extinction of 252 MA. But were there many *more* phyla in the Cambrian than now? Were there strange, fundamentally different kinds of animals in the Cambrian than now? That has been a very contentious issue, culminated in a late 1990s feud[16] of memorable bile between the late great evolutionist Stephen

Jay Gould and Cambridge University's Simon Conway Morris, who remains, essentially, Britain's paleontologist laureate.

In his *Wonderful Life*, Gould asserted that the Cambrian was full of "weird wonders," which he defined as body plans now longer present on the Earth. His view is that the Cambrian explosion was just that—an explosion of new body types, body plans, numbers of species. But to slightly mix metaphors, most explosions are deadly. In fact many of the new kinds of body plans—in Gould's view, new kinds of phyla—did not make it out of the Cambrian. Killed by the explosion, but not in the original sense. The effects of the vast increase in kinds of animals killed them by competition. With so many body plans, only some would stand the test of natural selection. Gould's view is that the diversification of body plans can be modeled by a pyramidal shape; the great diversification of body plans was fast, creating a fat base of the pyramid of numbers of body plans—also known as disparity (the diversity of body plans, not species). But as the Cambrian progressed, that base diminished, until there were far fewer phyla at the end of the Cambrian than soon after its start.

Many others disagreed that disparity has, in fact, increased since the Cambrian. Simon Conway Morris is the leading proponent of this point of view, one that is in direct contradiction to that of Stephen Gould. In Morris's view, the weird wonders were not separate phyla at all, just early and not yet recognizable members of well-known and still-living phyla. The consensus since this late-twentieth-century argument, one that was heated to unseemly levels between scientists, seems to be that Gould was wrong, and we can add little to this argument. But if this once-boiling scientific dispute has cooled to a low simmer, other aspects of the Cambrian explosion remain frontline science, the best science—controversial science.

NEW DATING OF THE CAMBRIAN EXPLOSION

The Cambrian explosion was obviously one of the most important and until recently least understood of major events in life's history as well. Much of the uncertainty came from dating—or lack thereof, at least in any sort of precision—and the older the rock, the greater the uncertainty. When he first defined the base of the Cambrian as the beds with the first trilobites within them, the early eighteenth century's Adam Sedgwick had no idea that actual age dating in years—rather than the relative appearance of fossils—would ever become available to his brethren (but we are sure he must have dreamed of the possibility). For almost two hundred years, in fact, an accurate date for the base of the Cambrian was a case in point. A major problem was that it had never really been defined either in biological terms or with respect to the actual rock record, and numerically dated calibration points were few and far between. Unlike a mass extinction event or other biological innovations, the Cambrian radiation did not have a specific obvious well-defined starting point. The global definition of the terms was chosen instead by a special committee of international specialists organized by UNESCO, under the auspices of the International Geological Correlation Program (coauthor Kirschvink was a voting member of this committee).

At issue was the actual position of whatever boundary was to be chosen, and how to date it. By the 1960s and 1970s, age guesses (for they were nearly that) for when the Cambrian explosion happened varied from over 600 million years ago to as young as 500 million years ago. It took the development of incredibly sensitive—and precise—radiometric dating techniques before progress could be made. The problem with dating was that in order to obtain a radiometric age date, volcanic rocks had to be interbedded with the sedimentary beds as ashes, for it is only the volcanic ashes—and only some of them—that contain the mineral zircon (which locks in uranium and lead ratios to form beautiful geological clocks). And almost none of these kinds of beds within beds were known from any Cambrian-aged rocks around the globe.

In an attempt to try something else, a prominent Australian geochronologist named William Compston (at the Australian National University in Canberra) developed a technique in the mid-1900s using rubidium-strontium isotopes in shale (a sedimentary, not volcanic rock) that gave age estimates of 610 million years for the first trilobites in China. We now know that his technique was totally wrong, and that techniques based on dating the mineral zircon with the uranium-lead are the way to go. Nevertheless, until the 1980s the "official" date for the base of the Cambrian was listed as 570 million years ago, and that date is occasionally still found in many compilations of the geological time scale online and in books.

But the second problem, not "when" so much as "what"—what first or last fossil occurrence should mark the base of the Cambrian—was more intransigent. As noted above, by the 1960s, paleontologists had improved their collecting methods and instrumentation, and it became increasingly clear that in fact a great deal of animal evolution, including animals with hard parts that could and did fossilize, predated the trilobites by great periods of time. The oldest hard-part fossils in strata beneath those with trilobites were tiny but recognizable parts of shells (the "small shelly fossils"). Some looked like tiny spines, some like small snail shells, some simply chunks of what looked like armor from some archaic mollusk or echinoderm. But at question were their actual ages of formation and existence.

International agreement was finally reached[17] in the early 1990s. Of the four-part appearance of animals known from the fossil record, the first, the Ediacarans, were kicked out of the Cambrian period altogether. Their time received its own name: the newly defined Ediacaran period of the Proterozoic era. The base of the Cambrian System was defined as strata containing the lowest, vertically burrowing trace fossils, thus predating the successive strata with small shelly fossils, which in turn underlay the strata with trilobites. The ability to burrow vertically through sediments is thought to imply the existence of a hydrostatic skeleton and the neuromuscular connections to control it, but this horizon was nearly 20 million years older than the actual Cambrian explosion (as recorded by the fossil record itself). Yet if finally sorted out, the dates when these strata were deposited was still unknown.

Without reliable radiometric dating, the extent of this interval—between the oldest recoverable animal fossils and the first appearance of trilobites—could (in some regions) be measured in tens of thousands of meters of strata between the Ediacarans and trilobites. This suggested that tens of millions of years separated them—but the 1980-era mass spectrometers (the instruments that can determine ages from rocks) needed large numbers of zircons to do the analyses properly.

However, technology advanced, and by the late 1980s, new, better instruments began to be used on the rare but crucial volcanic horizons that occasionally could be found in the sedimentary beds thought to be Cambrian in age. One such locality, discovered long after Sedgwick and all his contemporaries went to the great fossil record in the sky (or wherever paleontologists go), was located in the Anti-Atlas Mountains of Morocco. Here was the potential Rosetta stone for determining the age of the four acts of the Cambrian explosion.

AGE BREAKTHROUGH — AND AGE SURPRISE

It was in the late 1980s that coauthor Kirschvink collected samples of a volcanic ash from the Anti-Atlas Mountains of Morocco. This ash layer was stratigraphically located about fifty meters below the first occurrence of Cambrian trilobites in this great pile of sedimentary strata. But how long did it take for those critical fifty meters of underwater-derived strata to form? Unfortunately this volcanic ash produced only a tiny number of zircon grains, far too little to be dated using techniques that were conventional at that time. However, by that time Compston had developed an incredible instrument known as the super-high-resolution ion microprobe (SHRIMP), which was able to focus a collimated beam of cesium ions onto a small spot on a mineral grain. The plasma generated by this process was fed into a mass spectrometer, and with a few subtle manipulations they were able to produce an extremely high resolution uranium-lead date.

The result was stunning. The dates emerging for these Morocco samples were about 520 million years, rather than being older than 600 million years in age![18] Compston did everything he could to try and make the age older, but it did not work. There was at least an 80-million-year error in the age of the base of the Cambrian. This meant that the Cambrian explosion—at least the massive diversification of the animal phyla that is seen when the first shelly fossils appear—was more like a nuclear explosion, at least twenty-five times faster than supposed. Other groups at MIT (Sam Bowring) and elsewhere have since replicated these findings with additional volcanic ashes from Morocco, as well as others from exotic places like Namibia and the northern part of the Anabar Uplift in Siberia.[19] There was now a date for the appearance of the trilobites, and it was far younger than previously supposed. The paleontologists charged with selecting the formal base panicked when they thought the entire Cambrian would be only 10 million years long, so they abandoned the first trilobites as their guide and chose an older event—the first burrowing trace fossil—that ultimately was calibrated at about 542 MA.

It turns out that this unusual interval of the evolutionary activity and innovation has some other rather unusual features as well. Studies of the carbon isotopes across the Proterozoic-Cambrian boundary show that something rather strange was happening, with huge oscillations that lasted for hundreds of thousands to millions of years (these are now known as the Cambrian carbon cycles).[20] The magnitude of these is wild—the equivalent of grinding up and burning all of the existing biomass on Earth every few million years. Either that or something was causing extremely light carbon (which occurs in methane) to erupt into the atmosphere on a massive scale, with all of the associated greenhouse effects. Did the Earth go through a succession of short-term heating events? Mild heating can actually increase biological diversity by shortening generation times—an effect observed in the modern biota. Too much, of course, can be lethal!

Another oddity is that the Cambrian has long been known as having some extremely large apparent plate motion (plates are the enormous sheets of crust that compose the Earth's surface, and that move, diverge, or collide with other of these Earth tectonic plates). These motions can be tracked using the technique known as paleomagnetism, which can determine ancient latitudes of rocks as well as the directions of plate motions. It was using this tool that coauthor Kirschvink first proved the snowball Earth episodes of previous chapters. New paleomagnetic analyses coming out of multiple paleomagnetism laboratories were showing something seemingly impossible: that the continents were scooting across the surface of the global at great speed—or that the entire globe was rapidly moving under its poles of rotation. The north and south poles were staying where they always were: it was the globe beneath them that was moving.

This information came from samples taken from Australia, for example, indicating that while it straddled the equator, it underwent a nearly seventy-degree clockwise rotation between Early Cambrian and Late Cambrian time—in less than 10 million years, and perhaps much less time than this. However, because Australia was a part of the supercontinent of Gondwana, which included Antarctica, Greater India, Madagascar, Africa, and South America, this rotation must have involved well over half of the continental landmass at the time. Now data from virtually all over Gondwana tell a similar story—it was spinning counterclockwise precisely during the Cambrian explosion interval, 530 to 520 million years ago. Similar results from the large North American continent called Laurentia indicated that it moved from the chilly South Pole all the way up to the equator at essentially the same time.

At this point, the god of simplicity appeared; perhaps it was not a bunch of little tectonic plates moving around, but everything on the sphere moving together, relative to the spin axis. However, this would work only if Laurentia and Australia were roughly ninety degrees away from each other at the time (which—duh—had to be true if Australia was on the equator and Laurentia on the pole!). In fact, this single-rotation hypothesis makes very specific predictions about the relative orientation and configuration of all continental landmasses, an "absolute paleogeography." With apologies to Tolkien, "One motion to move them all, one rotation to spin them, one translation from the pole, and on the globe we'll find them!" One simple rotation of the entire solid Earth around the spin axis brought ~90 percent of the previously scattered paleomagnetic results into a clear focus.

Everything was happening at once. A big pulse of evolution, both in the number of species and in the body plans, a huge increase in biomineralization (the number and different kinds of outer skeletons evolved by many different phyla), the first predator-prey interactions among animals, huge swings in the organic carbon budgets, and wild oscillations in the positions of the continents, leaving scientists including Kirschvink and his students to ponder whether this was coincidence or cause and effect.

As more and more paleomagnetic evidence began to accrue, not only surprising but also downright impossible motions of the ancient plates (with their entombed continents fixed in oceanic crust) were detected. Uniformitarianism tells us to use the modern to understand the past, and today we can readily measure how fast plates are currently moving. In the Atlantic, where new oceanic crust is being created along the Mid-Atlantic Ridge, the rate at which the two plates bisecting the North and South Atlantic Oceans are slowly moving away from the axial origin line is only about one inch per year. These enormous plates, while created at the oceanic spreading centers, hold the continents in their stony embrace—so as (and where) the plates go, the continents go as well. The rates are varied. For instance, today there are much faster rates seen in the plates being created in the Pacific Ocean area, with speeds of three to five inches per year. The fastest possible rate is close to ten inches per year, but even this is theoretical and controversial. Yet paleomagnetic data were yielding speeds that measured multiple *feet* per year: This is impossible if only plate tectonics was involved. Yet the data are repeatable and stark. Something revolutionary took place, or at least something so different from modern processes as to cause enormous surprise in science. So much for the Principle of Uniformitarianism!

The first reaction to encountering this data suggesting such fast motion of the surface of the Earth was to doubt the reality of the data. Fair enough. As Carl Sagan once said, extraordinary (scientific) claims required extraordinary proof. Continental motions were so fast that normal plate tectonic motions noted above, typically at most a few inches per year, could not explain them. The new data, slowly but inexorably being produced by Kirschvink and a few others, were showing that the plates were moving too fast for the conventional theory of plate tectonics. To top it all: most of this motion was happening precisely coincident with the explosive increase in diversity from the animal phyla. If it was not plate tectonics, what could it have been? And how could this affect the evolution of animals?

The answer was a surprise—but should not have been, because a similar process is known to have occurred on Mars, the moon, and many satellites and minor planets for billions of years. Such bodies are capable of astounding changes in orientation. On Earth, the consequences for life may have been inestimable, and yet our dawning comprehension of this possibility is one of the great new revolutions brewing in our understanding of the history of life.

For over a century, geophysicists have known that the solid parts of a planet can move rather rapidly relative to the spin axis. The fundamental principle is that a spinning object would like to rotate around something called its maximum moment of inertia. A Frisbee is a good example: When thrown properly, it spins about the center point, and most of the mass at the edge of the disc keeps it rotating stably. But now put a small hunk of lead somewhere on the disk—but not at its center. The spin of the Frisbee will change as it tries to reorient itself to take this new mass situation into account, and the Frisbee will try to spin with the new heavy mass as far away from the rotational axis as possible: it wants to go to the equator. On a spinning planet, centrifugal and gravitational forces similarly tug on any anomalous mass. But on a spinning sphere, a much more orderly change takes place—the position of the spin axis will reorient so that the "weight," which may have been located, perhaps, two thirds of the way from the equator to the pole, will not be at the equator. The spinning ball has had its axis of rotation change position—because of the strange new weight that was added.

It is very well known that the moon and Mars have both realigned themselves in this way during their geological history. Both had new masses added to their surface that originally were *not* on their equators, but then somehow ended up on the equator. For example, the gigantic Tharsis province (a geological region on the Martian surface) is composed of an enormous quantity of heavy lava. In terms of geological time it was just like the weight we added to a Frisbee or a spinning ball; it was added after the formation of the planet. In fact, it is the largest positive gravity anomaly in the solar system and lies precisely on the Martian equator—now, that is. On the moon, the pre-*Apollo* surveys detected mass concentrations associated with the lunar mare basalts, also on the equator. These processes are fairly simple to understand on the moon and on Mars, because neither of these objects have plate tectonics. This realignment process is been termed true polar wander, or TPW. Prior to the discovery of plate tectonics back in 1966, all evidence for the poles being at different positions at earlier geological time was thought to have been a result of TPW.

A geologically rapid change of mass on a planet can happen in a number of ways, including the impact of a big asteroid or comet, or even internally through the eruption of magma from the deep interior of Earth to its surface. Similarly, big mass shifts can happen when one of the parts of the plate tectonic features, which are composed of spreading centers and, separately, subduction sites (where a plate dives down back into the deep Earth), either appear or disappear. Both of these are large enough to excite TPW as far as the Earth is concerned, as long as the masses involved are being maintained actively, not just floating buoyantly. But if they disappear, it will affect the orientation of the planet. Both subduction zones and spreading centers can disappear when one continent undergoing continental drift runs into another. Any offshore spreading centers or subduction zones between the converging continents are destroyed in the collision; only in this case it is the disappearance of a surface mass that causes reorientation, not the addition of a mass.

As it is rather unlikely that the observed biological changes associated with the Cambrian explosion were forcing the continents to move, a more plausible explanation was that the unusual burst of motion was somehow accelerating the pace of biological evolution. Several mechanisms have been discovered that might explain and connect some of these observations. First, when continents are at high latitudes they tend to build up large reservoirs of frozen methane known as clathrates, or gas hydrates, on the seafloor and in permafrost. As these areas move toward the equator, they will warm gradually and can episodically cause pulses of greenhouse gas emissions into the atmosphere, periodically warming the environment. Evolution and species diversification in particular tend to proceed faster in warmer environments, through a mechanism of accelerated metabolism.

At the time this was proposed in the literature, it was nicknamed the "methane fuse for the Cambrian explosion," and argued that thermal cycling of biological diversity may have been one of the major factors in promoting the proliferation of species. It is also a possible cause for the crazy oscillation in the carbon isotopes. It also turns out that geographically higher diversities exist naturally in the equatorial zone. When Ross Mitchell, a colleague of ours at Yale University, looked at the paleogeographic motions during this true polar wander event, he noticed that almost all of the newly evolving animal groups seemed to originate on the leading edges of the continents as they moved into the equatorial zone, with few to none originating as other areas moved in the high latitudes. This increase of diversity with latitude provides a stunningly simple explanation for the diversity increase, particularly if this happened when nature was experimenting with body plans via the Hox genes. It also implies that the paleontological record of this Cambrian explosion might be partially an artifact—because a side effect of true polar wander is to produce relative marine transgressions in areas moving onto the equator, and sea-level regressions in areas moving off. Sediments are preserved best during transgressions and removed during regressions. Hence, the rock record is biased during TPW events to preserve rocks that are recording a diversity increase.

Invoking true polar wander as a cause of events in the history of life is definitely a new field of research, unheard of in the twentieth century. Just as this mechanism is being used here as a new hypothesis for the Cambrian explosion, so too can TPW be used to try to explain the killing mechanisms in mass extinctions, one of which ended the Cambrian period and Cambrian explosion, killing off almost all the weird wonders described by Stephen Gould and Simon Conway Morris from the Burgess Shale. This mass extinction was given the unlikely name of SPICE.

ENDING THE CAMBRIAN—THE SPICE EVENT AND THE FIRST PHANEROZOIC MASS EXTINCTION

Any history recounting the Cambrian explosion can be overwhelmed by the sheer power and importance of animal body plan evolution—the radical change of the world's biota from the immobile, floating, and simple larger animals of the latest Precambrian world to the diverse exuberance of the animal cargo inhabiting the world's oceans at the end of the Cambrian. But why is there an "end of the Cambrian" at all? Here is a topic where a long-held understanding has been toppled.

Mass extinctions, the short-term periods of great mortality among both individuals and species, were variable in their severity. While the largest are included in the "big five" mass extinctions, when at least 50 percent of species died out, there were many more extinction events not as catastrophic (unless one was one of the victims, that is). One of the most celebrated of these occurred at the end of the Cambrian period.

The Late Cambrian mass extinction was actually three or four separate, smaller extinction events, mainly affecting trilobites and other marine invertebrates, especially brachiopods. It has long been accepted that these were caused by increases in warm, low-oxygen water masses affecting marine communities. Some of the earliest occurring of all trilobites, the olenellids, underwent total extinction, and in fact the entire nature of the trilobite faunas changed: trilobites of the Cambrian had many segments, primitive eyes, and no obvious defensive adaptations on the body (such as antipredatory spines), and were unable to do what modern-day pill bugs do when threatened—roll up into a tight ball. After the extinctions, and thus in the earliest times of the Ordovician period, newly evolving waves of trilobites had changed their entire body plan: now virtually all reduced the number of segments (many segments are easier for predators to break through during a predatory attack than fewer, thicker body segments) and had better eyes, defensive armor, and especially the ability to roll up into pill bug–like balls.

Warmth, low oxygen, and faunal change: that was the view of these late Cambrian extinctions. But then an entirely new series of data were recovered suggesting quite the opposite: evidence for cold water, not warmer; and evidence as well of a major burial of organic matter into the oceans—a process that caused oxygen levels to skyrocket. These changes are now named the SPICE event (after the

Steptoean positive carbon isotope excursion). But there is a huge contradiction with this new finding. It was first identified in the rock record not only because of a sudden extinction of species, but also as a major perturbation of the carbon isotope record, and hence carbon nutrient cycling. There is quite good evidence that a large percentage of trilobites died out in a succession of short-lived extinctions near the end of the Cambrian period.

One of the most interesting aspects of the SPICE event is that unlike most other mass extinctions, this one might have been accompanied not by a drop but by a short-term rise in oxygen. It is intriguing to speculate that a known volcanic eruption at about this time might have caused one of the short-term rapid continental movements mentioned above—a TPW, or True Polar Wander event. In this case more land area was moved *into* the tropics for a few million years, increasing carbon burial and spiking the atmospheric oxygen up to previously unheard-of levels. Something like *that* might have paved the way for the next major radiation of life following the Cambrian Explosion. One kind of ecosystem needs a great deal of oxygen. The coral reefs appeared soon after the SPICE event, starting in the next geological period, the Ordovician.

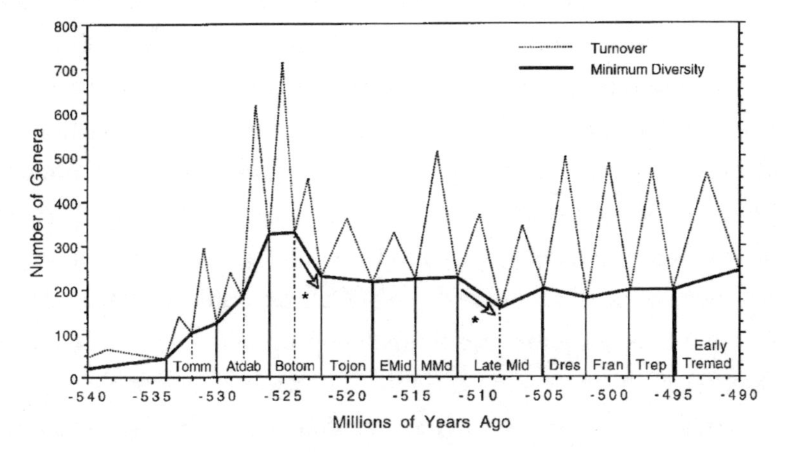

Biological turnover and genetic diversity during the Cambrian explosion. The classical Cambrian explosion interval spans the Tommotian, Atdabanian, and Botomian stages of the Siberian platform. The turnover shows the number of genera that either arose or disappeared during the particular stage. (From Bambach et al., "Origination, Extinction, and Mass Depletions of Marine Diversity," *Paleobiology* 30 (2004): 522–42)

CHAPTER IX

The Ordovician-Devonian Expansion of Animals: 500–360 MA

MODERN coral reefs have been called the "rainforests of the oceans" because they share with rainforests the trait of high species diversity and abundance in small areas, and that is often the shared first impression—that there is so much life. But from there the comparisons largely cease. In a rainforest, or any forest, most of the life to be found is plant life. Reefs, on the other hand, are composed almost exclusively of animals. On any reef there are indeed large numbers of leafy, bush-shaped forms that are plant*like*. Yet virtually all are formed by animals, from soft corals to sponges to lacy bryozoans. One could argue that the verdant green of photosynthesizing plants covering great swaths of our planet's continents are the most obvious evidence, if seen from space, that our planet is a world of life. But there is an entirely different kind of biological signal that can be seen from space—this one in the seas. It is the presence of tropical marine coral reefs, best illustrated by the Great Barrier Reef, which lines more than a thousand miles of eastern Australia's coastline. But there are many more reefs than the Great Barrier Reef, magnificent as it is. The equatorial seas are filled with numerous coral atolls, fringing reefs, and the vast pale-green lagoons that these wholly biological structures enclose. These reef systems are parts of a very ancient kind of ecosystem, one that predates forests and even land animal life of any kind. They remain one of the most diverse of all ecosystems, and are essentially long-lived superorganisms that pop up again after every mass extinction and planetary die-off of the last 540 million years.

A hallmark of the coral reef environment is the abundance of movement virtually everywhere, for the flitting and schooling of fish, to the never-ceasing wave action on the reef, to the waving and billowing soft corals, pulsing and undulating in the active water movement that characterizes the reef environment. Every coral reef is home to fish—many fish, of many sizes, shapes, and behaviors. Some school, some lurk, some swim in solitary splendor, and some—the omnipresent sharks—simply patrol. And it is not just the vertebrate members of these diverse communities that are seemingly always on the move. Closer inspections show that an amazing diversity of invertebrates is seemingly in constant motion as well—if usually slower than the fish. Smaller reef shrimp dance from coral to coral, while crabs large and small can be seen in their constant foraging. Snails, slower yet, perambulate according to some plan known only to them, and the gastropods to be found on any reef are diverse as well: there are large carnivorous species, such as the beautiful tritons, as well as equally large but herbivorous conch shells. Under the coral rubble, at least during the day, a cornucopia of the beautiful cowries huddle or slowly feed on tiny bits of algae, while the ferocious cone shells move among them, searching for most of their kind's normal prey—small worms. Some, however, such as the textile cones, are fish eaters, and use a highly modified tooth, shaped like a harpoon and dipped in poison, to spear fish and then consume them whole. Turgid sea cucumbers move on the sediment—or just beneath it, constantly ingesting massive amounts of sand at one end and constantly expelling large sandy pellets from the other. There they share the upper foot of white coral sand with heart urchins. Other echinoderms are there as well, from a variety of predatory starfish to the placidly perched—but also swimming—comatulid crinoids. Color—and especially motion by a great and diverse assemblage of species. The coral reef ecosystems of today are filled with motion and color and there is every reason to suspect they have always been so.

Reefs, in fact, are very ancient evolutionary inventions,[1] and their rise to prominence mirrors the rise to high diversity that followed the Cambrian explosion. In a way a hydrogen bomb is a good analogy. Thermonuclear fusion and the enormous explosion that ensues only take place in the immense heat of an atomic explosion. That is how a hydrogen bomb works: an atomic (fission) bomb with plutonium is ignited, which in turn creates sufficient heat and pressure to start the fusion reaction—and fusion explosion. In similar fashion the Cambrian explosion of diversity was the heat and fuel that led to the far greater

Ordovician diversification, and one of the most important of the products of this huge run-up in species numbers was the invention of the coral reef.

The first reefs—and by reef, we mean a wave-resistant, three-dimensional structure built by organisms—date all the way back to the earliest Cambrian period. They were not coral reefs, but composed of archaic, now long-extinct sponges called Archeocyathids.[2] Coral reefs are slightly younger; the first of these are found in the Ordovician period, and they really increase in distribution, size, and diversity by the Devonian period. They remained a rather constant and ecologically recognizable ecosystem until the end of the Permian—when the reefs and so much else were decimated by the Permian mass extinction.

Let us imagine that we were capable of going back into time and take a dive on a Paleozoic coral reef, one that is 400 million years in age. At first glance there is a surprising similarity to the reefs of today. Corals dominate the reefs; they are the bricks of the 3-D structures that are reefs, and like a brick house, are held together by many kinds of biological mortar, mostly encrusting species that serve to cement and bind the many heads and coral fronds into enormous and complex ramparts and foundations of limestone. But on closer view, the 400-million-year-old corals can be seen to be entirely different in basic appearance, and certainly in taxonomic composition. The massive coral heads are composed of a family that while building overall shapes similar to the corals of today are actually very different in their finer morphology: these are the tabulate corals, and these filled the same niches as are held today by scleractinian corals, the common corals of our modern reefs. Between these broadly branching and hemispherical tabulate coral colonies are other "framework builders," other bricks in the wall. Many of these are stromatoporoids, a strange, carbonate-producing sponge still living today, but never in the sizes or diversity of the Paleozoic era. Scattered among these two massive inhabitants are a second kind of coral, solitary in nature, called rugose corals, solitary species that look like the horns of a bull, but in this case the pointed end of the "horn," the calcium carbonate skeleton of these rugose corals, are cemented to the substrate, and the widest end, facing up, is the seat for a single broad sea-anemone-looking animal.

Like our modern scleractinian corals, no matter how large, and composed of how many of the small, tentacled bodies that are the basic body plan of a coral, the tabulates were a single "individual"—at least genetically. But in fact all corals, surely then as well as now, are vast *colonies* of tiny sea-anemone-like polyps, each a ring of poison-tipped tentacles surrounding a small central mouth. But unlike a seashore rock covered with a smattering of the small, common sea anemones (which are solitary polyps) found the world over, each of these tiny polyps linked to others around it by a thin sheet of tissue. Every part of these sometimes-vast colonies is genetically identical. But this is not just one animal. In fact, any coral supports a vast and diverse assemblage of plants within its tissues. Throughout both the coral's polyp-to-polyp connecting tissue, as well as in the polyp itself, are untold numbers of tiny plants—single-celled dinoflagellates that live in symbiotic bliss with the corals. It is a great deal for both: the tiny plants get the four things they most want: light, carbon dioxide, nutrients (phosphates and nitrates), and protection within the coral flesh, protection from the many organisms that would love to dine on a tasty if tiny plant.

ORDOVICIAN DIVERSIFICATION: BUILDING ON THE CAMBRIAN EXPLOSION

The Cambrian period came to an end because of a mass extinction, one that affected many of the more successful members of what has come to be known as the Cambrian fauna—sea life composed of such early animal entrants in the overall history of animal life as trilobites, brachiopods, and many of the very exotic arthropods of the Burgess Shale, such as *Anomalocaris* (although in 2010 a new fossil deposit of Ordovician age has yielded the youngest *Anomalocaris* of all, and thus perhaps the Late Cambrian mass extinction was kinder to some of the odd Burgess Shale fauna than previously considered). This particular extinction has long been known, but it is not listed as "major," in that less than 50 percent of marine forms died out. This acted like gasoline on the open fire of diversification, perhaps, in that those forms that were less adaptive died out, opening the way for new innovation and new species in the same manner that ridding a garden of weeds leads to a rapid proliferation of new growth from the nonweeded.

It was also as if the biological world discovered entirely new ways for animals and plants to make a living, as well as finding entirely new places to live: areas that were poorly populated in the Cambrian, such as brackish water and freshwater, as well as both deeper and shallower areas in the sea, right into the surf zones themselves, became ripe for colonization by animals. Many of these were still sedentary, spending their entire lives sitting in one place, filtering the ever-richer and more nutritious marine plankton. But species numbers and biomass numbers alike climbed.[3]

Many kinds of animals were present in the Ordovician that had not yet evolved in the Cambrian, and many of these appeared soon after the end of the Cambrian mass extinction. The result was an assemblage of animals that is markedly different than most of the Cambrian faunas. Trilobites are still there, but compared to the Cambrian oceans, when they may have been the most commonly encountered animal at most depths, they were overwhelmed in numbers as well as numbers of species of animals with shells—the brachiopods and more than a few mollusks as well. The biggest winners were animals that had evolved an entirely novel way of living—animals that were colonial. While colony formation was something that had been used by other biology far simpler in body plan, including many kinds of plants, microbes, and protozoa, in the Ordovician the leitmotif of colonial life dominated and drove the relentless diversification that is the hallmark of the Ordovician: corals, bryozoans, and new kinds of sponges among many others.

The reasons for this great diversification go back to oxygen.[4] Our view is that the true effects of oxygenation in the sea can be seen from this point. Here, then, we will make an interpretation that historians do, one that is as yet still new enough in science that it cannot be considered as hard truth, but one that has enormous explanatory power. It also lets us look, quite appropriately at this point in the book, at an overview of animal diversification. We will argue that it was oxygen levels more than any other factor that has left us with the diversity curve of animals through time, results that are hard science and accepted.

The Ordovician period can be regarded as the second part of the two-part initiation of animal diversity on Earth, with the first being the Cambrian explosion,[5] and in both cases, rising oxygen was the driver. Like the Cambrian, it was a time when new species as well as new kinds of body plans appeared at a faster rate than was characteristic of more recent times. This high rate of evolution and innovation was partly in response to filling up the world with animals for the first time. The history of life in the Cambrian was a filling of the seas with many experiments. The post-Cambrian history was one where many of these early and clearly primitive and inefficient evolutionary designs were replaced in what became a rocketing increase in biodiversity as competition ruthlessly killed off the less fit. Evolution became a means of exploring the engineering excellence of body plans.

THE HISTORY OF THE HISTORY OF BIODIVERSITY

The history of biodiversity, which can be thought of as the assembly and numbers of the various categories of organisms (especially animals, because they leave the most abundant and recognizable fossils), was first presented by the English geologist John Phillips, who is also credited with subdividing the geological time scale through the introduction of the concepts of Paleozoic, Mesozoic, and Cenozoic eras. Phillips, who published his monumental work in 1860 both defining these new eras and discerning the largest-scale pattern of evolutionary change that can be found in the fossil record, recognized that major mass extinctions in the past could be used to subdivide geological time, since the aftermath of each such event resulted in the appearance of a new fauna as recognized in the fossil record. But Phillips did far more than recognize the importance of past mass extinctions and define new geological time terms: he proposed that diversity in the past was far lower than in the modern day, and that the rise of biodiversity has been one of wholesale increases in the number of species, except during and immediately after the mass extinctions. His scheme recognized that mass extinctions slowed down diversity, but only temporarily. Phillips's view of the history of diversity was completely novel. Yet a century passed before the topic was again given scientific attention.

In the late 1960s, paleontologists Norman Newell and James Valentine again considered the problem of exactly when and at what rate the world became populated with species of animals and plants.[6] Both wondered if the real pattern of diversification was of a rapid increase in species following the so-called Cambrian explosion of about 530 to 520 million years ago (using the revised dates, not those favored in the 1960s), followed by an approximate steady state. Their arguments rested on the importance of preservation biases in older rocks. Perhaps the pattern of increasing diversification through time seen by Phillips was in reality the record of *preservation* through time, rather than the real evolutionary *pattern* of diversification. According to this argument, the change of species is reduced in ever-older rocks, so that sampling bias is the real agent producing the so-called diversification he saw. This view was soon after echoed by paleontologist David Raup in a series of papers[7] that forcefully argued that there are strong biases against older species being discovered and named by scientists, since older rocks experience more alteration through recrystallization, burial, and metamorphism; entire regions or biogeographic provinces have been lost through time (therefore reducing the record for older rocks); and there is simply more rock of younger age to be searched.

The argument as to whether diversity has shown a rapid increase through time or achieved a high level early on and has stayed approximately steady ever since dominated paleontological research for much of the latter part of the twentieth century. In the 1970s, massive data sets derived from library records began to be assembled by Raup and the late Jack Sepkoski[8] of the University of Chicago along with his colleagues and students. These data, compiling the record of marine invertebrates in the sea, as well as other data sets for both terrestrial plants and for vertebrate animals, seemed to vindicate Phillips's early view. In particular the curves discovered by Sepkoski showed a quite striking record, with three main pulses of diversification carried out by different assemblages of organisms.

The first was seen in the Cambrian (the so-called Cambrian fauna, composed of trilobites, brachiopods, and other archaic invertebrates), followed by a second in the Ordovician that led to an approximate steady state through the rest of the Paleozoic era (the Paleozoic fauna, composed of reef-building corals, articulate brachiopods, cephalopods, and archaic echinoderms), culminated by a rapid increase beginning in the Mesozoic and accelerating in the Cenozoic to produce the high levels of diversity seen in the world today through the evolution of the modern fauna, composed of gastropod and bivalve mollusk, most vertebrates, echinoids, and other groups.

The net view of biodiversity over the last 500 million years was thus about the same as that of John Phillips in 1860: there are more

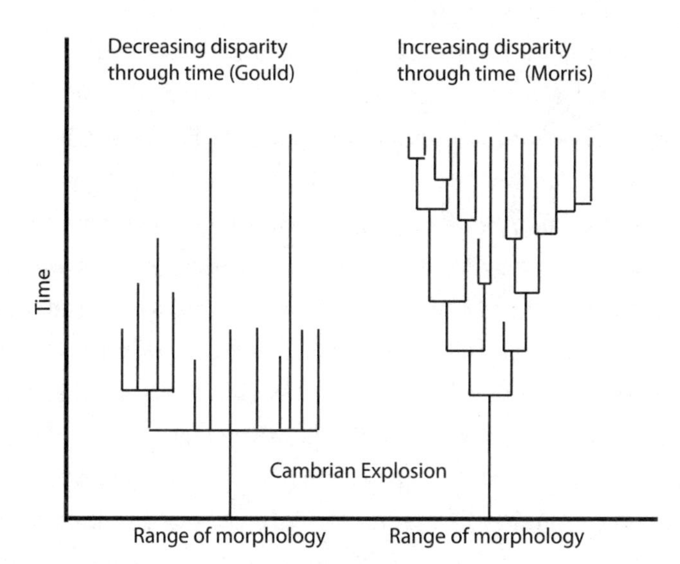

Competing hypotheses about disparity and the Cambrian Explosion. *Diversity* refers to the number of species, whereas *disparity* refers to the number of different kinds of anatomies, or body plans. Stephen Jay Gould thought that there were many more body plans (high disparity) during the Cambrian Explosion than now. He referred to many of the strange fossils from the Burgess Shale as "weird wonders" and thought that they were phyla now entirely extinct. The opposite view was held by Simon Conway Morris, who advocated that disparity has slowly increased through time.

species on the planet now than at any time in the past. Even more comforting, the trajectory of biodiversity seemed to show that the engine of diversification—the processes producing new species—was in high gear, suggesting that in the future the planet would continue to have ever more species. While not at the time viewed in any sort of astrobiological context, these findings certainly do not suggest that planet Earth is in any sort of planetary old age. All in all, the 130-year belief, from the time and work of John Phillips to that of John Sepkoski—that there are more species now than any time in the past—remained a comforting view. This long-held scientific belief suggested to many that we are in the best of biological times (at least in terms of global biodiversity), and there is every reason to believe that better times, an even more diverse and productive world, still lie ahead, even without weird contributions from biotechnology.

While the Sepkoski work seemed to show a world where runaway diversification is a hallmark of the Late Mesozoic into the modern day, worries about the very real sampling biases described by earlier workers persists, and a series of independent tests of diversity were conducted. Of most concern was a phenomenon dubbed the "pull of the recent"—that the methodology used by Sepkoski would undercount diversity in the deep past, making it look like there were ever more species in more recent times. Because of this very real concern, new tests were devised to examine biological diversity through time. In the early part of the new millennium the issue was reexamined[9] by a large team headed by Charles Marshall of Harvard (now at Berkeley) and John Alroy, then at the University of California at Santa Barbara. This team assembled a more comprehensive database, based on actual museum collections rather than using Sepkoski's method of simply tabulating the number of species recorded in the scientific literature for given intervals of past geologic time. To virtually everyone's surprise, the first results of this effort were radically different from the long-accepted view.

The analyses of the Marshall-Alroy group found that diversity in the Paleozoic was about the same as in the mid-Cenozoic. The dramatic run-up of species that had so long been postulated for diversity through time was not evident in this new study. The implications are stark: we may have reached a steady state of diversity hundreds of millions of years ago. It may be that diversity peaked early in the history of animals, and in contrast to all views since the time of Phillips, it has remained in an approximate steady state since, or perhaps may already be in decline. While many new innovations, such as the adaptation allowing the evolution of land plants and animals, surely caused there to be many new species added to the planet's biodiversity total, it may be that by late in Paleozoic time the number of species on the planet has been approximately constant.

Thus, following an initial burst of diversification in the lower Cambrian, animal diversity increased exponentially to reach equilibrium during the Paleozoic and then crashed at the end of the Permian, followed by an overall trend of increasing diversity, but importantly interrupted by short intervals of diversity decreases—the mass extinctions—of which five were particularly consequential. While these mass extinctions each led to substantial loss of taxa, all were followed by increased rates of species formation that led to levels not only equaling but in each case exceeding the original diversity prior to the extinction event.

This history suggests that a complex array of factors causing both diversification and extinction are responsible for the observed pattern of Phanerozoic diversity. Among the many potential factors that have been invoked to explain the observed increases in diversity are evolutionary innovation, the colonization of previously empty or unobtainable habitats, and the occurrences of new resources, while the major factors cited for diversity drops are changes in climate, reduction of resources or habitat area, new biological competition or predation, or external events such as asteroid impact.

Geochemists have long known that CO_2 levels and atmospheric oxygen levels show trends that are inversely related to one another: when oxygen levels rise, CO_2 is usually dropping. While it is difficult to understand how changing levels of CO_2 at concentrations that have little or no direct biological effect on individual organisms could somehow promote or inhibit diversification, it is quite plausible to

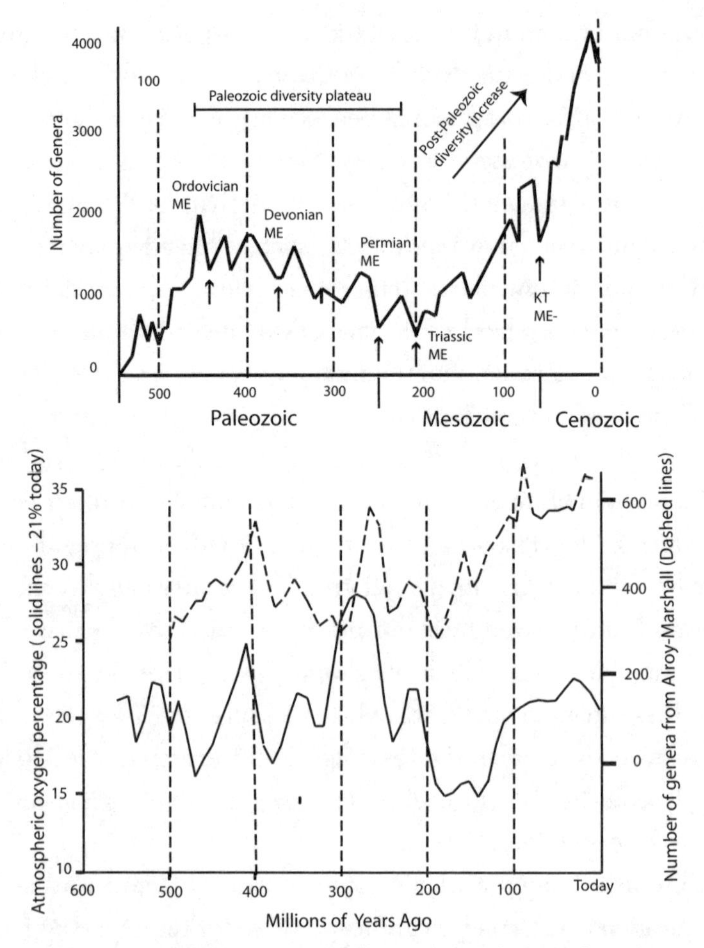

Top graph: The trajectory of marine invertebrate animal diversity from the Cambrian Period onward, as discovered by Jack Sepkoski. His findings, based on long and prodigious library research, indicated that the numbers of genera rose quickly in the Paleozoic and then plateaued, only to be cut down by the Permian Mass Extinction. Afterward, he saw a great rise in generic numbers to the present day. Bottom graph: Here we show the levels of oxygen, as modeled by Robert Berner, with newer (than Sepkoski) estimates of animal generic numbers published by John Alroy and others. Notice the strong correlation between peak oxygen levels (both high and low) and the trajectory of animal numbers. (From Peter Ward, unpublished results.)

suggest that it was not the changing CO_2 but in fact a combination of changing oxygen levels, with that change being affected by global temperature rates.

Cold water holds more oxygen than warm. In a cold world, with already high oxygen levels, life in the sea would rarely be compromised by too little oxygen. On the other hand, in a warm world where there was already relatively little oxygen, most bodies of nonflowing water would quickly go stagnant. And not just ponds to lakes—large oceans would succumb to this as well in a warm world, which is what a high CO_2 world is.

The data to date suggests that (at least for marine animals) overall, global taxonomic diversity is related to oxygen content, which might be expected, since all animals have a poor tolerance for anoxic conditions. What was *unexpected* was the finding that origination rates (measured either in species or genera, which are groups of related species all with the same common ancestor) appear to be *inversely* correlated to oxygen levels. The high levels of origination characterizing the period from 545 to roughly 500 MA (the Cambrian explosion) occurred during a protracted period of oxygen levels of between 14 and 16 percent, compared to 21 percent today. The dramatic rise in oxygen in the Silurian and again in the Carboniferous corresponds with the *lowest* levels of generic origination. The drop of oxygen in the Permian is correlated to a rise in origination rates but a drop in the total number of species. There seems to be a clear signal.

Times of high oxygen are like the boom times in a country's economy. There is little unemployment and businesses succeed and stay open—but not many new ones start up. Start-ups, it seems, are related to bad times. New ideas play out and new risks are taken in times of desperation. Yet while there are many new start-ups, few of them succeed, and at the same time many of the businesses that were succeeding during good times begin to fail at an ever-higher rate in the bad times.

Thus we see a dichotomy: more new businesses appear, but most of them quickly go bankrupt and disappear along with many of the previously successful. There is also less money circulating. The total number of businesses plummets. The same seems to be true of species. High oxygen means good times: large numbers of species, and nothing much new comes along. But when oxygen is low, species die out at a faster rate than they are replaced, even though the actual number of emerging species is higher than in the high-oxygen times.

There are many examples. One of the best: the long-term rise in oxygen beginning in the Jurassic and continuing to the modern day is accompanied by both a long-term drop in origination rates and a huge run-up in diversity. But what radical new designs appear? Birds, mammals, reptiles, amphibians—all of the Cenozoic forms are slight changes of body plans that originated in either Paleozoic or Mesozoic bad times, the low-oxygen times. There are no dinosaurs (the best examples of a radical innovation spawned directly from low oxygen) appearing in the Cenozoic.

The recognition that a combination of low oxygen and elevated carbon dioxide has stimulated species formation in the past by the formation of evolutionary novelty, while at the same time vastly increasing extinction rates, has a firm biological basis. The net effect is the reduction of species during the low-oxygen periods. Dropping oxygen levels with concomitant rising temperatures is the worst kind of one-two punch, as adapting to hotter, lower-oxygen environments is never a quick fix. More hair, more feathers, and more body fat can fix increasing cold. But staying cool is far more difficult and involves much more profound evolutionary change. This is even more true in animals trying to deal with staying alive in ever-lowering oxygen, since adaptation to lower oxygen levels is necessarily profound and must occur in multiple organ systems, ranging from blood pigments to more efficient circulatory systems, to better lungs or gills.

The most striking aspect of oxygen and its relationship to diversity was pointed out to us in 2009 by Bob Berner (of Yale), who alerted us to what he saw as a profound similarity between his latest oxygen-through-Phanerozoic time curve with the then-latest diversity curve of the John Alroy group. We show those two curves in the figure on page 158. While there is a slight direct correlation between oxygen levels and diversity when both are broken down into 10-million-year time bins, an absolutely amazing correlation is present between the change in atmospheric oxygen when plotted against the change in diversity for these same 10-million-year bins. For example, the correlation between the change of atmospheric oxygen as a percentage of total atmospheric gas content from 230 to 220 million years ago, plotted against the change in generic diversity during that same time interval, is highly significant—in other words, not by chance. The results are very strong indeed, from a statistical point of view.

A most interesting aspect of this is that since both were introduced, the model results from the Berner group (and others as well) estimating past oxygen and carbon dioxide levels have been controversial. Equally controversial are the various Alroy curves. Each set of results (one yielding oxygen and CO_2 values, the other the estimated number of animal genera through time) comes from models *that have completely different inputs*. None of the many values inputted into the GEOCARB and GEOCARBSULF models have anything to do with how many species there were at any given time. In similar fashion, the Alroy model is completely independent of the values used to model carbon dioxide and oxygen. Yet the near unbelievable correlation, while theoretically possible from chance, is hard to explain that way. There is no chance at work. It appears that oxygen and carbon dioxide levels (particularly oxygen) are the most important of all factors dictating animal diversity. The two independent curves in this fashion support each other in terms of that most important of scientific values: credibility.

INSECTS AND PLANT GROUPS

It is clear that the invasion of land unleashed the floodgates of both diversity and disparity. Our understanding of the overall diversification of life through time is that there are more kinds of life on Earth now—more kinds of species or any other way of tabulating diversity—than at any time in the past. But is this really true? What might the biases be?

All good science has a null hypothesis, and in this, it is that marine animal life on Earth reached present-day levels at the end of the Cambrian. This was the view of Stephen Jay Gould in the 1970s, and whether he really believed this is irrelevant: his take on this issue resulted in much stronger science.

The answer to this question, whether diversity rose rapidly or only slowly rose to present levels, relates to the relative preservation potential of modern-day organisms compared to those of the Cambrian. Today, about one out of three marine animals has hard parts that produce ready fossilization—anatomy such as shells, bones, and hard carapaces. But what if that number were one out of ten during the Cambrian? In such a case, there might be approximately equal numbers of animals during the Cambrian, compared to modern oceans. Support for this idea also came from the post-Sepkoski work of Marshall and Alroy, as their model results, while showing post-Cambrian increases in diversity, did not see the explosive runaway of post-Permian animal taxa found by Sepkoski.[10] The Alroy work has since gone through several iterations with new data.[11]

There are other sources of bias as well as perhaps dubious assumptions that have been made to arrive at models of diversity through time. For instance, what about the unequal sample sizes being studied? Critics of the entire diversity-through-time enterprises have noted that there is far more rock to sample of late Cenozoic or Pleistocene age than there is of Cambrian age. Furthermore, there are many more paleontologists studying late Cenozoic and Pleistocene fossils than there are professionals studying Cambrian-aged rocks and fossils. Andrew Smith of the British Museum[12] and independently Mike Benton[13] of the University of Bristol and Shanan Peters[14] of Wisconsin have all done remarkable work on this aspect.

It turned out that a quite simple test demonstrated that there has been an increase in marine animal taxa (be they species, genera, or families) since the Cambrian explosion. The test came from studying the number of trace fossils through time. Trace fossils are the results of animal activity, as we saw in the chapter on the Cambrian explosion, and each different trace found in strata had to come from a slightly different body plan. Their pattern of diversity mirrors the record from body fossils. It is now agreed that the overall pattern of diversity long recognized by invertebrate paleontologists has indeed given us a fairly accurate view of how life diversified on Earth.

By the end of the Devonian period, the major marine environments from the shallows to the deepest oceans were colonized. But this marine diversification was about to be overshadowed by a diversification that would ultimately prove far greater, creating the greatest pool of animal and plant species on Earth: the diversity of life on land.

THE ORDOVICIAN MASS EXTINCTION

The Ordovician period was also the time of the first of the so-called big five mass extinctions. All five involved animals and plants. There surely were mass extinctions before the Ordovician event, such as during the great oxygenation event and the various snowball Earth episodes. But animals were in the midst of differentiating at a rapid rate when something brought this increase in diversity to a halt. The best bet is that it happened when the Earth underwent a "little ice age" that turned the early coral reefs to dead piles of rubble because of sudden temperature drop. However, this is still a puzzle, as the extinction has two discrete steps, at either end of the last stage of Ordovician time, called the Hirnantian glaciation.

There are other more fanciful suggestions about the cause of the Ordovician mass extinction. The most interesting is that the Earth was hit by a giant blast of hard radiation coming from interstellar space, called a gamma-ray burst, during the Ordovician.[15] This is a most dramatic potential cause, but there is also not a single shred of evidence to support it, much as journalists have publicized it. Prior to 2011, the accepted cause for this mass extinction was that there was no accepted cause.[16] Most explanations opted for some kind of rapid cooling event. One prevalent idea is that perhaps volcanic outpourings caused the atmosphere to become obscured by sulfur aerosols[17] in a manner similar to that following the Krakatoa volcanic explosion of the 1800s, when Europe went through a "year without summer." Recently, however, geologists and geochemists at Caltech[18] attacked this late Ordovician glaciation problem from a superbly preserved sequence of rock on Anticosti Island, a remote Canadian island in the Gulf of St. Lawrence that was once located in the tropics. Using a new type of geochemical thermometer, they were able to measure both the relative ice volume and temperature, with unprecedented resolution. Lo and behold, they found that while ice volume changed only slowly before and after Hirnantian time, and the tropical temperature remained at a very hot but possible 32° to 37°C, there was a sharp shift at either end of it that was associated with the two steps of mass extinction. Tropical temperatures fell by ~5° to 10°C, the global ice volume peaked up to levels that equaled or topped those of the last (Pleistocene) glacial maximum, and carbon isotopes had a positive spike, suggesting a large perturbation of the global carbon cycle—in this case, presumably more organic carbon burial.

These new data narrow the actual kill mechanisms of these two extinction pulses to two possibilities, either a fast change in climate or a fast change in the level of the oceans all over the globe. In a follow-up paper, members of the same team[19] mined two enormous digital databases for North America, one showing the fossil distributions and another the volume of rock available for fossils to be found in (a necessary correction for the fossil discoveries!). Both processes were found to account for the extinctions—habitat loss from sea level drop and a sudden drop in temperature were both flagged as major elements of the die-offs. However, it is not clear if this is the entire story; the timing of the climatic perturbations, including the positive spike in carbon isotopes, is surprisingly similar to some of the events induced by the true polar wander mentioned in previous chapters. A short, sharp TPW (True Polar Wander) excursion could have triggered a short period of global cooling, producing, perhaps, a short-lived period of glaciation. This remains an enigmatic and still-to-be-researched topic. It is certainly not the traditional explanation. It is, in fact, new—the promise of our title.

CHAPTER X

Tiktaalik and the Invasion of the Land: 475–300 MA

A long point of contention between "evolutionists" and those opposed (creationists) to the knowledge that species evolve one from another has been the supposed dissimilarity between the first amphibian and its last-known fish ancestor: the fish fossils seemed too "fishy," and the first amphibians too "un-fish-like" to appease doubters. There indeed was merit to one aspect of this dispute: until recently, the oldest agreed-upon amphibian fossil, a Devonian-aged creature[1] named *Ichthyostega* (which means fish-amphibian), had a fish-like body (including a quite normal fish tail) and four legs. Its immediate ancestor appeared to be a creature with a similar-looking body—but without legs. This fish, which paleontologists have deemed as the true ancestor to *Ichthyostega* and the other early land vertebrates (or at least living some of their life on land), belongs to a group known as the sarcopterygians, which had fins with fleshy lobes around them.[2] These were the predecessors of limbs. The living fossil *Latimeria* (a coelacanth fish) is thought to be at least somewhat similar to the immediate ancestor of the eventual first amphibians including *Ichthyostega*. The critics asked: "Where are the missing links?" But a twenty-first-century fossil discovery changed all that—a fossil found in the frigid Devonian-aged strata of the high Arctic. It was named *Tiktaalik*, and is so transitional that its discoverers[3] dubbed it a "fishopod." This discovery is one of the most consequential of all revisions to what we call the history of life not only for filling in a large hole in our understanding (of the fossil record from water- to land-living vertebrates) but in helping solidify the entire theory of evolution.

This large fossil proved to be the perfect antidote to creationist doubters. It was unearthed in Arctic Canada by a team of international researchers led by Neil Shubin of the University of Chicago, and when finally (and painstakingly) removed from the sarcophagus-like coating of sedimentary rock holding its bones, the first *Tiktaalik* fossil was deemed to be a fish, complete with scales and gills. It also showed a flattened head and fins that had thin ray bones, the most familiar kind of fish fin. However, in this new fish's case, there were also the kind of sturdy interior bones necessary for an animal as large as this specimen (which would have been near three feet in length) to prop itself up in shallow water using its limb-like fins for support, just as four-legged animals do. With these strange fins and an amphibian (even crocodile-like) head, *Tiktaalik* has the combination of features that shows a perfect, step-by-step evolutionary transition between the fish and tetrapod body plans.[4]

The first appearance of vertebrates on land is the most dramatic event of what was nothing less than a succession of invasions of land by aquatic animals—and plants. Yet while most relevant to us, in fact we vertebrates were among the very last to climb out of the pool and join the roster of animals making the water to land transition. To tell the story in order, we begin with the first—plants.

THE INVASION OF LAND BY PLANTS

It can be argued that the greatest single event in all of life's history, save for the first formation of life itself, was life's invention of oxygen-releasing photosynthesis. It was this that allowed life to move from its dark and dank habitats as low standing biomass and fill the shallower waters of seas and freshwater bodies alike with the living by tapping the greatest energy source that our solar system has to offer, the sun. And in so doing, as an unintended by-product, our planet radically changed its atmosphere to one with such a high concentration of oxygen that a second unintended consequence became the greatest of all dangers to living plants—grazing animals. Yet as consequential as these changes were to life on Earth by aquatic plants, even more radical changes transformed the planet when plants evolved the means to break free of their watery shackles and colonize dry land. In a relative blink of an eye in terms of Earth history to date, in a period of less than 1 percent of the total age of life itself, this great invasion of land by plants changed all the rules—as well as the history of life on our planet.

As we saw in an earlier chapter, there is now abundant evidence that some kind of primitive photosynthesizing organisms found a way to grow on land surfaces hundreds of millions of years before the first animal, and in fact may have been a major cause of the last of the snowball Earth episodes between 700 and 600 million years ago. We have no idea what they were. Perhaps they were simply cyanobacteria, or perhaps they had real adaptions to land life, such as the ability to stay in place, obtain nutrients, reproduce, and get and then keep water. Candidates for this seem to be the still-extant single-celled green algae.

But even these plants from 700 million years ago may not have been first to get out of the sea, because an increasing number of geobiologists are concluding that there was land life far earlier: single-celled photosynthetic bacteria, making the water-to-land transition as much as 2.6 billion years ago. If so, these early colonists would have been long, long established when "higher" plants and animals finally climbed onto land as well.

What is known is that in less than 100 million years after the appearance of animals in the sea, some species of green algae, probably still living in freshwater, shed the shackles of a wholly aquatic lifestyle and migrated to the land, rapidly evolving from simple leafless twiglike plants not dissimilar to many moss species of today to true giants, thanks to one of evolution's great innovations: the leaf.

From about 475 million years ago, when the aquatic green algae began the numerous evolutionary changes that would allow them to attain nutrients—and most critically reproduce—in the combination of air and soil rather than entirely in water, to about 425 million years ago, when the fossil record shows the beautiful unmistakable remains of the first true vascular plants (those with roots and stems), the necessary changes were slow, step-by-step, and largely invisible to the fossil record. The evolution of these first small spiky and leafless plants to the first plant with true leaves took another 40 million years. But once the first leaves appeared, a great revolution of rapid change was unleashed. By around 370 to 360 million years ago, trees were up to twenty-five feet tall.

It took almost a hundred million years for the invading multicellular plants to change from small marine forms to the world-covering forests that were present by the end of the Devonian period. In one respect these plants had a far more significant effect on the land than the long-reigning microbes did, for the multicellular land-plant invasion utterly changed the nature of landforms and soil. It also changed the transparency of the atmosphere, for as more and more plants spread across the land, the restless sand dunes and dust bowls that had been the unceasing landforms of the Earth until that time were transformed. Roots began to hold the grit and dust of the land in place to a far greater extent than did the land-dwelling bacteria, which, as single cells or even thin sheets would have had little strength; as the primitive plants died and rotted in place, thicker and thicker soil began to form, and the ragged, rocky landscape that had always been Earth began to soften. From space the very air itself would have cleared; for the first time the edges of continents and seas, of large lakes and rivers would have become visible from short and great distances alike.

By the Late Devonian, forests had almost completely covered the land, changing the very way rivers moved across the landscape. And in so doing, plants ultimately caused atmospheric oxygen to climb far above the 21 percent found today, to levels as high as 30 to 35 percent—levels that allowed limbed lungless fish to crawl from the sea and survive the hundreds of thousands of years it would take to evolve an efficient, air-breathing lung. All of this conquest and change caused by land plants depended on a single great anatomical innovation—the evolution of the leaf.

LAND VS. SEA

Animal life emerged from the sea in a series of successive invasions, much like a succession of uncoordinated, ragtag, and poorly equipped and adapted armies might do—a few solders at a time, and most dying in the process. The standard explanation for this particular history is that these invasions took place because animals had finally evolved to a point where conquest of land was possible, with the driver being the presence of unexploited resources, less competition, and less predation (for a while, anyway). In other words, the evolutionary advances in arthropods, mollusks, annelids, and eventually vertebrates—the major animal phyla involved in the conquest of land—had finally and coincidentally arrived at levels of organization *allowing* them to climb out of the water and conquer the land. But our view is that the first conquest of land by animals took place as soon as atmospheric oxygen rose to levels allowing it.

Let us first look at what was required of both plants and animals to allow terrestrialization, the adaptations allowing life on land. Let us begin with plants, for without a food source on land, no animals would have made the effort to gain a terrestrial foothold.

By 600 million years ago, plant evolution had resulted in the diversification of many lineages of multicellular plants, some familiar to us still: the green, brown, and red algae that are familiar members of any seashore in our world.[5] But these were plants that had evolved in seawater. The needs of life—carbon dioxide and nutrients—were easily and readily available to them in the surrounding seawater. Reproduction was also mediated by the liquid environment. The move to land required substantial evolutionary change in the areas of carbon dioxide acquisition, nutrient acquisition, body support, and reproduction. Each required extensive modification to the existing body plans of the fully aquatic taxa. Much of this history is still disputed, especially with the understanding of how abundant and diverse various groups were in the Proterozoic era, even before the Proterozoic snowball Earth.[6] While the press loves anything that includes "oldest," "largest," or some other absolute, there is a disconnect between the rapid rate of discovery of the antiquity of land plants, their biological affiliations, and the need to more accurately date them. For instance, in 2010 the discovery of the "oldest" land plants was trumpeted based on new fossil discoveries from Argentina.[7] These fossils appear to be related to the common liverwort, and were dated at 472 million years in age. But the error on any such dating from such ancient rock is substantial. And besides, while these are indeed quite ancient "vascular" plants, kinds with complex internal transportation systems, in this case definitions of just what a plant is complicate the story. There were a lot of both body plans and species diversity of photosynthesizing organisms we can call plants well before 472 million years ago. Many paleobiologists suspect that a wide diversity of fungi as well as green photosynthesizing microbes to multicellular plants may have been on land earlier than is now considered, and that even a billion years ago there may have been a surprisingly vigorous and numerous assemblage of what collectively could be called plants, if we throw in lichens, fungi, and sheets of green microbes draping wetter landscapes and swamps.[8]

It was the green algal group, the Charophyceae, that ultimately gave rise to photosynthetic multicellular land plants that all can agree are true "plants," the kind of organism being described in most stories about oldest plants. Many obstacles had to be overcome; perhaps first among these was the problem of desiccation. A green alga washing ashore from its underwater habitat quickly degenerates and dies, as it rapidly desiccates in air, for there is no protective coating. But these green algae produce reproductive zygotes that have a resistant cuticle, and this same cuticle may have been used to coat the entire plant in the move onto land. But the evolution of this cuticle, which protected the liquid-filled plant cells inside, created a new problem: it cut off ready access to carbon dioxide. In the ocean, carbon in dissolved carbon dioxide was simply absorbed across the cell wall. So to accomplish this, in the newly evolved land plant, many small holes, called stomata, evolved as tiny portals for the entry of gaseous carbon dioxide.

The plant body must be anchored in place, and early land plants were probably anchored by fungal symbionts because there doesn't appear to be any differentiation in the higher forms. Additionally, this symbiotic relationship would provide for a means through which water could be recovered from the soil.

Moving onto land also created the problem of support. Plants need large surface areas facing sunlight. One solution is to simply lay flat on the ground, and the very first land plants probably did this. This kind of solution is still used by mosses, which grow as flat-lying carpets over soil. A visit to the Ordovician land probably would have been a visit to a moss world, where the world's tallest "tree" was all of a quarter inch tall. But this is a very limiting solution. Growing upright enables acquisition of much more light, especially in an ecosystem where there is competition between numerous low-growing plants, and various harder materials were incorporated by early plants to allow first stems and finally tree trunks. Concomitant would have been the evolution of a transport system from the newly evolved roots up to the newly evolved leaves. Finally, reproductive bodies that could withstand periods of desiccation evolved, ensuring reproduction in the terrestrial environment.

With these innovations, the colonization of land by plants was ensured, and with the formation of vast new amounts of organic carbon on land for the first time, animals were quick to follow. New resources spur new evolution. If the first terrestrial plants evolved from a small group of predominantly freshwater green algae, as is the most accepted opinion, they certainly did so without a lot of paleontological fanfare or evidence in the fossil record. They left behind a very fragmentary fossil record. Unearthing this fossil record (in both the literal and philosophical sense) required detective-like sleuthing of the first order.

The recovery of the fossil record of the earliest complex land plants began with a seminal 1937 paper, and for much of the discussion here, as well as the scientific history, we are indebted to our acerbic but brilliant colleague and friend, David Beerling of the University of Sheffield, who in his revolutionary book *The Emerald Planet* rather unapologetically complains that his field of Earth history, paleobotany, "gets no respect" in an almost comically Rodney Dangerfield way. But he is totally correct, in the sense that while dinosaurs and dinosaur hunters garner the lions' (raptors'?) share of scientific interest and glory, in fact, plants remain by far the most important group of organisms on Earth in terms of their effect on the history of life. A book about how our planet changed as a result of the "history" of life should have one chapter about animals and all the rest about plants. In any event, much of our take on the role of plants overtly comes from David's work, and especially his book.

The history of how land plants took over the terrestrial ecosystems, and in so doing changed the nature of life on Earth because of their effects on global temperature, ocean chemistry, and atmospheric inventory, can start with paleobotanist William Lander. Lander is the scientist who made these first discoveries and found the then-oldest-known land-plant fossil remains in 417-million-year-old rocks in Wales. (At the time these dates were completely unknown. In fact, the absolute age dates that we now use are a fairly new discovery.) While the 417-million-year-old fossils from Wales were thought to be the oldest record of land plants, soon other fossils began to appear in even older rocks, later dated as being 425 million years in age, also found in Wales.

This oldest plant was named as *Cooksonia*. From these early beginnings, land plants underwent a curiously long and much-delayed evolutionary radiation. Between 425 and 360 million years ago plants underwent their own version of the Cambrian explosion in animals; only this time it was an explosion of plants on land. But the newest view is that for at least 30 million years following the first appearance of land plants, not one of them had leaves. It now looks as if leafy plants were not firmly established until 360 million years ago.

There is indeed a mystery as to why leaves took so long. Even after the first appearance of leaves, it then took another 10 million years until they became widespread and distributed both in diversity and abundance throughout the planet. This extremely long period of time between the appearance of a land plant and that of a land plant with leaves can be compared with the much faster appearance of large and diverse mammals following the extinction of the dinosaurs, 65 million years ago. For the latter, it took no more than 10 million years for the major stocks of land mammals to appear, and appear not only in diversity but also in abundance and large size.

Once again we must look at the role of evo-devo and genes to understand this particular evolutionary history. Plants had to first evolve the genetic tool kit required to assemble leaves, but then they had to be able to use it, and the use seems to have been delayed. The best evidence to date indicates that plants with leaves had the genes necessary to build leaves, but then had to await changes within the environment in which they lived. In this particular case it was not a wait for the rise in oxygen—as it was for animals—but something entirely different: a wait for a drop in atmospheric carbon dioxide, as least according to the latest paleobotanical interpretations of the twenty-first century.

Here again is an example where the modern day can inform past history—our history of life. Experiments on living plants show that they are extremely susceptible to the level of carbon dioxide in which they live. All plants need carbon dioxide to undergo photosynthesis, but to do this, the plant has to absorb carbon dioxide out of the atmosphere around it. If there is a leaf, carbon dioxide has to enter through the otherwise impenetrable outer wall of the leaf. This is done through tiny holes called stomata. But there is a two-way street here. While carbon dioxide can enter through the stomata, water within the plant can also exit through the same holes. A theme that recurs over and over in the evolution of land animals and land plants is that desiccation remains one of the major obstacles to life. In high carbon dioxide settings, there are very few stomata. But when carbon dioxide is reduced, the number of stomata increases.

One would think that high levels of carbon dioxide would be the most optimal condition for any land plant. In terms of this physiology, in fact this is true. However, we know that carbon dioxide is one of the principal greenhouse gases. Times of high carbon dioxide are times of high heat on the surface of the Earth.

Plants have an exquisite signaling system, allowing fully grown and mature leaves to communicate with leaves just undergoing first growth and development. The larger leaves inform the smaller about the optimal number of stomata to produce for the environmental conditions they are all living in. If we go back in time to observe the levels of carbon dioxide in the atmosphere when land plants first began their evolutionary rise, over 400 million years ago, we see a period of extremely high carbon dioxide levels—and thus an extremely warm planet. So warm, in fact, that heat itself may have been a major brake on plant evolution and ecological success. The same stomata that let carbon dioxide in also allow the removal of water from the inside of the plant—and it is this process that actually cools the plant.

A little desiccation cools a plant, but a lot kills it, and as in so much, success comes from balance. In a very hot climate, a lot of cooling is needed. But in a high-CO_2 atmosphere, a plant needs very few stomata to handle its carbon dioxide needs. Yet the same number of stomata necessary for "ingestion" or carbon dioxide into the body of the plant might be too few to allow cooling—especially if the stomata are located on a large, flat surface—like a leaf. In such a case, a large leaf with few stomata will cause overheating to the point of death. This is the newest view of why it took so long for leaves to evolve. The genetic tool kit necessary to make them was in place. But the atmosphere had so much CO_2 in it that plants did not dare build leaves.

The new early twenty-first-century work of David Beerling and others suggests that it took a drop in carbon dioxide before leaves could be viable at all. Before this time any leaf would be a death sentence for the plant. Thus it was that it was only after 40 million years following the first appearance of *Cooksonia* that leaves as well as better internal plumbing systems within the plant (including new and deeper boring roots) first appeared. This latter, the ability to send roots to ever-greater depths, had two advantages for plants. First, deeper roots provided more stability. Second, deeper rooting gave greater access to both soil nutrients and water. The first plants have extremely shallow rooting systems. But once leaves evolved, roots also began to change and evolve to go ever deeper into the soil.

By the Devonian period, we see the evidence of roots that extended downward for as much as three feet. The new, deeper roots vastly increased the weathering of rocks beneath these early plants. As more plants lived in the soil, more and more of them died, adding organic material to the soil. At the same time, ever-deeper penetration by roots vastly increased both mechanical and chemical weathering of rocks beneath. This had important consequences for the makeup of the atmosphere as well as the temperature of the Earth.

We have seen that perhaps the most important driver removing carbon dioxide from the atmosphere is the weathering of silicate rocks, the granites, and sedimentary and metamorphic rocks with a granite-like chemical composition, a rock type rich in the element silicon. The reaction of chemically weathered silicate rocks on land is such that molecules of carbon dioxide are removed from the atmosphere. This is called the biotic enhancement of weathering, and it would have been taking place as soon as tree-rich forests began covering the land, about 380 to 360 million years ago. As roots went deeper into silicate rocks beneath, the granite and compositionally granite-like rocks of the continents began to weather much more quickly than the time before forests, and this caused carbon dioxide levels to plunge, and plunge quickly.

The lowering carbon dioxide levels allowed ice to appear on the continents, first only at the highest latitudes, but eventually at ever-lower latitudes. But the juggernaut of evolution favored taller trees, and with taller trees came deeper roots. Plants became taller, roots went deeper, and the planet became ever colder. The evolution of land plants with their ever-deeper rooting in fact plunged the planet into one of the longest-running ice ages ever in Earth history, one that began in the Carboniferous period. But before this happened, the world would have been warm, lush, and rich in plant-friendly levels of carbon dioxide. In short, the continents, newly green with vascular plants, would have been like a gigantic, stocked, but customer-free grocery store. Free food, if only you can get into the store. Or in this case, out of the sea and onto land—to stay.

THE FIRST LAND ANIMALS

The major problem facing any would-be terrestrial animal colonist was water loss. All living cells require liquid within them, and living in water does not provide any sort of desiccation problem. But living on land requires a tough coat to hold water in. The problem is that solutions that allow a reduction in surface desiccation are antagonistic to the needs of a respiratory membrane. So here we are twixt the devil and the deep blue sea: build an external coating that resists desiccation, an advantage, but at the same time risk death from suffocation. The alternative was to evolve a surface respiratory structure that allowed the diffusion of oxygen into the body, but caused increased risk of desiccation through this same structure. This dilemma had to be overcome by any land conqueror, and it was apparently so difficult that only a very small number of animal, plant, and protozoan phyla ever accomplished the move from water to land. Some of the largest and most important of current marine phyla certainly never made it: there are no terrestrial sponges, cnidarians, brachiopods, bryozoans, or echinoderms among many others, for instance.

The oldest fossil land animals all appear to have been small arthropods resembling modern-day spiders, scorpions, mites, isopods, and very primitive insects. It is unclear which of these quite different arthropod groups was first, but being first did not last long, as all of these groups are found in the fossil record in ancient deposits. Identifying these first land animals has necessarily relied on a fossil record that is notoriously inaccurate when it comes to small terrestrial arthropods. All of these groups have very weakly calcified exoskeletons, and thus are rarely preserved as fossils. By the Late Silurian or Early Devonian time intervals, however, or around 400 million years ago, the rise of land plants also brought ashore the vanguards of the animal invasion, and it is clear that multiple lines of arthropods independently evolved respiratory systems capable of dealing with air.

The respiratory systems in today's scorpions and spiders provide a key to understanding their successful transition from marine animals to successful terrestrial animals. Of all structures required to make this crucial jump, none was more important than respiratory structures. It also seems apparent that the earliest lungs used by the pioneering arthropods would have been transitional structures nowhere near as efficient as in later species. But in a very high oxygen atmosphere, air can diffuse across the body wall of very small land animals—and the first land animals all seemed to be small, as well as taking in oxygen by even their primitive lung structures.

Of the phyla that made it onto land, which included many kinds of arthropods, as well as mollusks, annelids, and chordates (along with some very small animals such as nematodes), the arthropods were preevolved to succeed, for their all-encompassing skeletal box was already fashioned to provide protection from desiccation. But they still had to overcome the problem of respiration. As we have seen, the outer skeleton of arthropods required the evolution of extensive and large gills on most segments to ensure survival in the low-oxygen Cambrian world where most arthropod higher taxa are first seen in the fossil record. But such external gills will not work in air. The solution among the first terrestrial arthropods, spiders, and scorpions was to produce a new kind of respiratory structure called a book lung, named after the resemblance of the inner parts of this lung to the pages of a book.

A series of flat plates within the body have blood flowing between the leaves. Air enters the book lungs through a series of openings in the carapace. This is a passive lung in that there is no current of air "inhaled" into these lungs. And because of this, they are dependent on some minimum oxygen content.

It is well known that some very small spiders are blown by winds at high altitudes and have been dubbed "aerial plankton." This would seemingly argue that the book lung system in spiders is capable of extracting sufficient oxygen in low-O_2 environments. But these spiders are invariably very small in size, so small that an appreciable fraction of their respiratory needs may be satisfied by passive diffusion across the body. Larger-bodied spiders are dependent on the book lungs.

Book gills may be more efficient at garnering oxygen than the insect respiratory system, which is composed of tubelike trachea. Like spiders and scorpions, the insect system is passive in that there is little or no pumping, although recent studies on insects suggest that some slight pumping may indeed be occurring, but at very low pressures. The book lung system of the arachnids has a much higher surface area than does the insect system, and thus should work at lower atmospheric oxygen concentrations.

The "when" of this first colonization of land is hampered by the small size and poorly fossilizable nature of the earliest scorpions and spiders. Present-day scorpions are more mineralized than spiders, and not surprisingly have a better fossil record. The earliest evidence of animal fragments is from late Silurian rocks in Wales, about 420 million years in age, near the end of the Silurian period—and a time when oxygen had already reached very high levels, the highest that had up to that time ever been evolved on Earth. These early fossils are rare and of low diversity, but identifications have been made: most of the material seems to have come from fossil millipedes.

A far richer assemblage is known from the famous Rhynie Chert of Scotland, which has been dated at 410 million years in age. This deposit has furnished fossils of very early plants, as well as the fossils of small arthropods. Most of these arthropods appear to be related to modern-day mites and springtails, which both eat plant debris and refuse, and thus would have been well adapted to living in the new land communities composed mainly of small, primitive plants. Mites are related to spiders. Springtails, however, are insects, presumably the most ancient of this largest of animal groups now on Earth today. It might be expected that once evolved, insects diversified into the most abundant and diverse of terrestrial animal life in our time. However, this was not the case, and in fact just the opposite appears to be true.

According to paleoentomologists, insects remained rare and marginal members of the land fauna until nearly the end of the Mississippian period, some 330 million years ago—when oxygen levels had reached modern-day levels, and in fact were on their way up to record levels, which climaxed in the Late Pennsylvanian period of some 310 million years ago. Insect flight also occurred well after the first appearance of the group, with undoubted flying insects occurring commonly in the record some 330 million years ago. Soon after this first development the insects undertook a fantastic evolutionary surge of new species, mainly flying forms. This was a classic adaptive radiation, where a new morphological breakthrough allows colonization of new ecological niches. But that radiation also took place at the oxygen high, and was surely in no small way aided and abetted by the high levels of atmospheric oxygen.

Insects were also not the first animals on land. That accolade may go to scorpions. In mid-Silurian time, some 430 million years ago, a lineage of proto-scorpions with water gills crawled out of the freshwater swamps and lakes that they were adapted to and moved onto and then about on land, perhaps scavenging on dead animals such as fish washed up onto beaches. Their gill regions remained wet, and the very high surface area of these gills may have allowed respiration of sorts. They certainly did not have functional lungs, only semi-serviceable gills.

Here is the timetable as we now know it: scorpions onto land about 430 million years ago (MA), but of a kind that may have been still tied to water for reproduction and perhaps even respiration; followed by millipedes at 420 MA, and insects at 410 MA. But common insects did not appear until 330 MA. How does this history relate to the atmospheric oxygen curve?

The newest estimates of atmospheric oxygen levels at this time indicate that a high oxygen peak occurred at about 410 million years ago, followed by a rapid fall, with a rise again from very low levels (12 percent) at the end of the Devonian to the highest levels in Earth history by somewhere in the Permian when it exceeded 30 percent (compared to 21 percent today). The Rhynie Chert, which yielded the first abundant insect-arachnid fauna, is right at the oxygen maximum in the Devonian. Insects are then rare in the record (according to paleontologists who study insect diversity) until the rise to near 20 percent in the Mississippian-Pennsylvanian, the time interval from 330 to 310 million years ago—the time of the diversification of winged insects.

The conquest of land by various vertebrate groups was seemingly enabled by a rise in atmospheric oxygen levels during the Ordovician-Silurian time interval. Had that not happened, it is possible that the history and kind of animals that did colonize land might have been much different—or it might never have happened at all; animals might never have colonized land. We also know that following this colonization, animals became seemingly rare, during the subsequent time of *low* oxygen.

There are three possibilities for this observed pattern of fossil abundances and diversity. First, this seeming pause in the colonization of land is not real at all; it is simply an artifact of a very poor fossil record for the time interval from 400 to about 370 million years ago. Second, the "pause" is real; because of very low oxygen there were indeed very few arthropods and especially insects on land. But the few that survived were able to diversify into a wave of new forms when oxygen again rose, some 30 million years later. Third, the first waves of attackers coming from the sea as part of the invasion of land were wiped out in the oxygen fall. Yes, here and there a few survivors held out. But the second wave was just that—coming from new stocks of invaders, again swarming onto the land under a curtain of oxygen. The colonization of land by animals (arthropods, and as we shall see, vertebrates as well) thus took place in two distinct waves: one from 430 to 410 million years ago, the other from 370 onward.

Arthropods are not the only colonists making a new life on land of course. Gastropod mollusks also made the evolutionary leap onto land, but did not make this transition until the Pennsylvanian (thus they were part of the second wave), when oxygen levels were even higher than at any time during the first wave. Another group that made shore were horseshoe crabs, at about the same time that the mollusks landed. But these are minor colonists compared to the group that most concerns this history of life—our group, the vertebrates. But amphibians did not just burst out of the sea. They were the culmination of a long evolutionary history, and before they emerge onto land in our narrative, let us look at the Devonian period, a time long called the Age of Fish. To do this we want to feature one of our favorite field areas, the Devonian-aged Canning Basin of Western Australia, where we two coauthors spent multiple field seasons in one of the most extraordinarily beautiful (if hot!) places on the planet. The Canning Basin preserves the world's best fossilized barrier reef system. It is as if the Great Barrier Reef were suddenly turned to stone and the water removed. While much of the work to date has been in studying that giant Devonian reef, in fact the rocks deposited in deeper water nearby during the Devonian Period have yielded some of the most extraordinary of all fossil studies that certainly need to be featured in any self-styled "new" history of life.

JOHN LONG AND THE GOGO FORMATION FISH

While common in salt water to freshwater and all salinities in between, in fact fish fossilize all too rarely. It usually takes a low oxygen sea bottom where a dead fish is rapidly buried for an entire fish to be preserved. Scavengers are all too efficient at tearing fish corpses apart. But here and there beautiful fish fossils can be preserved. Sometimes they appear in two-dimensional form, as from the Eocene-aged Green River shale of Colorado, perhaps the place where more fish fossils have been found than in any other locality. But other fish parts, especially big fish skulls, are sometimes preserved in large round balls of rock called concretions. These cannonball-like objects are often found in sedimentary rocks, and they can contain the most beautifully preserved of fossils. Such preservation is found in strata from northern Ohio of Devonian age, where gigantic fish skulls have been found for a century, including the skull of one of the iconic monsters, an ancient fish called Dunkleosteas, lately featured in the usually cheesy Discovery Channel programs about ancient predators. But such preservation is also found in a curiously named rock formation called the Gogo Formation, the same age rocks (but deeper water equivalents) of our own Devonian Age research. Among these cannonball concretions are some of the most important fossils ever found. They give us a window of the platform from which our amphibian ancestors ultimately emerged. To understand the conquest of land we first have to know the Devonian world of fish in all its diversity and complexity. In recent years Australian paleontologist John Long, a professor at Flinders University in Adelaide, Australia (but also with a long professional stint at the Los Angeles County Museum of Natural History), has taken new high-resolution-scanning technology to make breakthrough discoveries about the ancestry of all modern fish, as well as the lineages in deep time that are in our own DNA.

Long is a rarity in Australian academics in having a successful and thriving career in science outreach, and is the author of numerous books. But Long's "day job" has shown us that the evolution, morphology, diversity, and ecology of Devonian age fish was far more complex than is now portrayed in textbooks. By pioneering the use of imaging technology such as CT scanning, which bombards fossils with energy sufficient to produce 3-D slices of the fossils, Long has literally looked into the heads of the various fish groups.

The four "traditional" fish groups—today represented by lampreys and hagfish; sharks; the most diverse, the "bony" fish; and an entirely extinct group, the placoderms (the first jawed fish)—are far more complicated in all aspects than they have been long portrayed. Long's major discoveries from his field expeditions to the Gogo fossil sites included the first complete skull of one of the first bony fish, named *Gogonasus*, which showed that this species had large spiracles, or holes, previously unknown in fish on top of its head. But the most surprising discovery—beyond demonstrating a hitherto unknown diversity of other kinds of early fish, including new types of lungfish (closely related to the fish that ultimately crawled onto land) as well as strange fish called arthrodires—was the discovery of the first Devonian fishes showing embryos inside them. This latter discovery was the first time that reproduction by internal fertilization was demonstrated, as well as the oldest evidence for vertebrate viviparity yet discovered. One of his specimens was the only known fossil to show a mineralized umbilical structure linked to the unborn embryo. Long used his new high-tech methods to remarkably preserve 3-D muscle tissues, nerve cells, and microcapillaries, all new kinds of detail from fossil fish. But most important for understanding the move onto land, his soft tissue discoveries gave entirely new insight into how a fish could evolve ancestors that could walk—even upright on two legs.

THE EVOLUTION OF TERRESTRIAL VERTEBRATES

The transition of our own group from purely aquatic organisms to true terrestrial inhabitants began with the evolution of the first amphibians. The fossil record has given us a fair understanding of both the species involved in this transition and the time it happened. A group of Devonian period bony fish known as rhipidistians appears to have been the ancestors of the first amphibians. These fish were dominant predators, and most or all appear to have been freshwater animals. This in itself is interesting, and suggests that the bridge to land was first through freshwater. The same may have been true for the arthropods as well.

The rhipidistians were seemingly preadapted to evolving limbs capable of providing locomotion on land by having fleshy lobes on their fins. The still-living coelacanth provides a glorious example of both a living fossil and a model for envisioning the kind of animal that did give rise to the amphibians. But another group of lobe-finned fish, the lungfish, also are useful in understanding the transition, not in terms of locomotion, but in the all-important transition from gill to lung. The best limbs in the world are of no use if the amphibian-in-waiting could not breathe. There were thus two lineages of lobe-finned fishes, the crossopterygians (of which the coelacanth is a member) and the lungfish.

The split of the amphibian stocks from their ray-finned ancestors (in this case, the lobe fins) is dated at 450 million years ago, or at about the transition from the Ordovician period to the Silurian period. But this may have simply been the evolution of the stock of fish from which the amphibians ultimately came, not the amphibians themselves. Paleontologist Robert Carroll, whose specialty is in this transition, considers a fish genus known as *Osteolepis* the best candidate for the last fish ancestor of the first amphibian, and this fish genus did not appear until the early to middle part of the Devonian, or before about 400 million years ago.

The first land-dwelling amphibians may have evolved at this time, based on tantalizing evidence from footprints found in Ireland. A set of footprints from Valentia has been interpreted as being the oldest record of limbed animals leaving footprints, dated at about 400 million years in age. But there are no skeletons associated with this trackway, which is composed of about 150 individual footprints of an animal walking across ancient mud dragging a thick tail. This find has set off debate, since it predates the first undoubted tetrapod bones by 32 million years. Interestingly, however, the trackway dates to a time interval when oxygen levels either approached or exceeded current levels, and it is at this same time that the fossil record of insects, recounted above, yielded the first specimens of terrestrial insects and arachnids. Thus, just as the high oxygen aided the transition from water to land in insects, so too might it have allowed evolution of a first vertebrate land dweller.

The uncertainty about the age of the first vertebrate footprints on land was slightly alleviated by a discovery made in 2010, of a second set of tracks that was discovered to be 395 million years in age. They were preserved in marine sediments of the southern coast of (now) Poland. They were made during the Middle Devonian period. The tracks, some of which show digits, are thus 18 million years older than the oldest-known tetrapod body fossils. Additionally, the tracks show that the animal was capable of a type of arm and leg motion that would have been impossible in the more fish-like tetrapods and near tetrapods, such as the aforementioned *Tiktaalik* and its probable descendant, *Acanthostega*.

The animal that produced the tracks was large for the time: some estimates peg it at more than eight feet long. Perhaps this creature and its ilk were scavengers on the tidal flats, feeding on washed-up marine animals stranded by the tide, or the numerous land arthropods, including scorpions and spiders.

The first tetrapod bone fossils are not known until rocks about 360 million years in age, so the transition was in this interval between 400 and 360 MA. A rapid drop in oxygen characterizes this interval, and the first tetrapod fossils come from a time that shows minimal oxygen on the Berner curve. It is likely, however, that the actual transition from fish to amphibian must have happened much earlier, nearer the time of the Devonian high-oxygen peak but still in a period of dropping oxygen.

Most of our understanding about these crucial events comes from only a few localities, with the outcrops in Greenland being the most prolific in tetrapod remains. Although the genus *Ichthyostega* is given pride of place in most texts as being first, actually a different genus named *Ventastega* was first, at about 363 million years ago, followed in several million years by a modest radiation that included *Ichthyostega, Acanthostega*, and *Hynerpeton*.

Of these, *Ichthyostega* was the most renowned—until *Tiktaalik*, that is. Yet the new notoriety of *Tiktaalik* is a bit misplaced. It was a fish. *Ichthyostega* was something else. An amphibian, its bones were first recovered in the 1930s, but they were fragmentary, and it was not until the 1950s that detailed examination led to a reconstruction of the entire skeleton. The animal had well-developed legs, but it also had a fish-like tail. Later that further study showed that this inhabitant from 363 million years ago was probably incapable of walking on land. Newer studies of its foot and ankle seemed to suggest that it could not have supported its body without the flotation aid of being immersed in water.

The strata enclosing *Ichthyostega* and the other primitive tetrapods from Greenland came from a time interval that was soon after the devastating Late Devonian mass extinction, whose cause was most certainly a drop in atmospheric oxygen that created widespread anoxia in the seas. The appearance of *Ichthyostega* and its brethren may have been instigated by this extinction, since evolutionary novelty often follows mass extinction in response to filling empty ecological niches. But the success of *Ichthyostega* and its brethren was short-lived: the fossil record shows that within a few million years after its first appearance, it and the other pioneering tetrapods disappeared.

The appearance of *Ichthyostega* and its late Devonian brethren poses crucial questions. If these were indeed the first terrestrial vertebrates, why wasn't there a succeeding "adaptive radiation" of their descendants? But this did not happen. Instead there is a long gap before more amphibians appear. This gap has perplexed generations of paleontologists. In fact it came to be known as Romer's gap, after the early twentieth-century paleontologist Alfred Romer, who first brought attention to the mysterious gap between the first wave of vertebrates invading the land and the second. In fact, the expected evolutionary radiation of amphibians did not take place until about 340 to 330 million years ago, making Romer's gap at least 30 million years in length.

A 2004 summary by John Long and Malcolm Gordon similarly interpreted the tetrapods living in the 370- to 355-million-year-old interval, the time of a great oxygen drop, as entirely aquatic, essentially fish with legs, even though some of them had lost gills. Respiration was by gulping air in the manner of many current fish, and oxygen absorption through the skin. They were not amphibians as we know them today—species that can live for their entire adult lives on land. And it appears that none of the Devonian tetrapods had any sort of tadpole stage.

The long interval supposedly without amphibians was "plugged" in 2003 by Jenny Clack. While looking through old museum collections she came upon a fossil misinterpreted as a fully aquatic fish, but which she showed to be a tetrapod with five toes and the skeletal architecture that would have allowed land life. This fossil was given a new name, *Pederpes*, and it lived long after *Tiktaalik*. It indeed may have been the first true amphibian, and it did come from the time interval between 354 and 344 million years ago known as Romer's gap. But like so much about the past, sometimes fossils raise more questions than answers. It does tell us that somewhere in the middle of Romer's gap, a tetrapod did evolve the legs necessary for land life. However, it is still not known if it could breathe air or whether it could even emerge from the water for even for a few minutes.

Alfred Romer thought that the evolution of the first amphibians came about because of the effect of oxygen. Romer considered that lungfish or their Devonian equivalents were trapped in small pools that would seasonally desiccate. He thought that the lack of oxygen brought about by natural processes in these pools, as well as the drying, was the evolutionary impetus for the evolution of lungs—the amphibians-in-waiting were forced out of the pools and into the air. Gradually, those animals that could survive the times of emersion from water had an advantage. These fish still had gills, but the gills themselves allowed some adsorption of oxygen. It may be that the transitional forms had both gills and primitive lungs.

The transition from aquatic tetrapods such as *Ichthyostega* or, more probably, *Pederpes*, passed through the *Tiktaalik* grade of fish organization and involved changes in the wrists, ankle, backbone, and other portions of the axial skeleton that facilitate breathing and locomotion. Rib cages are important to house lungs, while the demands of supporting a heavy body in air, as compared to the near flotation of the same body in water, required extensive changes to the shoulder girdle, pelvic region, and the soft tissues that integrated them. The first forms that had made all of these changes can be thought of as the first terrestrial amphibians. Yet a great radiation of new amphibian species, which would be expected soon after the evolution of a respiratory system that could breathe air, not water, and limbs that could move a heavy body across land, did not occur until 340 to 330 million years ago. But when it did finally take off, it did so in spectacular fashion, and by the end of the Mississippian period (some 318 million years ago) there were numerous amphibians from localities all over the world.

The evidence at hand suggests that the evolution of the amphibian grade of organization, essentially a fish that came on land, may have taken place twice, or even three times, the first being some 400 million years ago as evidenced by the Valentia footprints as well as the *Tiktaalik* fossil discovery, and the second some 360 million years ago, and the last some 350 million years ago. *Ichthyostega*, long thought to mark the appearance of the first land vertebrate, may

Artist rendition of *Tiktaalik*, created for the Animal Planet program *Animal Armageddon*. (Art by Alfonse de la Torre in conjunction with Peter Ward, used with permission from Digital Ranch Productions, Rob Kirk)

have been far more fish-like than first thought, and the fact that it lost its gills is not evidence of a fully terrestrial habitat. In fact, we now know that over a hundred different kinds of modern fish use air breathing (as well as gills) of some sort. Air breathing has evolved independently in as many as sixty-eight of these extant fish, showing how readily this adaptation can take place. *Ichthyostega* may not even have been on the line leading to the rest of tetrapod lineages, but one that was evolving back into a fully aquatic lifestyle, forced off the land by its primitive lungs and the dropping oxygen levels of the Late Devonian.

It has long been assumed that the first amphibians were freshwater forms, and indeed this has been a major question in the history of life: was the route to land through freshwater first, or did some organisms evolve directly from salt water to air? However, new research has shown that early lobe-finned fish and lungfish—the immediate ancestors of the first tetrapods—were most often marine forms. Similarly, paleontologist Michel Laurin has noted several classical Carboniferous-aged localities that have yielded early amphibians and that have long been considered to represent freshwater deposits may in fact have been either marine or near marine deposits, such as intertidal or lagoonal environments. However, it seems equally sure that the famous *Tiktaalik* and some early amphibians such as *Ichthyostega* and *Acanthostega* have been interpreted as freshwater forms. It is thus likely that these first amphibians and near amphibians inhabited a wide variety of environments: salt water, freshwater, and terrestrial environments in the Late Paleozoic. This brings up an interesting point. Modern amphibians are intolerant of salt water; their skin, which takes in oxygen when immersed in water, cannot deal with the salt. This must be a trait evolved much later in their history.

In summary, colonization of the land came in two steps, each corresponding with a time of high oxygen. The time in between, the time of the Devonian mass extinction through the so-called Romer's gap, had little animal life on land. Thus Romer's gap should be expanded in concept to include arthropods as well as chordates.[9] It finally ended in the Carboniferous period (split in two in America, where we call it the Mississippian and Pennsylvanian periods), when oxygen levels rose in spectacular fashion, and in the last intervals of Carboniferous and then continuing into the successive Permian period, when the oxygen levels finally topped out at nearly 32 to 35 percent, creating a unique interval in Earth history. A time of giants.

CHAPTER XI

The Age of Arthropods: 350–300 MA

A staple of Hollywood in the immediate post–World War II interval, the dawn of the nuclear age, was the "giant-creature-produced-by-A-bomb-radiation" movie. Sometimes these monsters were examples of some kind of giant extinct life, often thawed out of some 70-million-year-old glacier. More often they were a familiar insect, scorpion, or spider of giant size. While easy to dismiss as "unscientific," these movie monsters do let us pose a legitimate question about the maximum size that can be obtained by any given animal body plan. Since large size is often a protection against predation, it seems that most animals grow as large as they can. What ultimately limits the size of animals? In the case of terrestrial arthropods (spiders, scorpions, millipedes, centipedes, and insects among a few other more minor groups) it is clear that two aspects of the arthropod body plan limited and still limit them from attaining large mammal-like size.

One of these is the exoskeleton. Because of scaling properties and strength of the material called chitin, the hard-part material that makes up most of the arthropod exoskeleton, a giant ant, spider, scorpion, or mantis of even human size would collapse, its walking legs snapping. The second aspect of arthropod design that limits size is respiration. Insects, spiders, and scorpions appear to be limited in size by the degree to which oxygen can diffuse into the innermost regions of their body. Today, no insect is bigger than about six inches in body length. In the past, however, much larger forms than this did exist, during the interval of the highest oxygen in Earth history.

THE CARBONIFEROUS-PERMIAN OXYGEN HIGH

While the various specialists modeling past atmospheric composition differ in values, their respective models suggest for past time intervals that there is unanimous agreement that oxygen reached extraordinarily high values in the time interval from about 320 to 260 million years ago, with maximal values occurring near the end of this interval. The Carboniferous period (again in North America subdivided into the Mississippian and Pennsylvanian periods) and the first half of the subsequent Permian period were the times of high oxygen, and the biota of the world at that time has left clear evidence of the high oxygen. Insects from the time present the best evidence.

The Carboniferous oxygen high (and much else as well) was well described by Nick Lane in his 2002 book *Oxygen*.[1] In a chapter titled "The Bolsover Dragonfly," Lane wrote about a fossil dragonfly discovered in 1979 that had a wingspan of some twenty inches. An even larger form, with a thirty-inch wingspan, is also known from fossils of this Carboniferous time, a beast aptly named *Meganeura*, yet another dragonfly. It was not only the wings that were large. The bodies of these giants were also proportionally larger, with a width of as much as an inch and a length of nearly a foot. This is about seagull size, and while seagulls are never linked in any sentence with the word "giant," an insect with a twenty-inch wingspan was indeed a veritable giant. In comparison, today's dragonflies may reach four inches in wingspan, but more commonly are smaller. Other giants of the time included mayflies with nineteen-inch wingspans, a spider with eighteen-inch legs, and two-yard-long (or longer) millipedes and scorpions. A three-foot-long scorpion could weigh fifty pounds, and would be a formidable predator of all land animals, including the amphibians. But, as we will see, the amphibians also evolved some giant species of their own.

In the case of insects, it is the nature and efficiency of the insect respiratory system in extracting oxygen and getting it into the most interior recesses of its body that dictates maximum size. All insects use a system of fine tubes, called trachea. Air actively ventilated into the tubes where it then diffused into the tissues. Air is pulled into the canals either by rhythmically expanding and contracting the abdominal region or by using the flapping of wings to create air currents around the tracheal opening. The tracheal system is thus made more efficient in either case. Flying insects achieve the highest metabolic rates of any animal, and experimental evidence shows that increasing oxygen to higher levels enables dragonflies to produce even higher metabolic rates. These studies showed that dragonflies are both metabolically and probably size limited as well by our current 21 percent oxygen levels.

Whether or not oxygen levels control arthropod size has been contentious. The best evidence that it does comes from studies of amphipods, small marine arthropods that are widely distributed in our world's oceans and lakes. Gauthier Chapelle and Lloyd Peck examined two thousand specimens from a wide variety of habitats and discovered that bodies of water with higher dissolved oxygen content had larger amphipods. More direct experiments were conducted by Robert Dudley of Arizona State University, who grew fruit flies in elevated oxygen conditions and discovered that each successive generation was larger than the preceding when raised at 23 percent oxygen. In insects, at least, higher oxygen very quickly promotes larger size.[2]

It was not only higher oxygen that allowed the existence of giant dragonflies. The actual air pressure is presumed to have been higher as well. Oxygen partial pressures rose, but not at the expense of other gases. The total gas pressure was higher than today, and the larger number of gas molecules in the atmosphere would have given more lift to the giants. There was clearly more oxygen in the air than now. The question is why.

Earlier we saw that oxygen levels are affected mainly by burial rates of reduced carbon and sulfur-bearing minerals like fool's gold (pyrite). When a great deal of organic matter is buried, oxygen levels go up. If this is true, it must mean that the Carboniferous period, the time of the Earth's highest oxygen content, must have been a time of rapid burial of large volumes of carbon and pyrite, and the evidence from the stratigraphic record confirms that this indeed happened—through the formation of coal deposits.

We are looking at a long interval of time: 70 million years, longer than the time between the last dinosaurs and the present day, in the 330–260-million-years-old time of high oxygen. It turns out that 90 percent of the Earth's coal deposits are found in rocks of that interval. The rate of coal burial was much higher than any other time in Earth history—six hundred times higher, in fact, according to Nick Lane in his book *Oxygen*. But the term "coal burial" is pretty inaccurate. Coal is the remains of ancient wood, and thus we see a time when enormous quantities of fallen wood were rapidly buried and only later through heat and pressure turned to coal. The Carboniferous period was the time of forest burial on a spectacular scale.

The burial of organic material during the Carboniferous was not restricted to land plants. There is much carbon in the oceans tied up in phyto- and zooplankton, the oceanic equivalents of the terrestrial forests, and here too large amounts of organic-rich sediments accumulated on sea bottoms. The ultimate cause of this unique buildup of carbon, leading to the unique maximum of oxygen levels, was the coincidence of several geological and biological events that culminated in the vast carbon deposit accumulations. First, the continents of the time coalesced into one single large continent by the closing of an ancient Atlantic Ocean. As Europe collided with North America and South America with Africa, a gigantic linear mountain chain arose along the suturing of these continental blocks.

On either side of this mountain chain great floodplains arose, and the configuration of the mountains also produced a wet climate over much of the Earth. Newly evolved trees colonized the vast swamps and their adjoining drying land areas that came into being. Many of these trees would appear fantastic to us in their strangeness, and one of their strangest traits was a very shallow root system. They grew tall and fell over quite easily. And there are lots of falling trees in our world, but nowhere near the accumulation of carbon. More was at work than a swampy world ideal for plant growth.

The forests that came into being some 375 million years ago were composed of the first true trees that used lignin and cellulose for skeletal support. Lignin is a very tough substance, and today it is broken down by a variety of bacteria. But even after nearly 400 million years, the bacteria that do this job take their own sweet time.[3] A fallen tree takes many years to "rot," and some of the harder woods, those with more lignin than the so-called soft woods like cedar and pine, take longer yet.

Decomposition of trees is accomplished by oxidation of much of the tree's carbon, so even if the end product is eventually buried, very little reduced carbon makes it into the geological record. Back in the Carboniferous, many or perhaps all of the bacteria that decompose wood were not yet present,[4] with the key to this the seeming inability of microbes to break down the main structural component of wood, the material lignin. Trees would fall and not decompose back then. Eventually sediment would cover the undecomposed trees, and reduced carbon was buried in the process. With all of these trees (and the plankton in the seas) producing oxygen through photosynthesis, and very little of this new oxygen being used to decompose the rapidly growing and falling forests, oxygen levels began to rise.

OXYGEN AND FOREST FIRES

The Carboniferous oxygen peak would have had consequences in addition to gigantism. Oxygen is combustible, and the more there is the bigger the fire; it facilitates fuel ignition, and the fuel in question was the huge and global forest of the coal age.

The Carboniferous period may have witnessed the largest forest fires ever to occur on Earth (at least until the dinosaur-killing and forest-igniting Chicxulub asteroid of 65 million years ago, that is). Like so much dealing with the change of oxygen over time, studies on the possibility of megafires provoked by high atmospheric oxygen have been controversial, but are becoming much less so as more and more evidence accumulates. Indeed, the forest fire controversy has been a major criticism of the entire theory that oxygen values have been different in the past (including higher). It was suggested that ancient forests would not have been able to survive the catastrophic fires, and since we have a long fossil record of the forest, the catastrophic fires did not take place.

Conditions of elevated oxygen at least theoretically should generate more rapid rates of flame spread, as well as lead to higher-intensity fires, and indeed large deposits of fossil charcoal in sedimentary rocks of Mississippian and Pennsylvanian age in North America[5] are evidence that there were forest fires back then: forest fires that were larger, more frequent, and more intense than those of today, although direct comparison suffers from the very different biological makeup of forests then and now.

If there were more and more intense forest fires, we would expect to see morphological adaptations to fire resistance over time. Plants evolved a well-known series of adaptations collectively known as fire-resistance traits, which include thicker bark, deeply embedded vascular tissue (Cambria), and sheathes of fibrous roots surrounding the stem.

One can also question why such high oxygen did not cause all Carboniferous forests to burn to the ground. While fires then do seem to have been more frequent, the presence of fire-resistant plants and the high moisture content both in the plants themselves and in the swampy terrain of large portions of the Earth's surface in the numerous coal swamps limited damage. Also important is the temperature of the "match" that started the forest fires. In recent studies[6] looking at oxygen levels and whether wood would burn, the investigators reported that plants will not burn at oxygen levels below about 11–12 percent. Yet they tried to start these fires with a lit match, rather than the far higher temperature than any lightning strike brings.

THE EFFECT OF HIGH OXYGEN ON PLANTS

Like animals, plants need oxygen for life. Oxygen is taken up within the cells during photorespiration. But the levels are far lower than those needed by animals for the most part. A second difference is that various parts of terrestrial plants, for instance, have different oxygen needs. Most plants live in two very different media—part in air, part in solid (the roots in soil). The very different environments of underground roots, surrounded by water, solids, and gas, required very different evolutionary needs. Leaves sit in air. They worry about losing water and getting enough light, not drowning in too much water (if they could worry, that is). But mostly the roots have a requirement that the leaves do not—the right level of oxygen. It is the root system that is most susceptible to damage or cell death from low oxygen, and this is all too familiar to those gardeners or house plant owners who water their plants too much. Roots live in the underground environment where low-oxygen conditions can occur, even at times of well-oxygenated air, especially if there is too much water in the soil. Roots can be smothered by groundwater with low oxygen values, for instance.

What about plants and high oxygen levels? Here there is far less data, but what is known suggests that elevated levels of O_2 are deleterious to plants. Higher levels of oxygen in air lead to increased rates of photorespiration, but a more serious consequence is that in higher oxygen levels there are more toxic chemicals called "OH radicals" that are dangerous to living cells. To further test these possibilities, David Beerling, a former student of Yale's Bob Berner, grew various plants in higher than current oxygen within closed tanks.[7] When oxygen levels were raised to 35 percent (thought to have been the highest levels of all time, occurring in the late Carboniferous or early Permian), the net primary productivity (a measure of plant growth) dropped by a fifth. It may be that the higher oxygen of the Carboniferous through early Permian caused a reduction in plant life to some degree, although this is not observable in the fossil record by any dramatic change or mass extinction during this interval.

OXYGEN AND LAND ANIMALS

The conquest of the land by the chordates, our lineage, required many major adaptations. Most pressing was a way of reproducing that allowed development of the embryo in an egg out of water. The amphibians of the Pennsylvanian and Permian presumably still laid eggs in water, and thus could not exploit the resources of land regions that were without lakes or rivers. The evolution of what is termed the amniotic egg solved this. Presumably, it was this egg that ensured the existence of a stock of vertebrates now known as reptiles. The evolution of the amniotic egg differentiates the reptiles, birds, and mammals from their ancestral group, the amphibians.

The fossil record suggests that the amniotes are monophyletic: that is, they have but one common ancestor, rather than this condition arising more than once. That ancestor, an amphibian, lived some time in the Mississippian, and thus this crucial transition took place as oxygen levels were rising. The first amniotic eggs were probably produced at oxygen levels equal or even higher than that of today.

Reptiles are also considered to be monophyletic, a single species stock that diverged from amphibian ancestors perhaps some time in the Mississippian period of more than 320 million years ago. As we have seen, this was a time of rising oxygen, and a time as well of a major diversification of land- and water-dwelling amphibians. But while genetic evidence of this divergence can be dated back to as long ago as 340 million years, fossils that are ascribed to the first reptiles (instead of terrestrial amphibians) have been recovered from several localities globally. Fossils of small reptiles named *Hylonomus* and *Paleothyris* have been found interred in fossilized tree stumps of early Pennsylvanian age, and it may be that the fossil record of this later appearance is more valid than the assumption of a Mississippian evolution of the group. In either case, these first reptiles were very small, usually only about four to six inches long.

The skulls of these first reptiles had no tympanum (eardrum) and thus they could not hear well, or at all, and unlike the labyrinthodont amphibians, they lacked the large pair of fangs that was found in most of the larger carnivorous amphibians. Compared to these huge amphibians, the first true reptiles had a postcranial skeleton adapted to provide better and surely faster locomotion. They had very long tails relative to their body.

That these forms laid the first amniotic eggs is still speculation. There are no fossil eggs in the stratigraphic record until the lower Permian, and this single find remains controversial. But the pathway to the amniotic condition probably passed through an amphibian-like egg (without a membrane that would reduce desiccation), but lay in a moist place on land. It would have been the evolution of a series of membranes surrounding the embryo (the chorion and amnion) covered by either a leathery or calcareous but porous egg that was required for fully terrestrial reproduction. One possibility seemingly never mentioned is that these first tetrapods evolved live birth, so that the embryos were not born until substantial development within the female had taken place.

Eggs capable of successfully producing viable offspring eventually were produced on land, and it was to these new amniotic eggs that the level of oxygen and heat must have played a part. There is a huge trade-off in reproduction for any land animal using an egg-laying strategy. Moisture must be conserved, so the openings of the egg must be few and small. But reducing permeability of the egg to water moving from inside to outside also reduces the movement of oxygen into the egg by diffusion.[8]

Without oxygen the egg cannot develop. It may be no accident that the evolution of the first amniotes occurred during a time of high oxygen. It seems inescapable that this reproductive strategy was and remains in those animals living at varying altitude, affected by atmospheric oxygen content, with higher oxygen contents producing more rapid embryonic development. High oxygen may have allowed live birth. Some biologists have suggested that live birth could not take place because, at least in mammals, the placenta delivers lower levels of oxygen than is present even in arterial blood in the same mother. But this generalization is for mammals only, where so much development takes place within an environment that can be regulated for its oxyen levels, temperature, and amount of liquid. Reptiles have a very different reproductive anatomy. It may be that low oxygen even favors live birth. Evidence to support this comes from three lines of evidence. First, it is well known that birds (egg layers) living in high-altitude habitats routinely feed at higher altitude than the maximum altitude where they can reproduce.

The maximum altitudes of birds' nests of many mountainous species repeatedly show this pattern. The highest nests are at eighteen thousand feet, and higher than this the embryos will not successfully develop.[9] While at least three factors may be involved in this limit (lowered oxygen content with altitude, desiccation because of air dryness at altitude, and relatively low temperatures), it may be oxygen content that is most important.

Second, recent experiments by John VandenBrooks from Yale University have shown that alligator eggs taken from natural clutches and then raised in artificially higher oxygen levels showed dramatically faster than normal development rates. The embryos grew some 25 percent faster than the controls held at normal atmosphere oxygen levels. Increased oxygen clearly influences growth rates, at least in American alligators. Finally, Ray Huey of the University of Washington maintains that a higher proportion of reptiles at high altitude use live birth than do those at lower altitude.

As four-legged vertebrates emerged from their piscine ancestors, many new anatomical challenges had to be overcome. No longer was there water to support the animal's body; in air, both support and locomotion had to be accomplished by the four legs. An entirely new shoulder and pelvic girdle design had to evolve, along with the muscles necessary to allow locomotion. Equally daunting was the problem of acquiring sufficient oxygen to allow sustained exercise. Early tetrapods apparently used the same set of muscles for motion and taking a breath, and they could not do both at the same time. Fish seem to have no problem with sustained exercise or with respiring during activity, suggesting that oxygen is not a limiting factor in daily activity. For land tetrapods, this is not the case. The body plan of the earliest land tetrapods provided for a sprawling posture with legs splayed out to the sides of the body trunk. In walking or running with such a body plan, the trunk is twisted first to one side and then to the other in a sinuous fashion. As the left leg moves forward, the right side of the chest and the lungs within are compressed. This is reversed with the next step.

The distortion of the chest during this kind of locomotion makes "normal" breathing impossible; each breath must be taken between steps. But this process makes it impossible for the animal to take a breath when running. Thus, modern amphibians and reptiles cannot run and breathe at the same time, and it is a good bet that their Paleozoic ancestors were similarly impaired. Because of this there are no reptilian sprinters. This is why reptiles and amphibians are ambush predators. They do not run their prey down. The best of the modern reptiles in terms of running is the Komodo dragon, which will sprint for no more than thirty feet while attacking prey. This is called Carrier's constraint, after its discoverer, physiologist David Carrier.

The dilemma of not being able to breathe and rapidly move at the same time was a huge obstacle to colonizing land. The first land tetrapods would have been at a huge disadvantage to even the land arthropods, such as the scorpions, for the vertebrates would have been slow and would have needed to constantly stop to take a breath. This is why we contend that oxygen levels would have been critical: only under high oxygen conditions would the first land vertebrates have had any chance of making a successful living on land.

One consequence of this problem was that the early amphibians and reptiles evolved a three-chambered heart. This kind of heart is found in most modern amphibians and reptiles, and is adaptive for creatures that have the problem of inferior respiration while moving. While a lizard is chasing prey it is not breathing, and thus the fourth chamber of the heart, which would be pumping blood to the lungs, is superfluous. The three chambers are used to pump blood throughout the body, but the price that must be paid is that it takes the lizard longer to reoxygenate the blood when activity ceases.

OXYGEN AND TEMPERATURE, REPRODUCTION AND THERMOREGULATION

At this point we can summarize and discuss the variables in land animal reproduction and try to relate these to generalizations about both oxygen levels and temperature. There are two possible strategies, as we have seen: egg laying or live birth. In the egg case, the eggs are either covered with a calcareous shell cover or a softer, more leathery shell cover. Today, all birds utilize calcareous eggs, while all living reptiles that lay eggs use the leathery covering. Unfortunately, there is little information about the relative oxygen diffusion rates for leathery—or parchment—eggs compared to calcareous eggs.

The utilization of egg laying or live birth has important consequences for land animals. The embryos developed by the live-birth method are not endangered by temperature change, desiccation, or oxygen deprivation. But the cost is the added volume of the parent, which must invariably make her more vulnerable to predation in addition to needing more food than would be necessary for the adult alone. Egg layers are not burdened with this problem, but have the trade-off of a less safe environment—the interior of an egg outside of the body—that leads to enhanced embryonic death rate through predation or lethal conditions of the external environment.

Before the end of the Mississippian period three great stocks of reptiles had diverged from one another to become independent groups: one that gave rise to mammals, a second to turtles, and a third to the other reptilian groups—and to the birds. The fossil record shows that there are many individual species making up these three. A relatively rich fossil record has delineated the evolutionary pathway of these groups. It has also required a reevaluation of just what a "reptile" is. As customarily defined, the class Reptilia includes the living turtles, lizards, and crocodiles. Technically, reptiles can now be defined by what they are not: they are amniotes that lack the specialized characters of birds and mammals. Less appreciated is that all three of these lineages originated in a world with extensive glaciation and very high oxygen. It is the assumption here that coming from a cold but high-oxygen world would have affected many aspects of the biology of these animals. Let us look at some of these characteristics.

One of the enduring questions about the history of life concerns the history of thermoregulation in animals. There are three distinct kinds: endothermy (warm-blooded), ectothermy (cold-blooded), and a third category (homeothermic) that is essentially neither of the others, and is associated with very large size. The evolution of each of these has long attracted scientific scrutiny, with thermoregulation pathways—most important, the question of whether or not dinosaurs were warm-blooded—being the most discussed and controversial of all. The fact that each of these characteristics is primarily either physiological or involved body parts that only rarely leave any fossil record (such as fur) is in large part responsible for the controversies.

We know that all living mammals and birds are warm-blooded, with the former having hair and the latter feathers, just as we know that all living reptiles are cold-blooded, with neither hair nor feathers. The status of extinct forms remains controversial. Of interest here is whether or not oxygen concentration and/or characteristic global temperatures affected thermoregulation or characteristic body covering in the various stocks in the past.

REPTILE DIFFERENTIATION

The number of openings in the skull is a convenient way of differentiating the three major stocks of "reptiles."[10] Anapsids (ancestors of the turtles) had no major openings or fenestra in their skulls; synapsids (ancestors of the mammals) had one; and diapsids (dinosaurs, crocodiles, lizards, and snakes) had two. The fossil record suggests that all three arose at a time of high atmospheric oxygen.[11] The earliest member of the latter group, the diapsids, is known from latest Pennsylvanian rocks, and it was small in size, about twenty centimeters in total length. From the time of their origin until the beginning of the fall of oxygen, which probably began in earnest some 260 million years ago, in the middle and late part of the Permian period, this group did little in the way of diversification or specialization. They remained small in size, and while the split to the various diapsid groups may have happened in the latest Pennsylvanian through early Permian (the time of highest oxygen), the animals themselves remained small and lizard-like. They gave no indication that they would be the ancestors of the largest land animals ever to appear on Earth, in the form of the Mesozoic dinosaurs. If the time of highest oxygen stimulated insects to their greatest size, the same cannot be said of the diapsids.

The most pressing questions are whether or not this group was warm-blooded and how it reproduced. No unequivocal Permian eggs are known at all from any group, so we cannot know how they bred. It is presumed that they laid primitive amniotic eggs with a leathery covering on land, but we cannot rule out the possibility of live birth. It was not until the latest Permian—well into the oxygen crisis that was to culminate in the greatest of all mass extinctions—that the diapsids were stimulated into the diversifications they would become famous for. After all, they gave rise to dinosaurs.

The diapsids evolved shapes allowing movement. They were fleet carnivores. One of the other reptile groups, the anapsids, took another direction. No one would accuse a turtle of being fleet afoot, and that is what the anapsids became: turtles, and before that, huge slow-lumbering monsters known as pareiasaurs, one of the largest of all skeletonized reptiles known from the late Permian world.

Based on their earliest members, however, it would have been hard to predict that the anapsids would become so slow and lumbering and hiding inside armor. They were initially smaller, faster, and very successful during the Late Pennsylvanian, but less so into the Permian. As the glaciers receded from the long ice age spanning the first half of the Permian period, they evolved into giant forms, including cotylosaurs and the even larger pareiasaurs. These were armored giants, surely slow moving, and herbivores that lived right until the end of the Permian. It is very likely that the gigantic size of the earlier Permian anapsids was allowed by high oxygen.

The last major reptilian group was the synapsids, and these were our own ancestors. If diapsids did little during the Pennsylvanian through the early Permian oxygen high, the same cannot be said of the third group of amniotes from this time, the synapsids, or mammal-like reptiles. Like the diapsids, the most primitive are known from Pennsylvanian rocks, and also like the diapsids of this time, these ancestors of the mammals had a small, lizard-like shape and mode of life in all probability. It is assumed that like the diapsids (and the amphibians that they came from), these early synapsids were cold-blooded. They, in turn, gave rise to two great stocks: the pelycosaurs, like early Permian *Dimetrodon*, and their successors, the therapsids, the lineage giving rise to the mammals. It is this latter group that is also called the mammal-like reptiles.

Unlike the diapsids, the synapsids diversified during the oxygen high and at the peak of oxygen became the largest of all land vertebrates. In the latter part of the Pennsylvanian, the pelycosaurs probably looked and acted like large monitor lizards, or even the iguanas of today, with splayed limbs. By the end of the Pennsylvanian some attained the size of the Komodo dragon of today, and they may have been fearsome predators. By the beginning of the Permian period, some 300 million years ago, they made up at least 70 percent of the land vertebrate fauna. And they diversified in terms of feeding as well. Three groups were found: fish eaters, meat eaters, and the first large herbivores.

Both predators and prey could attain a size of close to fifteen feet in length, and some, such as *Dimetrodon*, had a large sail on the back that would have made them appear even larger. They also either partially or totally solved the reptilian problem of not being able to breathe while running by changing their stance. The synapsids show an evolutionary trend of moving their legs into a position so that they were increasingly under the trunk of the body, rather than splayed out to the side, as in modern lizards. This created a more upright posture, and removed or at least greatly decreased the lung compression that accompanies the sinuous gait of lizards and salamanders. While there was still some splay of the limbs to the sides of the trunk, it was certainly less than in the first tetrapods. With the evolution of the therapsids in the Middle Permian, the stance became even more upright.

The sail present on both carnivores and herbivores of the Late Pennsylvanian and early Permian is a vital clue to the metabolism of the pelycosaurs; it was a device used to rapidly heat up the animal in the morning hours. By positioning the sail so as to catch the morning sun, both predators and prey could rapidly warm their large bodies, allowing rapid movement. The animal first attaining warm internal temperature would have been the winner in the game of predation or escape, and hence natural selection would have worked on this. But the larger clue from this is that during the oxygen high, the ancestors of the mammals had not yet evolved endothermia, or "warm-bloodedness." So when did this trait first appear? That revolutionary breakthrough must have happened among the successors to the pelycosaurs, the therapsids. We must note as well that this period, while a time of oxygen high, was a period of low temperatures. There was a great glaciation known from this interval, and a sizable portion of the polar regions of both hemispheres would have been covered in ice, both continental and sea ice.

While much of our understanding of pelycosaurs' evolution comes from fossils found in North America, younger beds in this region have few vertebrate fossils. The transition to the therapsids is best seen in Europe and Russia, but even here the transition is poorly known because of few fossiliferous deposits of the critical age. This gap in our knowledge of the synapsid fossil record extends from perhaps 285 million years ago to around 270 million years ago. Two main regions tell us about the history of this group: the Russian area around the Ural Mountains, and the Karoo region of South Africa. The record in the Karoo begins with glacial deposits perhaps as much as 270 million years in age, and then there is a continuous record right into the Jurassic, giving an unparalleled understanding of this lineage of animals.

The therapsids split into two groups: a predominantly carnivorous group and an herbivorous group. By about 260 million years ago the ice was gone in South Africa, but we can assume that the relatively high latitude of this part of the supercontinent Pangaea (about 60 degrees south latitude) remained cool. It was still a time of high oxygen, certainly higher than now, but that was changing. As the Permian period progressed, oxygen levels were dropping. Seemingly two great radiations of forms occurred, among both carnivores and herbivores. From perhaps 270 to 260 million years ago the dominant land animals were the dinocephalians, and these great bulky beasts reached astounding size: not dinosaur sized, but certainly approaching any land mammal today save, perhaps, elephants, and some of the largest of the dinocephalians certainly must have weighed as much as elephants. *Moschops*, for instance, a common and well-known genus from South Africa, was five meters high, with an enormous head and front legs longer than the back. It was hunted by a group of similarly sized carnivores.

The dinocephalians and their carnivores were hit by a great extinction, still very poorly understood, that occurred some 260 million years ago. There is still little range data for both the dinocephalians and their immediate successors in terrestrial dominance, the earliest

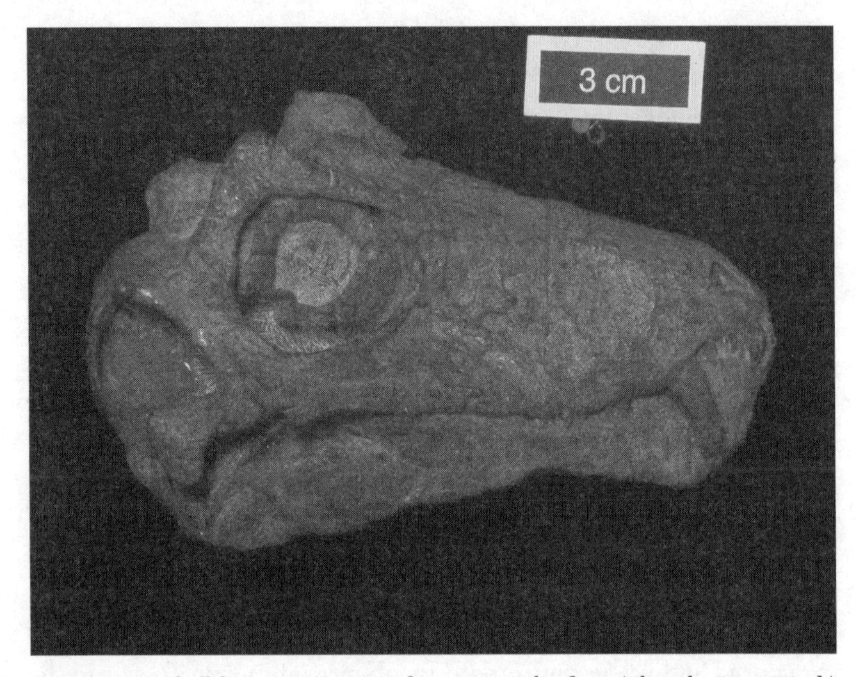

Gorgonopsian skull from Late Permian deposits, South Africa. (Photo by Peter Ward.)

dicynodonts and their predators. Until new fossils from South Africa and Russia are obtained, this uncertainty will remain. Sadly, there are few fossils of this age and fewer paleontologists studying them, so we may not know for generations, assuming that future generations continue to hunt fossils.

The dicynodonts were the dominant herbivores of the time from 260 to 250 million years ago. They were almost eliminated from the planet in the Permian extinction, which we will describe in more detail in the next chapter. They were hunted by three groups of carnivores: the gorgonopsians, which died out at the end of the Permian, the slightly more diverse therocephalians, and the cynodonts, which ultimately evolved into mammals during the Triassic.

ANIMAL SIZE AND OXYGEN LEVELS

The rise of atmospheric oxygen to unprecedented values of over 30 percent was accompanied by the evolution of insects of unprecedented size. The giant dragonflies and others of the late Carboniferous through the early Permian were the largest insects in Earth history. Perhaps it is just coincidence, but most specialists agree that the high oxygen would have enabled insects to grow larger, since the insect respiratory system requires diffusion of oxygen through tubes into the interior of the body, and in times of higher oxygen, more of this vital gas could penetrate into ever larger-bodied insects. So if insects got larger as oxygen rose, what about vertebrates? New data indicates that this is true as well.

In 2006, paleontologist Michel Laurin measured fossil skull lengths and body lengths of various reptiles ranging from the Carboniferous through the Permian, from about 320 million years ago until about 250 million years ago. Both of the size descriptors closely tracked oxygen levels. As O_2 levels rose in the Late Carboniferous, so too did the size of the reptiles increase, and, as O_2 began to drop in the mid Permian, size began to trend downward. As we will return to in the chapter on Cenozoic mammals, study on (much) later mammals, by Paul Falkowski and his colleagues, demonstrated a very similar phenomenon during the Early Cenozoic, when oxygen levels have been modeled to have risen significantly, while at the same time, the mean size of mammal species also increased.

This trend of changing size also occurred among the mammal-like reptiles as the Permian Period came to a close. The largest therapsids of all time, the dinocephalians of the Middle Permian, evolved at the peak of oxygen abundance. As oxygen began to drop in the mid-Permian, successive taxa assigned to various therapsid groups, and most important the dicynodonts, showed a trend toward smaller skull sizes. While some relatively large forms still lived in the latest Permian—the genus *Dicynodon* and even the carnivorous gorgonopsians come to mind—by this time many of the dicynodonts were smaller. The latest Permian taxa *Cistecephalus, Diictodon*, and a few others were very small. Research in 2007 showed that the Late Permian through Early Triassic genus *Lystrosaurus* was smaller in the Triassic than it was in the Permian, and the various cynodonts of the late Permian and early Triassic, as oxygen levels were precipitously falling, were all small in size. There are exceptions—a few giants in the Triassic named *Kannemeyeria* and *Tritylodon* are examples—but in general the therapsids of the Triassic are much smaller than those of the Permian. A recent paper by our colleague (and now at the University of Washington) Christian Sidor has confirmed the drop in size. Thus there is a strong correlation between terrestrial animal size and oxygen levels from the latest Permian into the Triassic. In high oxygen, tetrapods grew large, and then they grew smaller as oxygen levels diminished.

Antonio Lazcano, origin of life specialist and humanist, in Galapagos Islands contemplating a "lower" life form. (Photo by Peter Ward.)

THE FIRST AGE OF MAMMALS

Yale University's great Peabody Museum is home to one of the largest collections of fossils in the world. It also is home to the greatest paleontological paintings ever done.

There are two gigantic murals gracing an immense wall in the Peabody Museum of Natural History on the Yale University campus. For generations of Americans these two murals—*The Age of Reptiles*, painted over three years (1943–1947), and *The Age of Mammals*, painted over six years (1961–1967)—have been the iconic views of land life's journey through time.

The first, *The Age of Reptiles*, begins in dark swamps and ends with exploding volcanoes towering over *T. rex*. The second also begins in the jungle, but one with very different and very familiar vegetation. Combined, they tell us that amphibians begat reptiles, which begat mammals. But our view would now require two very different murals to correctly show the vertebrate assemblages in these deep time periods pictorially represented. In fact, we advocate here that there were three separate "ages" of mammals (knowing, of course, that an "age" is nothing more than informal labeling and categorizing, without scientific validity).

The first age of mammals was in the Permian period, the heyday of the therapsids and their ancestral synapsids. *Technically* they are not yet mammals. But they were close. It was a species-rich as well as numerically abundant assemblage. In South Africa, there were as many as fifty genera at a time (and since a normal genus normally contains several [to many] species, the actual diversity at the species level was higher yet; perhaps 150 species is a conservative estimate).

South Africa today is not so different latitudinally and perhaps even climatically from the South Africa of southern Gondwanaland, some 255 million years ago. Today there are 299 species; we can imagine the African veldt of today, but stocked with *Dicynodon* instead of the large herbivores, and many kinds of carnivores from the lion-sized gorgonopsians to the weasel-sized theriodonts. Vast herds grazing not on grass but the low, bushy *Glossopteris* and ferns. Africa of the first age of mammals.

The second age of mammals can be thought of as the time between the late Triassic and the end of the Cretaceous: mammals chained. Held in check by the dinosaur overlords. Living in the ecological cracks: at night, in burrows, in trees. Never bigger than a house cat, and usually far smaller.

Finally, the third age of mammals. Zallinger's age of mammals. The post-K-T mass extinction outpouring of species filling the families so well known to us today. This is the story most obvious to us: from ratlike survivors of the Chicxulub asteroid's wrath to the early giants such as titanotheres and uintatheres (rhino-like beasts) to the mammalian panoply we know so familiarly today.

Until about 2000 we knew the first age of mammals largely from the South African Karoo desert. But in the twenty-first century vast new collections have been made from north central Africa by Christian Sidor, and in Russia another gigantic assemblage is now known, thanks to paleontologist Michael Benton's work. In this second age, the mammals remained very small. It would not be until the Paleogene that mammals would finally gain ascendancy, and like some long-denied heir, would finally get an "age" named after them.

One could almost believe that the whole age of dinosaurs was a big mistake. That but for one huge flood of basalt there might have been a quite different history. Human intelligence 250 million years ago? It did not take long to go from apes to something more advanced not so long ago.

CHAPTER XII

The Great Dying—Anoxia and Global Stagnation: 252–250 MA

THE Karoo desert of central South Africa can be a bit of a disappointment to its first-time visitors. When the two words "Africa" and "desert" are found in the same phrase, there is often an image of the Sahara desert, Africa's most famous dry place, or the Kalahari, another vast wasteland with little life because of its shifting sands and harsh conditions of blazing heat by day and freezing temperatures each night. With animals and plants having such a hard time, and existing at such low standing diversity and abundance, it is no wonder that the human populations in the Sahara and Kalahari are limited in size as well. Very little in the way of plant or animal crops can be farmed there.

Unlike these two African deserts, the Great Karoo desert has no shifting sand dunes; it is mainly rock that is often well vegetated, and there seems no place in its vastness where one cannot find sheep dung, evidence for the ubiquity of this introduced species. There are no elephants or giraffes, no hippos or crocodiles or water buffaloes or rhinos; it has animal life, and in places lots of it, but its species are not ones redolent in the memory of Africa. There are also quite a few people, on large ranches. Thus the Karoo is not the place for desert-seeking tourists. What it does have, however, is a hundred-million-year-long accumulation of sedimentary rocks deposited in a time interval from about 270 million years ago to perhaps 175 million years ago.

In the middle of this vast rock heap is the world's best record of large terrestrial animal life living both before and after that most consequential of all mass extinctions, the Permian-Triassic mass extinction. Generations of paleontologists going back to the middle 1800s have searched the ancient riverbeds and river-valley floors that the Karoo strata were created by and in. Animals often are carried into rivers after death, or are in waterholes where they may have been attacked long ago, leaving bones to fall into mud and become preserved there. This region was the prime record for this period until very recently, with new work in both eastern Russia by our colleague Mike Benton of Bristol and the north central part of Africa, in the country of Niger, where another of our colleagues, Christian Sidor of the University of Washington, have unearthed important new records.[1] Yet even these new regions cannot compare the richness and temporal resolution that the Karoo rocks have given us—if "given" should be used at all. In fact, the Karoo has given up its vast store of information about one of the most critical times in life's history on Earth very grudgingly. It has to be taken, and while the work to do this seems glamorous (who does not dream of being a paleontologist finding a giant leering skull of some ancient predator, such as *T. rex*), it is at best difficult on the humans who pursue this passion.

A drive from Cape Town into the center of the Karoo is an all-day affair. But because the rocks are slightly tilted, while the landscape inexorably rises in altitude as one travels north and east into the Karoo, the entire book of strata that the Karoo holds can be read from its ancient mid-Permian-period cover to its Jurassic, dinosaur-bearing last chapter. It is not only time that changes as one goes upward through the many thousands of aggregate feet that is the Karoo sedimentary record. One starts in a time of ice and icebergs and ends in what may have been one of the hottest times in Earth history, as well as passing through an interval tens of millions of years in length when atmospheric oxygen receded to its lowest level since animals first occurred at all, nearly 600 million years ago. Yet if much can be understood by reading this entire record, there is one interval of rock, representing time, that has been more studied than any other.

These are the several hundreds of meters of strata deposited between 252 and 248 million years ago—rocks deposited in the last millennia of the Permian period (and thus the Paleozoic era, which ends with the end of the Permian) and first few millions of years following the vast mass extinction of 252 million years ago.

For decades now, geoscientists have been asking several principal questions of these rocks, and their rare but often exquisitely preserved skulls and body skeletons: First is the question of how long the mass extinction took, from the start of extinction rates first exceeding the normal "background" extinction rate, which has been calculated to have been about one extinction each five years. Second, we want to know if the catastrophic extinction on land took place simultaneously with the Permian marine mass extinction. Third, and perhaps most interesting, is the question of what caused the mass extinction. Finally, it is important to discover how quickly terrestrial ecosystems recovered, because these latter clues might give us useful information for surviving any future Permian-like mass extinction, a prospect far more probable than our species seems to realize.

To paraphrase one of the great twentieth-century paleontologists, David Raup of the University of Chicago: Were the surviving species gifted with good genes—or simply good luck?

RESULTS OF THE PERMIAN EXTINCTION

If intense controversy still exists about the cause or causes of the Permian extinction, on one aspect of that time interval are all in agreement: in the aftermath of the extinction ecosystems were profoundly affected, and extinction recovery was long delayed. It is this latter evidence that readily distinguishes the Permian extinction from the later Cretaceous-Tertiary event. While both caused more than half of the species on Earth to disappear, the world recovered relatively quickly after the "K-T" event. This may have been due to different causes for the two. Asteroid impact on the Earth and the environmental destruction accruing from the impact have for more than a decade been accepted as the cause of the K-T event. But the killing conditions following the impact soon dissipated. This was not the case after the Permian event. As we have seen above, while some Earth scientists believe that the Permian as well as the K-T events were caused by large-body impacts on the Earth, it seems as if the environmental conditions causing the Permian extinction persisted for millions of years after the onset of the extinction. It is not until the Middle Triassic, some 245 million years ago, that some semblance of recovery seemed to be under way.

These results would be expected if some part of the Permian mass extinction were directly or indirectly caused by the reduction in oxygen at the end of the Permian. The newest Berner curves show that oxygen stayed low into the Triassic, and there is even some indication that the oxygen levels did not bottom out and begin rising until near the end of the lower Triassic, which might account for the long delay in the recovery. This evidence suggests that the environmental events producing extinction just kept persisting. If so, and if animals were capable of any sort of adaptation in the face of these deleterious conditions, we would predict that the Triassic would show a host of new species not only in response to the many empty ecological niches brought about by the mass extinction, but might also show new species arising in response to the longer-term environmental effects of the prolonged extinction event itself. This is the pattern that is observed during the Triassic; the world was refilled with many species looking and acting like some of those going extinct (therefore an ecological replacement), but a host of novel creatures also appeared, especially on land. In the next chapter we will postulate that many of the latter new species evolved to counter the continued low oxygen that had begun near the end of the Triassic, but that continued right into the Jurassic, a period of more than 50 million years. The Triassic was truly the crossroads of animals adapted to two different worlds, one of higher oxygen and one of low.

THE CONTROVERSY: IMPACT VS. GREENHOUSE

With the end of the twentieth century and the arrival of the twenty-first, ever more attention was indeed being paid to the Permian extinction, largely because it was the most devastating of all, with the now oft-repeated estimate that as many as 90 percent of all species disappeared. But how fast, which is a clue to how, began to be best appreciated with the work of paleontologists from China and the United States in extensive studies of a thick Permian and Triassic limestone cropping out near Meishan, China.[2] Geologists worked to plot the thickness and identity relative to each other of every sedimentary layer. Then fossils were collected from the beds that had been so meticulously measured.

Each fossil was carefully identified, and its collection level in the piles of strata noted. The paleontologists made use of Charles Marshall's new statistical method, called confidence interval methodology,[3] which allowed estimates of the ultimate time range a given fossil might have had. The geologists in China had a great advantage going for them. In China there were scattered ash layers that could be dated using sensitive machines to measure uranium/lead isotope ratios, and this was done most recently on samples by MIT's Sam Bowring.[4] The newest work by this group now puts the extinction as lasting no more than sixty thousand years, which is amazing resolution in rocks a quarter billion years in age.

The Chinese effort combined results from five different stratigraphic sections in the Meishan locality, with sampling intervals made every thirty to fifty centimeters. A total of 333 species of marine life were ultimately found in these rocks, belonging to such varied sea creatures as corals, bivalve and brachiopod shellfish, snails, cephalopods, and trilobites among others. Nowhere at any stratigraphic horizon at any time has so thorough a collecting effort, or so rich a fauna, been documented with such precision.

The various environmental conditions in the seas at the end of the Permian included widespread evidence of oceanic anoxia, or low oxygenation in both the shallow and the deep sea. This was worked out beautifully in 1996 by Yukio Isozaki of Tokyo University, who located the boundary in deep-sea bedded cherts that had been thrust onto the Japanese mainland. Precisely around the mass extinction event, the normally red charts turned a deep black, just as everything died. The anoxia was apparently of such magnitude that many marine organisms were rather suddenly killed off, just as they are today in modern red tides. There is also evidence of global warming at the time of the extinction, and the coincidence of the Siberian lava eruptions at the same time as the mass extinction.

There have been various suspects as to the cause of this extinction. First is the possibility that the Siberian flood basalts introduced large volumes of gas into the atmosphere, triggering large-scale climate change and acid rain, as suggested by Berkeley geolochronologist

Paul Renne and others. With new information from disparate sources, a sudden methane release into the atmosphere became a viable candidate for the killer. But in spite of no evidence to support impact, the understanding that impact could cause extinction was still on everyone's mind. The new evidence from China argued for some sort of "quick strike." Among potential causes of mass extinction, only asteroid impact was thought to be capable of such mass death in so short a time.

At the turn of the century, Earth historians were enamored with large-body extraterrestrial impact as the cause of most, if not all, mass extinctions. In 2000, the Permian extinction looked like nothing known: it was still suspected to be some sort of impact extinction by the geological fraternity, but one seemingly different from the dinosaur-killing K-T event that had made sensational news in 1980. Perhaps the Permian extinction was many impacts, or a single large impact superimposed on some other kind of extinction mechanism. The most puzzling thing was that search as they might, none of the investigators looking at the Chinese rocks in the late twentieth and early twenty-first century could find the well-known clues associated with the impact extinction ending the Cretaceous already by then so well studied at the many K-T boundary sites, such as iridium, glassy spherules, and shocked quartz grains.

In 2001, and then over the next several years, a team led by geochemist Luann Becker reported[5] the discovery of high levels of complex carbon molecules given the ridiculous name Buckminsterfullerenes, mercifully shortened to Buckyballs. They used this evidence to argue that like the mass extinction of the end of the Cretaceous, the Permian extinction was also the result of the collision of a large asteroid with the Earth. Only this one hit 251 million years ago.

The Buckyballs described by this team are large molecules that contain at least sixty carbon atoms, and because they have a structure resembling a soccer ball or a geodesic dome, they were named for architect Buckminster Fuller, the inventor of the geodesic dome. The hypothesis is that geodesic-dome-like carbon molecules trapped the gases helium and argon inside their cage structures, and that these new indicators of impact exist in strata of latest Permian age at three different geographic sites scattered around the world. The Becker et al. team interpreted these particular Buckyballs as extraterrestrial in origin, and therefore like iridium (which, pointedly, was not found), because the noble gases trapped inside have an unusual ratio of isotopes. For instance, terrestrial helium is mostly helium-4 and contains only a small amount of helium-3, while extraterrestrial helium, the kind found in these fullerenes, is mostly helium-3. According to the authors, all this star stuff could have been brought to Earth only by a comet impacting the Earth at the end of the Permian (more correctly, it ended the Permian).

The researchers announced that the comet or asteroid was six to twelve kilometers across, or about the size of the K-T asteroid that left the huge Chicxulub crater near what is now the town of Progreso on Mexico's Yucatán Peninsula 65 million years ago. But such a large Permian impactor would be expected to have left a monstrous crater, just as the later Chicxulub impact did, and so the Becker team began an earnest search of potentially overlooked or buried impact craters.

Two years later, in 2003, they announced that they had found a giant buried crater in the seabeds off Australia.[6] The case for an impact cause for the Permian extinction seemed made. But then problems arose, in both the interpretations of the Buckyballs and the probability that the large underwater structure named the Bedout crater was an impact crater at all.

Science is about replication and prediction (among other things), and on both points the Permian extinction Buckyball hypothesis ultimately collapsed (although, curiously enough, impact and Buckyballs were, in 2012, still the first response spit out by Googling "Permian extinction"). But we workers searching for the causes of this mass extinction had our doubts early, and a certainty that the hypothesis could not be correct.

The original Becket et al. study was based on samples taken in China, Japan, and elsewhere. Later work could not replicate the results from China, and our friend Yukio Isozaki had shown several years earlier that the critical boundary interval that Becker had sampled near Osaka in Japan had actually been removed by low-angle faulting right at the boundary interval—three entire conodont zones on either side of the boundary were missing. Yet they reported that the helium-3 anomaly was there, just where they had been told (erroneously) the boundary should be. Something was fishy. Eventually our colleagues at Caltech demonstrated that helium-3 leaks out of a Fullerene cage in fewer than one million years, so none should have been left after 252 million years. Furthermore, the deep structure interpreted to be the crater that gave rise to all the Buckyballs, helium-3, and death to the world's biota turned out to be a great emplacement of volcanic rocks unrelated to any sort of asteroid or comet impact.

A group of geologists and organic chemists teamed together and used a fairly new tool to look at latest Permian and early Triassic marine strata. Rather than looking for body fossils, they extracted organic residues from the strata[7] in search of chemical fossils, which, if found, are known as biomarkers. The biomarker recovered can come only from a photosynthesizing purple bacteria species that can exist only in shallow water devoid of oxygen and saturated with toxic hydrogen sulfide. Apparently a great biomass of H_2S-producing microbes filled the oceans—not just a small area like the Black Sea of today, but most or even all of the world's oceans—based on newer studies by teams from MIT, who by 2009 had discovered the same biomarker in more than a dozen localities of latest Permian age scattered across the globe.[8]

A possible solution to the enigma of the cause of the largest of all mass extinctions came from a team of geochemists from Penn State in 2005. Led by Lee Kump of Penn State, one of the world's foremost experts on the chemistry of the oceans and especially its carbon cycle, along with his longtime colleague Mike Arthur (also of Penn State), their paper suggested the H_2S present at the end of the Permian, produced in the sea by microbes (a different species from the purple sulfur bacteria, to be accurate), was directly involved in the extinctions both on land and in the sea.[9]

THE KUMP HYPOTHESIS—AND THE DAWN OF THE GREENHOUSE EXTINCTION THEORY

The Kump et al. scenario is as follows. If deepwater H_2S concentrations increased beyond a critical threshold during oceanic anoxic intervals (times when the ocean bottom and perhaps even its surface regions lose oxygen), then the oceanic conditions (such as those in the modern Black Sea) separating sulfur-rich deep waters from oxygenated surface waters could have risen abruptly to the ocean surface. The horrific result would be great bubbles of highly poisonous H_2S gas rising into the atmosphere. This new entry into planetary killing provides a link from the marine to the terrestrial extinctions, because H_2S accumulates in the troposphere to lethal levels for plants and animals under relatively modest fluxes of H_2S from the ocean. This proposal relates not only to the end of the Permian, but may have occurred at other times in Earth history, and thus was perhaps a dominant perturbation causing mass extinctions.[10]

Kump and his team did some rough calculations and were astounded to conclude that the amount of H_2S gas entering the late Permian atmosphere would be more than two thousand times greater than the small modern flux (this is the toxic killer coming from volcanoes). Enough would have entered the atmosphere to most likely lead to toxic levels.

Moreover, the ozone shield, a layer that protects life from dangerous levels of ultraviolet rays, would also have been destroyed. Indeed, there is evidence that this happened at the end of the Permian, for fossil spores from the extinction interval in Greenland sediments show evidence of the mutation expected from extended exposure to high UV fluxes attendant on the loss of the ozone layer.

Today we see an ozone hole in the atmosphere over Antarctica, under which the biomass of phytoplankton rapidly decreases. If the base of the food chain is destroyed, it is not long until the organisms higher up are perturbed as well. The complete loss of our ozone layer has even been invoked as a way to cause a major mass extinction if the Earth was hit by particles from a nearby supernova, which would also destroy the ozone layer. Finally, an abrupt increase in methane concentrations significantly amplifies greenhouse warming from an associated CO_2 buildup and methane levels that would have risen to >100 ppm. As the H_2S goes into the atmosphere, at the same time destroying the ozone layer, greenhouse gases also do their work in making the planet hotter. It turns out that the lethality of H_2S increases with temperature. Thus a new and plausible alternative to impact was put forth. The extinctions would have been drawn out, or in pulses—a succession of short-term events, killing each time.

Up until now we have looked at evidence from the rocks themselves. But there is a second way to unravel past events, and that is to use some of this data to model what past atmospheres have been like. There are many kinds of such models, and many are relevant to trying to predict what our Earth's future atmosphere and heat level may be like. For the Permian, levels of oxygen and carbon dioxide, as well as potential global temperatures, have been modeled. First, changes in atmospheric CO_2 and O_2 have been calculated by Yale's Bob Berner. He and others have found that there must have been a pronounced spike in CO_2 levels accompanied by plunging oxygen levels at the end of the Permian. Second, Lee Kump's group undertook the difficult job of looking at the potential distribution of H_2S emission around the globe. For this they used a global circulation model, or GCM.

These models were originally developed to understand modern-day weather and climate patterns. But because the positions of the continents are known for the critical period at the end of the Permian and into the Triassic, as well as temperatures and levels of oxygen and carbon dioxide in the atmosphere and oceans, the method could be applied to the Permian. Kump and his team reasoned that the critical element to track would be phosphorus. This is a prime component of fertilizer, and if oceanic phosphorus levels were observed to rapidly rise at the end of the Permian, the amount of hydrogen sulfide gas could be calculated as well due to the beneficiaries of the raised phophorus levels—the sulfur microbes.

The emergence of H_2S did not happen once; it occurred over and over, as succession of burps clustered around the time that the

Permian-Triassic boundary strata were being deposited around the world. Kump finished with the most ominous note. Not only did the model show where the H_2S would emerge from the sea into the air, but he also showed new calculations that completely corroborated his earlier 2005 estimates of how much H_2S would have eventually gone into the atmosphere. The results: there would have been more than enough to kill off most land life, and as the nasty stuff also dissolves in seawater, it would have been greatly lethal in shallow marine settings as well, especially among shallow-water organisms that secreted calcium carbonate skeletons, such as corals, clams, brachiopods, and bryozoans—all invertebrate victims of the greatest extinction.

Since the introduction of the Kump et al. interpretation, others including Tom Algeo of the University of Cincinnati have greatly increased our understanding of the chemical aspects of this particular mass extinction, through numerous references.[11]

ALTITUDINAL COMPRESSION

The study of past mass extinctions is not new; in fact it is one of the very first kinds of research that could actually be called "science" when geology was first stirring as a discipline, in the first years of the nineteenth century. What is new about it is our understanding of the role of microbes in causing one or more of the so-called big five mass extinctions of the Phanerozoic.

Yet if extinctions themselves are not a new topic, the opposite side of the coin—the aftermath to mass extinction—has emerged in the past decade as a major new subdiscipline of evolutionary biology and paleobiology alike. We have learned that the more devastating the mass extinction, the more different the world coming after, not only in the immediate aftermath—the first few hundred thousand to million years later—but for subsequent tens of millions of years, and for some biological lineages, for all time.

A previously unrecognized aspect of the changing oxygen levels would be their effect on species migration and gene flow. Mountain ranges in our world are often barriers to gene exchange, producing

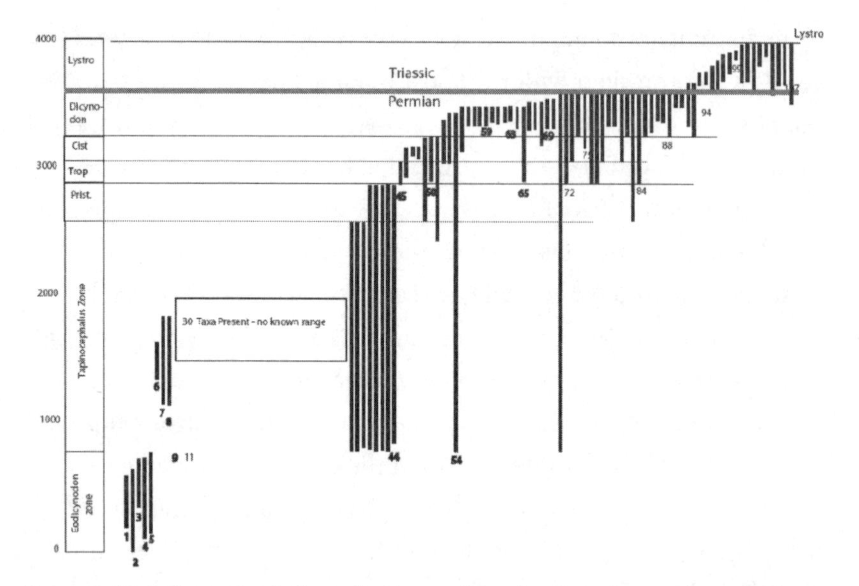

Ranges of vertebrate fossils from the Karoo of South Africa, from approximately 260 to 250 million years ago. Each vertical line represents a genus of vertebrate animal (based on fossils recovered from the strata). While most extinctions took place over a fairly narrow interval, this pattern is nothing like that seen at the end of the Cretaceous. The "smeared" appearance of extinctions shown here is characteristic of "greenhouse mass extinctions"—not a single level of extinction, but successive extinctions.

different biota on either side of the range. At the end of the Permian, just living at sea level would have been equivalent to breathing at five thousand meters, a height that is greater than that found atop Mount Rainier in Washington State. Thus even low altitudes during the Permian would have exacerbated this, so that even a modest set of hills would have isolated all but the most altitude- or low-oxygen-tolerant animals. The result would be a world composed of numerous endemic centers hugging the sea level coastlines.

The high plateaus of many continents may have been without animal life save for the most altitude tolerant. This goes against expectation based on continental position: because the continents at this 250-million-year-old time were all merged into one gigantic supercontinent (named Pangaea), we would expect a world where there were very few terrestrial biotic provinces, since animals would be able to walk from one side of the continent to the other without an Atlantic Ocean in the way. But altitude became the new barrier to migration, and new studies of various vertebrate faunas appear to show a world of many separate biotic provinces, at least on land.

The late twentieth- and early twenty-first-century work of Roger Smith, Jennifer Botha, and coauthor Ward in the Karoo desert, Mike Benton in Russia, and Christian Sidor in Niger[12] showed that each of these separate localities in Africa had distinct and largely nonoverlapping faunas. Thus, during times of low oxygen, altitude would have created significant barriers to migration and gene flow.[13] Low-oxygen times therefore should have had many separate biotic provinces, at least on land. The opposite occurred during high-oxygen times: there would be relatively few biotic provinces and a worldwide fauna.

The drop in oxygen did more than make mountain ranges barriers to migration. It made most areas higher than a thousand meters uninhabitable during the late Permian through Triassic time interval. This effect, called altitudinal compression, could have had a major impact on Triassic land life in the time of lowest oxygen. The removal of habitat because of altitudinal compression would have caused species from the highlands to migrate toward sea level or die out. Doing so would have increased competition for space and resources, and perhaps would have introduced new predators, parasites, or diseases in the previously populated lowlands, causing some number of species to go extinct. We calculated that by the end of the Permian more than 50 percent of the Earth's land surface would no longer have been habitable because of altitudinal compression. There may even have been extinction caused by the effects modeled long ago by Robert MacArthur and E. O. Wilson in *The Theory of Island Biogeography*. Those two scientists noted that diversity is related to habitat area, and that when islands or reserves of some sort became smaller, species died out. Altitudinal compression would accomplish the same by making the continental landmasses functionally lower in usable area.

PERMIAN EXTINCTION REDUX

A final aspect of the Permian extinction comes from research not yet published, but because it comes from coauthor Ward it will be reported here, as it is most pertinent to the topic of the Permian mass extinction. One of Ward's graduate students, Frederick Dooley, combined with Lee Kump to produce an unexpected discovery. Dooley studies the effect of hydrogen sulfide on plants and some animals; Kump has been modeling ocean conditions at the end of the Permian, including estimated amounts of hydrogen sulfide in the global oceans' surfaces. Kump arrived at a value that Dooley then used in actual experiments on single-celled oceanic phytoplankton, as well as the most important oceanic zooplankton, the tiny shrimp-like creatures called copepods. The levels were not sufficient to kill the algae, and surprisingly actually made them grow faster. The copepods, on the other hand, died almost instantly. Without copepods to feed on phytoplankton and keep it in check, these tiny plants sink to the sea bottom and rot, removing any last vestige of oxygen. This would produce a great oscillation in the carbon isotope pattern, as well as kill off every marine animal species that has an early life history in the upper water column as temporary plankton. The result would be a planet choked in rotting plants but nearly without animals. This is exactly what happened at the end of the Permian—in the oceans, anyway. On land it would have been very much like some combination of World War I and II combined. Roger Smith of South Africa now has very credible evidence of an extraordinary period of drying and sudden heat in the South Africa of 252 million years ago, while our own work on the vertebrates in the Karoo, published in 2005, remains the best record of land animal extinction across that boundary.[14] Roger Smith thinks that the drought and heat alone can account for the extinction of most vertebrates. We maintain that the world war analogy is apt: great armies dying in the desert, and in World War I, killed by poisonous chlorine gas. Long ago it was death in the desert from poisonous hydrogen sulfide in the air and sea.

CHAPTER XIII

The Triassic Explosion: 252–200 MA

ONE of the great joys of academics is the sense of community among the faculty, be it a community college or the most high-powered research institution in the land. Much of this comes from the very nature of the American university system, which requires a six- or seven-year trial period, followed by tenure. Permanence. Perhaps more than in any other profession, university faculties have a high stability, and compared to most other professions, a relatively low rate of turnover. The result is that relationships can literally last for appreciable parts of one's lifetime. In this the university faculty systems are indeed much like the system they were spawned from, the cloistered seminaries where monks would start as young men and then pass through life with others of their kind. And as was true in the old abbeys, with age and wisdom one learns to respect those with even greater experience—and listen to them.

In the year 2000 or thereabouts, the authors of this book were at lunch with several of the eldest of the science faculty of the California Institute of Technology. One of these elder statesmen was the great Sam Epstein, one of the most distinguished professors of geochemistry, perhaps of all time. Sam was present in the halcyon days at the University of Chicago, when Nobel laureate chemist Harold Urey discovered a way to measure the temperature at which ancient carbonate rocks were formed by comparing the isotopes of oxygen found in the precipitated carbonate rock. The ratio of the isotope O^{16} varied with the much more rare isotope O^{18} in proportion to the temperature of formation.

Sam eventually moved to Caltech and spent his career making high-precision measurements of many kinds of samples, using many different methodologies. But his first love seemed to be ancient temperatures. After a wonderful lunch, he took Kirschvink and Ward to his downstairs lab, which was in the process of being dismantled.

The geochemical equipment of the 1950s and 1960s, Sam's heyday, was mainly composed of handmade and hand-blown glassware, walls of thin tubes spiraling, crisscrossing, making spider webs of glass interrupted by strange flask-like shapes, rubber tubing coming and going, greased glass stopcocks of exquisite manufacture—everything custom made by the artisans who kept science in those days going, the skilled technicians now banished by budget cuts and the new generations of solid-state technology.

We walked through the lab, and conversation moved on to a topic then of our keen interest: the Permian mass extinction and its possible causes. At that moment the impact hypothesis was still viewed as the probable cause. Sam, however, would have none of it. He turned to us with a smile and told us the following short story. In his younger days he had taken samples of marine limestone that dated to the earliest Triassic, samples that had probably been formed in a very shallow seaway somewhere near the Permian equator in what is now Iran. On a whim, or because that is what he loved most, Sam began analyzing these samples for their ancient temperatures. He was stunned, he said, to find that all had been formed in temperatures above 40°C, with some of the temperatures exceeding 50°C—from 104°F to over 120°F! The samples had come from ancient corals, creatures that need water of normal salinity.

Such temperatures can be found in the stagnant pools and lagoons. But brachiopods do not live in such places. The temperatures found by Sam Epstein could not have been formed anywhere on our Earth. They spoke of a postextinction world of unreal water temperature in the main ocean.

Sam, then in his eighties and with only another year ultimately to live, smiled a sad smile. He told us that he never had the guts to publish these data. Any paleotemperature analysis requires really pristine samples to be accurate, and quite often samples looking as if they had not been reheated or exposed to groundwater or chemically changed in any obvious way had, in fact, had their oxygen isotope temperatures "reset," and such resets were normally to produce what looked like abnormally high temperatures. This process becomes ever more common the older the sample. But Sam was quite convinced that he had proof of ocean water temperatures above 100°F in the first million years after the Permian extinction—in the first million years of the Triassic.

Several years later, in analyzing paleotemperatures from a different, lower Triassic site, we too found what looked like one-hundred-degree-plus water. This time the depth was even greater than the estimated ancient water depths where Sam Epstein's Triassic brachiopods had grown so long ago. Like Sam Epstein, we did not publish these results.

The prize never goes to the faint of heart. In 2012 a joint Chinese-American research team, trying to understand why it took so long for life to recover in the seas after the Permian extinction, published an amazing paper.[1] Their findings: water temperature of 104°F in the sea, and a blistering 140°F on land! Unlike the work of Epstein, this study involved the analysis of over fifteen thousand samples, making it the most detailed and painstaking look yet at the environmental conditions in the aftermath of the Permian extinction.

The scientists completing this study allowed themselves to speculate about what that ancient hot world would have been like. Most marine organisms die above the plus-100°F level found by the investigators; in fact photosynthesis essentially stops at temperatures much above this. In that world, the entire zone of the topics would have been devoid of animals, and complex life would have hung on only at high latitudes. Land animals would have been rare even in the mid-latitudes. In such heat there would have been enormous volumes of moisture in the air, and the topics would have been wet year-round. But it might have been a wet desert, with no plant life at all.

Ever better geochronology now shows that this time of high temperature extended at least for the first 3 million years of the Triassic, and indeed may have been climbing ever higher during that time, with a maximum temperature occurring during a time interval known as the Smithian stage (a million-year time interval of around 247 million years ago) having the highest of all known temperatures since the time when animals first occurred. Sam Epstein was right.

Our data from Opal Creek[2] were right. We were wrong in not publishing those data.

The Permian extinction was clearly one of the most fundamentally catastrophic of all events—if, that is, one was a multicellular plant or animal. From a microbe's point of view—especially one of the sulfur-loving, oxygen-hating microbes that made up the majority of all life on Earth from its very inception right up to the first evolution of animals—that event was like a return to paradise. Seen from our vantage point so long after, the Permian extinction was a repeat of what happened at the end of the Devonian, itself the first of what we now call greenhouse extinctions. Many more were destined to come at the end of the Triassic, multiple times in the Jurassic and Cretaceous, and ending with the last-known greenhouse extinction at the end of the Paleocene epoch, some 60 million years ago. But none were ever to be so great as the Permian event, or to unleash a more diverse assemblage of animals in the aftermath of extinction.

The Permian extinction gave the world many new creatures, but for us, two entirely new lineages, both thriving and evolving by the end Triassic period. In no small way the Permian extinction brought to life mammals, and brought about the means that would create our long-term nemesis, the dinosaurs. Yet while being among the most important of all land animals (few animal groups are awarded an "age of . . ." before their names), the Triassic dinosaurs and mammals were late arrivals in the Triassic explosion, and remained both relatively small of stature (especially the mammals, which rarely exceed rat size) and small in both absolute abundance and species diversity. The age of dinosaurs was not to start until the successive Jurassic period, while the still-running age of mammals had to await the Cenozoic era.

Long before the late (Triassic) arrival on the evolutionary stage of dinosaurs and mammals, the other animals and plants of the Triassic period make up a most interesting assemblage of organismal characters, cast with new versions of already long-running taxa mixed with entirely new entrants, new designs arising yet radically different from the actual survivors of the Paleozoic era. It is this mix that makes the Triassic appear to be a veritable crossroads in time. In some ways it was not unlike the Cambrian explosion—a slew of newly invented body plans filling up an empty world, just as the first animals rapidly evolved into the cornucopia of body plans that filled the seas after the extinction of the first animals, the Ediacarans. And like the great Cambrian explosion, many of the body plans of novelty turned out to be but short-term experiments, to be pushed into extinction by the competition and/or predation of better-designed organisms. There is no time period other than the Cambrian and Triassic in which such a diversity of new forms appeared. Two reasons seem paramount: The Permian extinction emptied the world to such a degree that virtually any new design would work, for a while at least. But there is a second, new view of the Triassic that may be just as (or more) important than this.

Just coming out of the most devastating of all mass extinctions, this early Triassic world was very, very empty of life. At the same time, all modeling suggests that a long interval of the Triassic was a time of oxygen levels lower than those today. Earlier we suggested that times of low oxygen, especially following mass extinction, foster disparity: the diversity of new body plans. These two factors combined to create the largest number of new body plans seen since the Cambrian, and here we propose that it is to that seminal Cambrian time that we most accurately compare the Triassic. We call this time and its biotic consequences the Triassic explosion.

The Triassic was a time of amazing disparity on land and in the sea. In the latter, new stocks of bivalve mollusks took the place of the many extinct brachiopods, while a great diversification of ammonoids and nautiloids refilled the oceans with active predators. Fully a quarter of all the ammonites that ever lived have been found in Triassic rocks, a time interval that is only 10 percent of their total time existence on Earth. The oceans filled with their kind, in shapes and patterns completely new compared to their Paleozoic ancestors, and why not, for, as shown above, this kind of animal was the preeminent, low-oxygen adaptation among all invertebrates. A new kind of coral, the scleractinians, began to build reefs,[3] and many land reptiles returned to the sea. But it is on land that the most sweeping changes in terms of body plan replacements and body plan experimentation took place. Never before and never since has the world seen such a diverse group of different anatomies on land. Some were familiar Permian types: the therapsids that survived the Permian extinction diversified and competed with archosaurs for dominance of the land early in the Triassic, but this ascendance was short lived. The many kinds of reptiles were locked in a competitive struggle with them, and with each other, for land dominance. From mammal-like reptiles to lizards, earliest mammals to true, the Triassic was a huge experiment in animal design.

On the face of it, the mammals should have come out competitively "ahead" of the pure reptiles. After all, most of the mammal-like reptiles by this time were warm-blooded, probably capable (as now) of far more parental care than the presumably egg-laying dinosaurs; the mammalian teeth, one of the main reasons that mammals eventually did dominate the world, in their endlessly malleable tooth morphologies allowed all kinds of food acquisition, from small seeds to grass to meat of many kinds. Yet they did not win. Their extinction closed out the first age of mammals and gave rise to the second—composed of a very different group of mammals.

One of the major changes that has and continues to allow entirely new kinds of study of all groups of extinct animals is the great revolution in communication, morphological characterization and image analyses, and profound literature search skills that the computer revolution has allowed. Now large databases can be produced and then searched and analyzed in lightning blazes of microprocessor skill.

No longer does each fossil have to be laboriously measured by hand with micrometers, and no longer is it a single investigator traveling from museum to museum to do the work. Almost every new study that brings change to our history of life comes from large teams of investigators, ultimately inputting huge numbers of numbers to be crunched. Now the machines do much of this for us. And the results can produce new insights.

One such study by paleontologists Roland Sookias and Ludwig Maximilian, of the University of Munich, looked at the sizes of Triassic vertebrates that lived on land.

In this and subsequent work by this group it was found that only two major body plans emerged in the Early Triassic amid the emptiness left behind by the Permian mass extinction: those with four legs (quadrupeds), and those that used only two (bipeds). As the nearly 50-million-year-long Triassic period progressed into the Jurassic, with its own 50-million-year-long time interval, they found that the saurians diversified into a far larger number of species and shapes (and absolutely bigger in size, one measure of disparity) than did the mammal-like reptiles. While paleontologists long intuited this by perusing collections, here were numbers for the first time to substantiate this.

Their study also confirmed that the saurian grew faster, reaching adulthood and large size faster than did the other group. This "time to breeding" difference might be the most important metric of all. Faster growth and breeding meant that the saurians quickly adapted to the ecological roles of large herbivores and big predators before the smaller, slower-growing therapsids had a chance to evolve into these anatomical forms and ecological niches.

Questions remain. During the Late Triassic, when dinosaurs were well established, it would be expected that they would have immediately grown large, Jurassic large, and would have been common as well. According to Chicago paleontologist Paul Sereno, who has done more than any other to bring the earliest times of the dinosaur hegemony to light, neither was true. For almost 20 million years, from their first appearance some 221 million years ago until the end of the Triassic, about 201 million years ago, dinosaurs and therapsids alike remained both relatively rare and small in size.[4] There may have been more of them than the therapsids during this time, but the overall picture is that neither group was doing very well. Our own take on this is that nothing on land was doing very well at all, and that, in fact, it was perhaps far more advantageous for the four-legged land animals to return to the sea, which they did in higher numbers during the Triassic than at any other time in Earth history.

The conventional answer for the reason for the Triassic explosion is that the Permian extinction removed so many of the dominant land animals that it opened the way for more innovation than any nonextinction time, or perhaps any other mass extinction time as well. Perhaps, as well, it was simply that many terrestrial animal body plans finally came to an evolutionary point of really working efficiently. Even as late as the end of the Permian and into the Triassic, groups as evolutionarily mature as the mammal-like reptiles (the groups dicynodonts and cynodonts, by this time) were still trying to attain the most efficient kind of upright posture, rather than the less efficient, splayed-leg orientation of the land reptiles, with all of the ramifications and penalties that this entailed.

Body plans were being evolutionarily modified by intense selective pressures, and dominant among these was the need to access sufficient oxygen to feed, breed, and compete in a low-oxygen world. There is an old adage about nothing sharpening the mind faster than imminent death. The same might be said about evolutionary forces when faced with the most pressing of all selective pressures, which was attaining the oxygen necessary for the high levels of animal activity that had been evolutionarily attained in the high-oxygen world of the Permian, when nothing was easier to extract from the atmosphere. The two-thirds drop in atmospheric oxygen certainly lit the fuse to an evolutionary bomb, which exploded in the Triassic. Thus the diversity of Triassic animal plans is analogous to the diversity of marine body plans that resulted from the Cambrian explosion. As we have earlier recounted, the Cambrian explosion followed a mass extinction (of the Ediacaran fauna), and it was a time of lower oxygen than today. The latter stimulated much new design.

TRIASSIC REBOUND

The officially designated early Triassic time interval was from 250 to about 245 million years ago, and during this time there is little in the way of recovery from the mass extinction. The oxygen story for the Triassic is stunning. Oxygen dropped to minimal levels of between 10 and 15 percent, and then stayed there for at least 5 million years, from 245 to 240 million years ago. There is also a very curious record of large-scale oscillation in carbon isotopes from this time, indicating that the very carbon cycle was being perturbed in what looks like either a succession of methane gas entering the oceans and atmosphere or a succession of small-scale extinctions taking place. Again, the similarity to the early Cambrian is striking.

All evidence certainly paints a picture of a stark and environmentally challenging world for animal life. Microbes may have thrived, especially those that fixed sulfur, but animals had a long period of difficult times. However, difficult times are what best drive the engines of evolution and innovation, and from this trough in oxygen on planet Earth emerged new kinds of animals, most sporting respiratory systems better able to cope with the extended oxygen crisis. On land two new groups were to emerge from the wreckage: mammals and dinosaurs. The former would become understudies while the latter would take over the world.

As we saw in the last chapter, the Permian extinction annihilated almost all land life. The therapsids were hard hit. Much less is known about the archosauromorphs (reptiles with a somewhat crocodile-like anatomy), for at the end of the Permian they are a rare and little-seen group in areas such as the Karoo or Russia that have yielded rich deposits with abundant dicynodont (mammal-like reptile) faunas. In the Karoo desert, at least, very few well-preserved archosauromorphs have come from uppermost Permian study sections worked on by coauthors Ward and Kirschvink in the company of South Africa's Roger Smith.

If we are still poorly informed about their Permian ancestry, there is no ambiguity about the success of the earliest Triassic archosauromorphs. In the Karoo, in only a few meters of the strata that seem to mark the transition from Permian to Triassic there are relatively common remains of a fairly large reptile known as *Proterosuchus* (also known as *Chasmatosaurus*). This was definitely a land animal with a very impressive set of sharply pointed teeth. It was also definitely a predator, but like those of a crocodile, its legs were splayed to the sides (if somewhat more upright than the crocodilian condition). But this condition was to rapidly change in the archosauromorphs to a more upright orientation as the Triassic progressed, and more gracile and rapid predators soon replaced the early archosauromorphs such as *Proterosuchus*.

While the need for speed was surely a driver toward this better locomotor posture, just as important may have been the need to be able to breathe while walking. Like a lizard, *Proterosuchus* may still have had a back and forth sway to its body as it walked, and as we have seen previously, this sort of locomotion causes compression on the lung area due to what is known as Carrier's constraint,[5] the concept that quadrupeds with splayed-out legs cannot breathe while they run, because their sinuous side-to-side swaying of the body impinges on the lungs and rib cage, inhibiting inspiration. For this reason lizards and salamanders cannot breathe while walking, and *Proterosuchus* may have had something of this effect, although not as pronounced as in modern-day salamanders or lizards.

A solution is to put the legs beneath, but this is only a partial solution.[6] To truly be free of the constraint that breathing put on posture, extensive modification to the respiratory system as well as the locomotor system had to be made. The lineage that led to dinosaurs and birds found an effective and novel adaptation to overcome this breathing problem: bipedalism. By removing the quadruped stance, they were freed of the constraints of motion and lung function. The ancestors of the mammals also made new innovations, including a secondary palate (which allows simultaneous eating and breathing) as well as a complete upright (but still quadruped) stance. But this was still not satisfactory, and a new kind of breathing system evolved. A powerful set of muscles known as the diaphragm allowed a much more forceful system for inspiring and then exhaling air.

There are other clues than dinosaur bones to the nature of life on Earth, and the challenges it faced during the low-oxygen times of the Triassic. Part of the Triassic explosion was a diversification of reptiles returning to the sea. Many separate lineages did this, and the reasons why this happened may be tied up in the problems posed by the hot, low-oxygen Triassic world.

Oxygen is necessary to run metabolic reactions in animals; it enables the chemical reactions that are life itself. But as in a chemistry experiment, several factors control the reactions themselves. One of the most important is temperature. Metabolic rate is the pace at which energy is used by an organism. It is far higher in endotherms than in ectoderms. But even in the same organism, the metabolic rate is directly and importantly influenced by temperature to a surprising degree. Recent studies have shown that as much as one third to one half of *all* energy expenditure by an animal is used for simply staying alive through activities such as protein turnover, ion pumping, blood circulation, and breathing. Other required activities, such as movement, reproduction, feeding, and other behavior come on top of this, and the rate that "fuel" is used goes up with rising temperature.[7] But as metabolic rate goes up, so too does the need for oxygen, for the chemical reactions of life are oxygen dependent. The key finding is that metabolic rates double to triple with each ten-degree rise in temperature. The consequences of this in a world that has less oxygen availability than now, but warmer average temperatures, would be major.

There is no direct link between oxygen levels in the atmosphere and temperature. But there is a direct link between temperature and CO_2, the well-known greenhouse effect. And as we saw in chapter 3, levels of oxygen and atmospheric CO_2 are roughly inverse: when oxygen is high, CO_2 is low, and the converse. Many periods in the past with low oxygen had high CO_2, and thus were hot. In a low-O_2 world that is hot, the animal loses. We have already seen many solutions to dealing with low oxygen. One of them is obviously the simple solution of staying cool. Some solutions to staying cool—or cool enough—are physiological; some are behavioral.

One of these is all at once morphological, physiological, and behavioral. It is to return to the sea, the cool sea, for even in the hottest world of the past, the ocean would be essentially cooler in terms of physiology. And for this reason, perhaps, many Mesozoic land animals traded feet for flippers or fins and returned to the sea at a prodigious rate.

As noted earlier in this chapter, in this time of higher global temperature (perhaps 30°F warmer, in fact, on a global average) and only half the atmospheric oxygen found today, an increasing proportion of tetrapod diversity was composed of animals that re-evolved a marine lifestyle. Never before and never since have so many lineages given up the land for the sea. Today we celebrate the many kinds of whale, seal, and penguin families, the three groups coming from land dwellers that now show the greatest marine adaptations. Yet whales and seals combined make up only 2 percent of all mammal genera, and penguins but 1 percent of birds. But the Triassic oceans had many more kinds of such changed creatures, animals adapted to land that had revolved a body plan for life in the sea. In the Triassic there were giant ichthyosaurs, as well as seagoing tetrapods such as placodonts (the latter were like large seals, but unlike seals, had blunt teeth expressly evolved to crack shellfish); in the Jurassic the ichthyosaurs remained and were joined by a host of long- or short-necked plesiosaurs; and in the Cretaceous the ichthyosaurs disappeared to be replaced by large Mosasaurs. But all had a common theme: back to the ocean.

The existence of so many marine tetrapods was confirmed with the important research of marine reptile expert Nathalie Bardet, who in 1994 published a review[8] of all known marine reptile families of the Mesozoic. The surprise was that proportionately there were so many in the Triassic period. But why would so many animals evolve a marine lifestyle?

The two dominant environmental factors of those days would have been the low oxygen and the high global temperature of our planet. Ray Huey, a reptile specialist at the University of Washington, suggested too that the high heat of the Early Triassic through Jurassic would have been an evolutionary incentive for some number of reptiles to go back into the sea. In fact, in 2006, coauthor Ward showed that there was a very interesting and inverse correlation between Mesozoic oxygen levels and the number of marine reptiles. When oxygen was low, the percentage of marine reptiles was high. But as oxygen rose, the proportion of tetrapod families fully aquatic markedly dropped. This may not be that the absolute number of marine forms decreased as much as it was that the number of terrestrial dinosaurs markedly increased. Yet it marks an unusual and new view of the greenhouse planet that was Mesozoic Earth.

TRIASSIC-JURASSIC MASS EXTINCTION

One of the striking new findings of the oxygen-through-time results has been the level of Triassic oxygen. Only several years ago, the minimum oxygen levels of the past 300 million years was rather universally pegged at the Permian-Triassic boundary of 252 million years ago. But that time of oxygen low has been substantially moved, and now may correspond much more closely to the Triassic-Jurassic (T-J) boundary of 200 million years ago than previously thought. Thus, rather than the Triassic being a time of oxygen rise, or even a time with two downturns—one at the end of the Permian, one at the end of the Triassic—we are confronted with the possibility that oxygen was lower in the Late Triassic than in the early part of the period, perhaps as low as 10 percent of the atmosphere at sea level, or about half the modern-day levels. This time corresponds to one of the major changes of the Triassic, the winnowing out of most land vertebrates, with the exception of the first dinosaurs.

The cause of this mass extinction, like the others, has been long debated. What is clear is that, like the Permian mass extinction, the Triassic-Jurassic mass extinction occurred in a dead heat (literally and figuratively) with the emplacement of one of the largest flood basalt episodes in the history of Earth, one second only to the Siberian Traps event of the Late Permian. Back-to-back mass extinctions, 50 million years apart, both temporally linked to large flood basalts, events well known to rapidly increase carbon dioxide levels in both air and sea to many times the starting values. Some estimates place the peak CO_2 levels in the atmosphere as from 2,000 to 3,000 ppm, compared to our own 400 (2014) ppm (but rising fast!).

The utter destruction of plant life makes a dent in the carbon cycle and changes the relative proportion of carbon 12 to carbon 13. The use of this comparison, the carbon isotope analyses discussed at many other points in this book, seems to be a fixture of the mass extinctions. But it was not until a report by Ward and others in 2001, from T-J interval strata nestled along a shoreline fronting an old-growth, cold-temperature rainforest located on one of British Columbia's Queen Charlotte Islands,[9] that this carbon isotope perturbation was found. Just as with the Devonian and Permian greenhouse extinction before it, the newly found signal is characterized by oscillating changes in the ratio of C_{13} to C_{12}, brought about by changes in the abundance, kind, and burial history of diverse kinds of life on the planet.

As for the Devonian and Permian events, this signal seemed to indicate that this extinction as well as the others were caused by something other than impact. The conclusion that the T-J was yet another in the "family" of greenhouse extinctions was briefly challenged by another kind of discovery soon after the first carbon isotope shift was reported on. Paul Olsen of Columbia University and colleagues announced to great press effect that the T-J was caused, in fact, by large-body impact with the Earth. This seemed to provide a nice symmetry—an asteroid ending the age of dinosaurs, and another, 135 million years earlier, seemingly started that same age of dinosaurs. Or so it seemed. Olsen's evidence of impact had been found at a site in the Newark, New Jersey, region, home to the most diverse assemblage of late Triassic and early Jurassic dinosaur footprints on the planet. It was the association of dinosaurs and mass death that whetted the journalist's appetite for extensive press coverage.

Olsen and his colleagues reported an iridium anomaly from continental T-J boundary beds in New Jersey. It was just such an anomaly that had first alerted the Alvarez team (in 1980) to the possibility of impact at the end of the Cretaceous; iridium had become the gold standard of impact evidence. But here the two studies wildly diverged. Where the Alvarez group followed the physical and geochemical evidence from their Italian boundary section with data confirming mass extinction of small ocean life at the same time as the impact, the Olsen paper for the Triassic event followed their physical and geochemical evidence with just the opposite: they found that rather than eliminating most life in their section, instead the impact seemed to have acted like a biotic fertilizer, leading to both more and bigger life!

The Olsen group was sampling strata deposited on land (or more correctly, in streams and shallow lakes on land), and the "fossils" they studied were footprints, not the remains of body parts. But in spite of these rather startling differences, the Olsen et al. conclusion was the same: that a great asteroid had hit the Earth (this time about 200 million years ago, the age of the Triassic-Jurassic boundary), and that like the K-T event, the dinosaurs were affected. But the argument was that the impact killed off competitors of the dinosaurs, leading to a rise in diversity and animal size. And unlike the secrecy surrounding Luann Becker's work and methods dealing with the Permian extinction, Paul Olsen brought all who cared to look to his urban outcrops. Plenty of the many specialists working on mass extinctions at the time made the trip.

Olsen's samples had yielded iridium, and unlike the Becker work, various labs confirmed his findings. But a finding of iridium alone may not have propelled this work into *Science*, the prestigious flagship of scientific publishing. Olsen and his colleagues had pulled another and totally different array of evidence out of their New Jersey rocks. At numerous outcrops equal in age to that yielding the iridium, Olsen and crew had noticed that a significant change was observable among the footprints. The beautiful three-toed footprints, known to residents of this area for more than two centuries, increased in number, size, and diversity of shape.

One would think that the footprints found in strata deposited after the T-J mass extinction would be fewer in number (number of animals around), fewer in kind (a lesser species diversity), and smaller in size, since one lesson that we do know from the asteroid-caused K-T extinction is that it was disproportionably lethal to larger-sized animals. While no dinosaur or any of the many kinds of reptiles and mammal-like reptiles matched size with the biggest dinosaurs going extinct at the end of the Cretaceous, many were equal in size to dinosaurs that did go extinct as a result of the K-T asteroid. Thus, fewer, fewer kinds of, and smaller-sized footprints would be expected in earliest Jurassic rocks, if the Triassic's end was caused by an impact. Yet just the opposite was observed in all three of these evidence lines: there were more footprints, of more different kinds, and many were larger, much larger, than the largest of the Triassic footprints. It was this evidence as much as the iridium finding that convinced *Science* that this research article was important enough to publish in their journal.

Just as in the case of Luann Becker's work of a year prior to Olsen's publication, the *Science* paper by Olsen et al.[10] was scrutinized in painstaking detail. Two experts on interpreting impact deposits, Frank Kyte of UCLA and David Kring of Arizona, were both of the opinion that the iridium finding was certainly indicative of an impact about that time; both also pointed out that the amount of iridium reported from the various sites of the Olsen group was at least an order of magnitude less than that found at virtually every K-T boundary site. Something fell to Earth, all right, but it was small—probably too small to cause the amount of extinction at the end of the Triassic. Thus, while evidence for impact at the end of the Triassic was much more believable than at the end of the Permian, it was still hard to believe based on this new evidence that the Triassic extinction was a K-T-like impact extinction.

There is indeed a large crater in Quebec. It is one of the biggest craters visible on the planet, named Manicouagan Crater—with a diameter of about 100 km (in comparison, the Chicxulub crater is 180–200 km in diameter). It had long been thought to be of the right age, too—somewhere near 210 million years in age, which was about the age of the Triassic-Jurassic boundary. The radioactive decay measures indicated that the Triassic came to an end about 199 million years ago. In 2005 this date was slightly changed to 201 million years ago. And not only did the T-J get younger, but the age of the Manicouagan crater got older. Better dating placed its age at 214 million years ago.

Our own work on the Queen Charlotte Islands was designed to look at the T-J extinction, but also to search for any possible fossil die-off prior to it—in rocks we could age as being around 214 million years in age. The "kill curve" estimates of the late twentieth century predicted that any impact event leaving a crater the size of Manicouagan would easily kill off between a quarter and third of all species on Earth—and we found nothing! Perhaps we have overestimated the lethality of asteroid impacts?

TRIASSIC BLACKNESS

By the early years of the new century, geochemist Robert Berner of Yale University had greatly increased the resolution of his complicated computer models that estimated the amount of oxygen and carbon dioxide for any 10-million-year interval of the past 560 million years. His results showed a startling match between the times of lowest oxygen levels or the most rapidly dropping oxygen levels and mass extinction events.

All three of the mass extinction events with problematical causes showed strata indicating deposition in low-oxygen conditions.[11] Under such circumstances, strata usually turn black (because they contain the mineral pyrite and other sulfur compounds that are said to be reduced in that they were produced by chemical reactions that can occur only in the absence of oxygen). A second clue came from the fact that the rocks of these ages were thinly bedded or even laminated, often showing delicate sedimentary structures with the strata. Because so many animals burrow, most strata deposited in the sea since the Cambrian are what is called bioturbated by the vast number of invertebrates that ingest sediment at the bottom of a body of water in order to strain out any organic material. The presence of the fine bedding could occur only in environments with no or only rare animals. Through these three avenues—modeling, rock mineralogy (dictating color), and sedimentary bedding—it was clear that the Permian, Triassic, and Paleocene extinctions took place in a low-oxygen world.

Other evidence discovered in the late 1990s and early part of the new century showed that while oxygen may have been low, another constituent of the Earth's atmosphere was at the same time high: carbon dioxide. Like the evidence for low oxygen, the CO_2 evidence came from Berner's models as well as from evidence preserved in the rock record, or more accurately in this case, in the fossil record. Unfortunately, there is no way of actually measuring the exact volume of CO_2 that was present at any time in the past. Carbon dioxide does not color rocks or affect bedding. But some very clever work on fossil leaves resulted in an important breakthrough that allowed a relative measure of CO_2. Using this method, for instance, a paleobotanist could determine whether carbon dioxide levels were rising, falling, or staying the same over million-year intervals, and furthermore the method allowed estimates on how many times higher or lower the levels were from some base-level observation.

The CO_2 measure turns out to be both clever and simple, as so often it turns out with wonderful breakthroughs. Botanists looking at modern-day plant leaves had done experiments in which they grew plant species in closed systems where the amount of CO_2 could be raised or lowered relative to the level found in our atmosphere (about 360 ppm when these experiments were first conducted). Plants, it turns out, are highly sensitive to carbon dioxide levels, since even the small amounts of CO_2 in the atmosphere must serve as the source for their carbon, the major building block of life. They acquire this mainly in their leaves through tiny portals to the outside world called stomata. When grown in high levels of CO_2, the plants produced a small number of stomata, as even just a few sufficed in high CO_2 levels. The investigator then eagerly turned to the fossil record; leaf stomata are readily observable in leaf fossils. The results confirmed Bob Berner's model results.

At the end of the Permian and during the early Triassic, the fossil leaves showed only a few stomata. Carbon dioxide was spectacularly high at all three times. Moreover, not only was it high, but the rise in CO_2 happened quickly, on the order of thousands, not millions of years.

These two results give an entirely new view of mass extinctions. Each occurred in a world quickly warmed by the short-term rise in carbon dioxide (and perhaps methane as well, based on yet another line of evidence). And in addition to being hot, it was a place also low in oxygen. High-temperature, low-oxygen conditions coincided with major mass extinction. While modern-day greenhouses are not places of low oxygen (just the opposite by photosynthesis), they are places that heat very quickly due to the greenhouse properties of the glass panes covering the whole structure. Sunlight comes through the windowpanes, but when sunlight is radiated back in the form of light waves and heat, the glass panes trap the energy, which then warms the air, much like carbon dioxide, methane, and water vapor molecules do.

Heat is dangerous to any animal. The highest temperature that any animal can withstand is not even halfway to the temperature that boils water. At 40°C most animals die off, and the last holdouts die at 45°C. As is all too tragically known from the many sad cases of kids left in cars on a sunny day, rapid heating can be lethal. And the two aspects of this physiological system—the amount of oxygen available and the amount of heat energy—combine to make things even more lethal: animals need more oxygen as heat increases.

Of the three extinctions, the data for the Triassic-Jurassic CO_2 rise is particularly stunning. University of Chicago paleobotanist Jenny McElwain, collecting rocks in the dangerous and frigid outcrops amid the ice of Greenland in the last years of the twentieth century, showed without doubt that the end of the Triassic was ushered in by a sudden rise in CO_2 in an already low-oxygen world.

Increasingly, the Triassic began to look like an event similar to that at the end of the Permian. What it did *not* look like was the K-T extinction event, in which the extinctions were sudden and spread across every animal and plant group. But it was as if none of them "saw" it coming from an ecological or evolutionary sense. At the end of the Triassic, on the other hand, every group except the saurischian dinosaurs were undergoing size reduction (or at best, maintaining roughly equal diversity) in the time intervals leading up to and after the Triassic-Jurassic mass extinction, as if they knew bad times were coming, and small size would be more adaptive.

The groups with the simplest lungs (amphibians and the early-evolved reptiles) fared the worst, and many groups that had been very successful early in the Triassic, such as the phytosaurs, underwent complete extinction. Both amphibians and archosauromorphs probably had very simple lungs inflated by rib musculature only. Mammals and advanced therapsids of this time, probably both having diaphragm-inflated lungs, did better, but crocodiles, presumably with abdominal pumps, did poorly. The success of the saurischians may have been due to a multitude of factors (food acquisition, temperature tolerance, avoidance of predators, reproductive success), but our conclusion is that this group was unique in possessing a highly septate lung (one with many tiny flaps to increase surface area) that was more efficient than the lungs of any other lineage, and that in the very low oxygen world that occurred both before and after the Triassic-Jurassic mass extinction, this respiratory system conveyed great competitive advantages. Under this scenario, the saurischian dinosaurs took over the Earth at the end of the Triassic and maintained that dominance well into the Jurassic because of superior activity levels.

We now know that alone among the many kinds of reptilian body plans of the mid to late Triassic, the saurischian dinosaurs diversified in the face of either static or, more commonly, falling numbers in the other groups. We also know that oxygen reached its lowest levels of the past 500 million years in the Late Triassic. Something about saurischians enhanced their survival in a low-oxygen world. The ground truth suggests that a long and slow drop in oxygen culminated in the Triassic mass extinction, but that this extinction was really a double event, separated by a range of 3 to 7 million years.

There are few places on land where this time interval with abundant vertebrate fossils can be found. We really do not know the pattern of vertebrate extinction as well as we do the extinction in the sea. We do not know how rapidly the prominent victims of the mass extinction—the phytosaurs, aetosaurs, primitive archosauromorphs, tritylodont therapsids, and other large animals—disappeared. But by the time the gaudy Jurassic ammonites appeared in the seas in abundance, leaving behind an exuberant record of renewal in early Jurassic rocks, the dinosaurs had won the world. What kind of lungs did they have? There is no certainty but one: they had lungs and a respiratory system that could deal with the greatest oxygen crisis the world was to know in the time of animals on Earth.

A new view of things is that saurischian dinosaurs had a lower extinction rate than any other terrestrial vertebrate group because of a competitively superior respiration system—the first air-sac system. The fact that saurischians were actually *expanding* in number across this mass extinction boundary is the most striking aspect of all.

CHAPTER XIV

Dinosaur Hegemony in a Low-Oxygen World: 230–180 MA

THE word "Jurassic" is now irrevocably linked to dinosaurs and dinosaur parks, thanks to the *Jurassic Park* franchise of movies. In fact the real Jurassic was a world looking nothing like the cinematic view seen in those three progressively idiotic movies. Those movies were filled with plants that had not yet evolved in the Jurassic: the angiosperms, or our familiar flowering plants. In fact, it is impossible to even characterize a "Jurassic" world, because it was a world that was utterly changed from its earliest Jurassic appearance (201 million years ago to its latest iteration, of about 135 million years ago). At the beginning, it was a shattered world: a world again coming out of mass extinction; a world without coral reefs; a world where the dinosaurs were still few in number, species, and size; a world of such low oxygen that insects could barely fly, but to no matter, as no vertebrate flier could have caught them anyway. But things were to change in the relatively short time interval (in terms of geological periods, anyway).

By the end of the Jurassic the largest land animals of all time were common: dinosaurs were lords of all creation; tiny primitive birds and tinier primitive mammals hid in the lowest-rent districts in town. At the beginning the seas were so bare that stromatolites had made a comeback, and the larger fish and predators were few indeed.

By its end there was a veritable cornucopia of the most spectacular marine denizens to have ever populated the sea: long-necked reptilian plesiosaurs, dolphin-like ichthyosaurs, and splendid primitive fish—similar to the modern-day gar and sturgeon (both with strange body armor)—schooled among vast coral reefs and an ocean filled to exuberance with all manner of ammonites and their more squid-like relatives, the belemnites. The ammonites came in all varieties, from smooth to ribbed, and varied in shape from planispiral to, at the end of the period, peculiar, gently arcuate cones. The largest of all ammonites came from Jurassic rocks, the giant from Fernie, British Columbia, that in life was close to eight feet in diameter and surely would have weighed a half ton. Yet an odd thing has happened to the scientists studying this most iconic of prehistoric periods: most are dying off and not being replaced.

It is fair to say that geology as a modern discipline came into being because of the Jurassic. It was Jurassic beds that were first mapped by William "Strata" Smith, in the earliest 1800s, and it was Jurassic strata that demonstrated that fossils could be used to correlate beds in far-off lands. It was ammonites from Jurassic strata that provided Darwin with the then-best-known examples of evolutionary change. (References to this period are found in chapter 1, and as always, the historical works by the British historian of science Martin S. Rudwick are recommended.)

The Jurassic period showed the same pattern of short-term evolutionary explosion common to all of the post-extinction time intervals. They are called recovery periods. Each began with a low diversity of survivors of the mass extinction, but ended only after a 5- to 10-million-year time interval. Yet after this post-extinction hangover, diversity would always then again rise. The new animals and plants were always composed of a largely different assemblage of species. In most cases these species are newly evolved during the recovery interval, but in some cases they are taxa that lived a precarious, low-abundance existence prior to the extinction, but then exploded in numbers and ecological success in the new world. The early Jurassic was no different, and from the seeds of recovery a great new assemblage of marine creatures evolved into existence, composed in large measure of new kinds of mollusks, marine reptiles, and many new kinds of bony fish. But it is not the marine fauna that the Jurassic (or the succeeding Cretaceous period) is known for. No one has made three blockbuster movies featuring marine life with the name Jurassic in the title. The public wanted, and still wants, but one thing.

DINOSAURS

It is impossible to write a history of life without dwelling at length on dinosaurs. Yet at the outset it seems a losing prospect given the warrant of this book: that the histories narrated here have an element of the "new." So much is written about these antediluvian saurians (the Victorians' view of them) that bringing anything fresh to the table seemed impossible at the outset of this book's writing. It was thus a very welcome surprise to find that in fact new findings clutter the record of twenty-first-century science. Any nonspecialist summary of dinosaurs normally dwells on three issues: whether they were warm-blooded, how they reproduced and what is known of their behavior of nesting, and how they ultimately died out. But there are other interesting questions as well, including perhaps the most interesting of all: why were there dinosaurs at all, or at least the dinosaurs' body plan? This, in turn, is related to the question of how they respired. A second question we will look at here is also related to respiration in its own way: what is new about the dinosaur-to-bird story? A great deal, it turns out, most coming from new Chinese discoveries (but new Antarctic discoveries as well, which the authors of this book were witness to). Finally, our new century has given information about two of the most fundamental aspects of dinosaur physiology, a definitive answer to the long-running mystery of whether dinosaurs were warm-blooded, as well as new discoveries about the characteristic growth rate of dinosaurs. And here too one of the fascinating directions coming from these new data bring us right back to the differences between dinosaurs and "true" birds, not just avian dinosaurs, but species with all the traits we now associate with being a bird.

WHY WERE THERE DINOSAURS?

To tell the new history of dinosaurs, we have to dip back in time to a few million years prior to the Triassic-Jurassic mass extinction, the topic ending the last chapter. Dinosaurs were really Jurassic and Cretaceous dominants. During the Triassic they were just another rare small low-diversity and low-abundance vertebrate trying to survive in a low-oxygen world. Over and over, however, it really looks like a dominant theme in the history of life is that times of crisis promote new innovation. Diversity stays low, but disparity—the measure of the number of different, and in the dinosaur's case, radically different body plans and anatomies—skyrockets. An analogy comes from Tom Wolfe's wonderful book *The Right Stuff*. In it he describes the often short and violent ends of test pilots in the late 1950s, when giant new jet planes were being developed. Sooner or later any test pilot would find himself in a death dive. But Wolfe describes the reaction of the pilots: very coolly going through the progressions—trying method A, no, try B, try C, try . . . In the latest Triassic world, so many organisms were those crashing jets, with evolution as the pilot, trying this morphology, then another, then another. To use this analogy, it was dinosaurs that pulled out of the death spiral that the low-oxygen, late Triassic biosphere had become by evolving the most sophisticated and efficient set of lungs that the world has ever seen.

Some 200 million years ago, only 50 million after the great paroxysm of the Permian extinction, the Triassic period came to an end in another bloodletting. As we saw in the last chapter, of the many lineages of land life that suffered through this extinction, it was only the saurischian dinosaurs that came through unscathed. The mass extinction ending the Triassic period was not just a phenomenon on land: it also wiped out most stocks of chambered cephalopods, but in the lower Jurassic they diversified in three great lineages: nautiloids, ammonites, and coleoids. Scleractinian reefs flourished once again, and large numbers of flat clams colonized the seafloor. Marine reptiles belonging to the ichthyosaur and the new plesiosaur stocks again were top carnivores.

On land the dinosaurs flourished, and mammals retreated in size and numbers to become a minor aspect of the land fauna, but showed a significant radiation into the many modern orders near the end of the Cretaceous. Birds evolved from dinosaurs in the latter parts of the Jurassic. This is all known, and not the topic of the sort of revisionist history that is the goal of this book. Instead, let us look at the record of oxygen during the Jurassic, and compare that to the numbers and kinds of dinosaurs in the ancient Jurassic Park.

Because of its general interest and rather sensational aspects, perhaps the most commonly asked question about dinosaurs is the manner of their extinction. The 1980 hypotheses by the Alvarez group that the Earth was hit 65 million years ago by an asteroid, and that the environmental effects of that asteroid rather suddenly caused the Cretaceous-Tertiary mass extinction, in which the dinosaurs were the most prominent victims, keeps this question paramount in people's minds. This controversy is rekindled every several years by some new finding that brings it to the surface once again. Thus its preeminence, even superseding the question as to whether or not the dinosaurs were warm-blooded. Way down on the list of questions about dinosaurs is the inverse of the dinosaur extinction question, not why did they die out, but why did they evolve in the first place? We certainly know when they first appeared in the second third of the Triassic period (some 235 million years ago), and we certainly know what these earliest dinosaurs looked like: most were like smaller versions of the later and iconic *T. rex* and *Allosaurus*. Bipedal forms that quickly became large. What has not been largely known or even considered among those who do know about it is the new understanding that 230 million years ago was the time when oxygen may have been nearing its lowest level since the Cambrian period.

Why were there dinosaurs? This question can now be answered in multiple ways. There were dinosaurs because there had been a Permian mass extinction, opening the way for new forms. There were dinosaurs because they had a body plan that was highly successful for the Earth during its Triassic period. But perhaps these generalizations do not cut to the heart of the matter. Chicago paleontologist Paul Sereno, who has unearthed some of the oldest dinosaurs and has made their ascendancy a major part of his study, looks at the appearance of dinosaurs in another way. In his 1999 review "The Evolution of Dinosaurs," he noted: "The ascendancy of dinosaurs on land near the close of the Triassic now appears to have been as accidental and opportunistic as their demise and replacement by therian mammals at the end of the Cretaceous." Sereno goes on to suggest that the evolutionary radiation following the evolution of the first dinosaurs was slow and took place at very low diversity. This is quite unlike the usual pattern seen in evolution when a new and obviously successful kind of body plan first appears: usually there is some kind of explosive appearance of many new species utilizing the new morphology of evolutionary invention in a short period of time. Not so with the dinosaurs. Sereno further noted: "The dinosaurian radiation, launched by 1-meter-long bipeds, was slower in tempo and more restricted in adaptive scope than that of therian mammals."

For millions of years, then, dinosaurs as well as other land vertebrates remained at a relatively low diversity, a finding that Sereno and others continue to find perplexing. But now this question can be answered. The history of animal life on Earth repeatedly showed a correlation between atmospheric oxygen and animal diversity as well as body size: times of low oxygen saw, on average, lower diversity and smaller body sizes than times with higher oxygen. It appears that these same relationships held for dinosaurs. The 2006 work by Ward, *Out of Thin Air*, was the first source explicitly relating dinosaur body plan and later giantism: the oxygen levels.

If dinosaur diversity was indeed dependent on atmospheric oxygen levels, the extremely low atmospheric oxygen content of the Late Triassic readily explains the long period of low dinosaur diversity after their first appearance in the Triassic.

Low-oxygen times killed off species (while at the same time stimulating experimentation with new body plans to deal with the bad times). Support for this thesis can found by comparing the latest atmospheric oxygen estimates for the Triassic through the Cretaceous with the most complete compilation of dinosaur diversity through the same sampling period. The latter, published in 2005 by paleontologist and sedimentologist David Fastovsky and colleagues, showed that dinosaur genera stayed roughly constant from the time of the first dinosaurs in the latter half of the Triassic and the first half of the Jurassic. It is not until the Late Jurassic that dinosaur numbers started to rise significantly, and this trend then continued right up to the end of the Cretaceous, with the only (and slight) pause in this rise coming in the early part of the late Cretaceous. By the end of the Cretaceous (in the Campanian age, 84 to 72 million years ago) there are hundreds of times more dinosaurs than during the Triassic to late Jurassic. So what was the cause of this great increase? The relationship suggests that oxygen levels played a role in dictating dinosaur diversity. Through the late Triassic and first half of Jurassic, dinosaur numbers were both stable and low, as was atmospheric oxygen when compared to today's values. Gradually, oxygen rose in the Jurassic, hitting 15 to 20 percent in the latter part of the period. It is only then that the numbers of dinosaurs really began to increase. Oxygen levels steadily climbed through the Cretaceous, and so too did dinosaur numbers, with a great rise in dinosaur numbers found in the late Cretaceous, the true dinosaur heyday. The dramatic rise in oxygen at the end of the Jurassic was also the time that the *sizes* of dinosaurs increased, culminating in the largest-known dinosaurs appearing from the late Jurassic through the Cretaceous.

There were surely many other reasons for this Cretaceous rise. For instance, in mid-Cretaceous times the appearance of angiosperms caused a floral revolution, and by the end of the Cretaceous period the flowering plants had largely displaced the conifers that had been the Jurassic dominants. The rise of angiosperms created more plants and sparked an insect diversification. More resources were available in all ecosystems, and this may have been a trigger for diversity as well. Yet the relationship between oxygen and diversity and between oxygen and body size has played out over and over in many different groups of animals, from insects to fish to reptiles to mammals. Why not in dinosaurs as well?

Dinosaurs evolved during or immediately before the late Triassic oxygen low (between 10 and 12 percent, equivalent to an altitude of fifteen thousand feet today), a time when oxygen was at its lowest value of the last 500 million years. We have already seen that many other animals changed body plans in response to extremes of oxygen, and so too with the dinosaurs. The dinosaur body plan is radically different from earlier reptilian body plans, and appeared in virtually a dead heat (and in great global heat) with the oxygen minimum. Perhaps this is a coincidence. But because many of the aspects of "dinosaurness" can be explained in terms of adaptations for life in low oxygen, that seems unlikely. The initial dinosaur body plan (evolved by the first saurischian dinosaurs such as *Staurikosaurus* and the somewhat younger, dinosaur-looking beast named *Herrerasaurus*) was in some part in response to the low-oxygen conditions of the time, and from that it can be concluded that the initial dinosaur body plan of bipedalism evolved as a response to low oxygen in the Middle Triassic. With a bipedal stance, the first dinosaurs overcame the respiratory limitations imposed by Carrier's constraint. The Triassic oxygen low thus triggered the origin of dinosaurs through formation of this new body plan.

It is totally misunderstood what it means to inhabit a world where there is only 10 percent oxygen at sea level, as it would have been for the last 20 million years of the Triassic period. That is the level of oxygen found atop Mount Rainier, in Washington State. It is the amount of air atop the highest Hawaiian volcanoes, where the giant Keck Observatory peers out into space, a place where astronomers learn very quickly about the loss of energy and mental acuity that comes from such low oxygen levels. The principle of uniformitarianism fails us in this, because the use of high altitude to better understand low oxygen levels fails on so many levels, because at altitude, it is not just the oxygen that is lower in concentration, but all gases. One of these gases is water vapor, and this has a real effect on bird eggs at altitude. There had to have been major adaptations to such low oxygen, however, as the most important constraint on land animal evolution at this time. And indeed there were. One such we have named "dinosaur," for the first dinosaurs, all bipedal with new kind of lungs and respiration, became the most efficient land animals of all time in low-oxygen settings. Those that have survived, the creatures we call birds, maintain this excellence.

The fossil record shows that the earliest true dinosaurs were bipedal and came from more primitive bipedal archosauromorphs slightly earlier in the Triassic. These archosauromorphs were the ancestors of the lineage giving rise to the crocodiles as well, and may have been either warm-blooded or heading in that way. We see bipedalism as a recurring body plan in this group, and there were even bipedal crocodiles early on. Why bipedalism, and how could it have been an adaptation to low oxygen?

Even modern-day lizards cannot breathe while they run (and they have had hundreds of millions of years to potentially get this

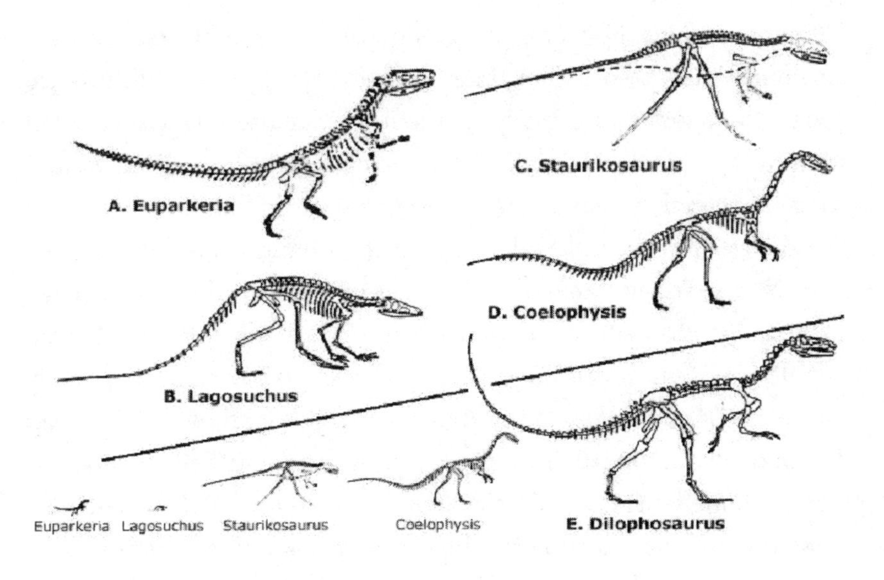

right). This is due to their sprawling gait. Modern-day mammals show a distinct rhythm by synchronizing breath taking with limb movement. Horses, jackrabbits, and cheetahs (among many other mammals) take one breath per stride. Their limbs are located directly beneath the mass of the body, and to allow this, the backbone in these quadruped mammals has been enormously stiffened compared to the backbones of the sprawling reptiles. The mammalian backbone slightly bows downward and then straightens out with running, and this slight up-and-down bowing is coordinated with air inspiration and exhalation. But this system did not appear until true mammals appeared, in the Triassic. Even the most advanced cynodonts of the Triassic were not yet fully upright, and thus would have suffered somewhat when trying to run and breathe.

If a creature runs on two legs instead of four, the lungs and rib cage are not affected. Breathing can be disassociated from locomotion; the bipeds can take as many breaths as they need to in a high-speed chase. At a time of low oxygen but high predation, any slight advantage either in chasing down prey or running from predators—even in the amount of time looking for food and how food is looked for—would have surely increased survival. The sprawling predators of the Late

Permian, such as the fearsome gorgonopsians, were, like most predators during and before their time, ambush predators, as all lizards are today. Predators that actively seek out prey require high speed and endurance. So what must it have been like for the animals of the Triassic when they found that for the first time the predators were out searching for them, rather than hiding and waiting?

In the Triassic, the crocodile lineage and the dinosaur lineage shared a quadrupedal common ancestor. This beast may have been a reptile from South Africa named *Euparkeria*. This group is technically called the Ornithodira, and from its earliest members this group began to evolve toward bipedalism. This is shown by their anklebones, which simplified into a simple hinge joint from the more complex system found in quadrupeds. This, accompanied by a lengthening of the hind limbs relative to the forelimbs, is also evidence of this, as is the neck, which elongates and forms a slight S shape. These early Ornithodira themselves split into two distinct lineages. One took to the air. These were the pterosaurs, and the late Triassic Ornithodira named *Scleromochlus* might be the very first of its kind; a still-terrestrial form looks like a fast runner that perhaps began gliding between long steps using arms with skin flaps. The oldest undoubted flying pterosaur was *Eudimorphodon*, also of the Late Triassic.

While these ornithodires edge toward flight, their terrestrial sister group headed toward the first dinosaur morphology. The Triassic *Lagosuchus* was a transitional form, being between a bipedal runner and a quadruped. It probably moved slowly on all fours, but reared up on its hind legs for bursts of speed—the bursts necessary to bring down prey, for this was a predator. But it still had forelimbs and hands that had not yet attained the dinosaur type of morphology, so it is not classified as a dinosaur. Its successor, the Triassic *Herrerasaurus*, meets all the requirements and is classified as a dinosaur—the first. But, as we shall see below, it may have lacked one attribute that its immediate descendants would rectify: a new kind of respiratory system that could handle the still-lowering oxygen content of Earth's atmosphere.

This first dinosaur was fully bipedal,[1] and it could grasp objects with its hands, having a thumb like we do. This five-fingered hand was distinct from the functionally three-toed foot (there were five actual toes, but two were so vestigial that only three toes touched the ground while running or walking). Because it was not totally bipedal, evolution no longer had to worry about maintaining a hand that had to touch the ground for locomotion. So with a free appendage no longer necessary for locomotion, what to do with it? The much later and more famous *T. rex* reduced the size of the forearm to the point that some have suggested it was nonfunctional. Not so for these first dinosaurs, however. While their posture was that of the later carnivorous dinosaurs so familiar to us, their hands were obviously used—probably for catching and holding prey while on the run.

So this is the body plan of the first dinosaurs, from which all the rest evolved; bipedal, elongated neck, grasping hands with a functional thumb, and a large and distinctive pelvis for the massive muscles and necessary large surface area for these muscles used in walking and running. These early bipeds were relatively small, and before the end of the Triassic they again split into two groups, which remained the most fundamental split of the entire dinosaur clan. A species of these bipedal, Triassic dinosaurs modified its hipbones to incorporate a back-turned pubis, compared to the forward-facing pubis of the first dinosaurs. As any schoolboy knows, this change in pelvic structure marks the division of the dinosaurs into the two great divisions: the ancestral saurischians, and their derived descendants, with whom they would share the world for about the next 170 million years: the ornithischians.

Of interest here, of course, is how dinosaurs breathed.[2] It has been discovered that their respiratory system was quite different from the cold-blooded reptiles we find today, but very similar to that of the warm-blooded birds. The lungs of modern-day amniotes (reptiles, birds, and mammals) are of two basic types (although we will see that there are more than two respiratory systems, which include lungs, circulatory system, and blood pigment type). Both kinds of lungs can be reasonably derived from some single kind of Carboniferous reptilian ancestor that had simple saclike lungs. Extant mammals all have alveolar lungs, while extant turtles, lizards, birds, and crocodiles—all the rest—have septate lungs. Alveolar lungs consist of millions of highly vascularized, spherical sacs called alveoli. Air flows in and out of the sacs. It is therefore bidirectional.

We mammals use this system, and our familiar breathing—in, out, in, out—is quite typical. Our air has to be pulled into these sacs, and then expelled again as oxygen switches place with carbon dioxide. We do this by a combination of rib cage expansion (powered by muscles, of course) and contraction of the large suite of muscles collectively called the diaphragm. Somewhat paradoxically, contraction of the diaphragm causes the volume of the lungs to increase. These two activities—the interacting rib expansion/diaphragm contraction—create a reduction in air pressure within the lung volume, and air flows in. Exhaling is partially accomplished by elastic rebound of the individual alveoli: when they inflate they enlarge, and soon after they naturally contract due to the elastic properties of their tissue. The many alveoli used in this kind of lung allow for a very efficient oxygen acquisition system, which we warm-blooded mammals very much need in order to maintain our active, movement-rich lifestyles. But the fact that air goes in and out of the same tube is very inefficient and reduces oxygen uptake relative to the amount of energy expended to get it.[3]

In contrast to the mammalian lung, the septate lung found in reptiles and birds is like one giant alveolus. To break it into smaller pockets that increases surface area for respiratory exchange, a large number of blade-like sheets of tissue extend into the sac. These partitioning elements are the septa, which give these kinds of lungs their name. There are many variations on this basic lung design among the many different kinds of animals that use it. Some kinds of septate lungs are partitioned into small chambers; others have secondary sacs that rest outside the lung, but are connected to it by tubes. As in the alveolar lung, airflow is bidirectional in most—but, as recently discovered, not all, and the exceptions to this rule, recently found, have profoundly changed our understanding of not only the paleobiology of early reptiles but their fate during the Permian mass extinction.

Septate lungs are not elastic and thus do not naturally contract in size following inhalation. Lung ventilation also varies across groups with the septate lung. Lizards and snakes use rib movement to draw air in, but as we have seen, locomotion in lizards inhibits complete expansion of the lung cavity, and thus lizards do not breathe while moving.

The variety of modifications of the septate lung makes this system more diverse than the alveolar system. For instance, crocodiles have both a septate lung and a diaphragm—an organ not found in the snakes, lizards, or birds. But the crocodile diaphragm is also somewhat different from that in mammals: it is not muscular, but is attached to the liver, and movement of this liver-diaphragm acts like a piston to inflate the lungs, with muscles attaching to the pelvis. The mammalian (including human) diaphragm pulls the liver in just the same way a crocodilian one does, creating a visceral piston, but the way this is accomplished differs in crocs and mammals.

Until recently, the septate lungs of crocodile and alligator lungs were considered relatively primitive and therefore inefficient. But a radical new finding not only makes us reassess the respiratory ability of the extant forms, but also puts an entirely new view on the reptilian success across the Permian extinction and in the Triassic.

The most inefficient way to breathe is the mammalian way: inhalation and exhalation through the same tube into the lungs. The inefficiency comes from the disorder of the gas molecules as one exhalation finishes and one inhalation starts. In any sort of more rapid breathing, there is a chaotic collision of exhaled air trying to get out before inhalation begins—and quite often the same gas molecules, including volumes of air with more CO_2 and less O_2, are sucked back in. It has long been thought that the crocs have this problem as well. But a study in 2010 showed that in fact the crocodilians use a separate one-way path that is similar to that in birds and dinosaurs. The revelation is that the ancient Permian- and Triassic-aged stem reptiles, the groups ultimately giving rise to the modern crocodiles and birds, and to the extinct dinosaurs, were also more efficient in their breathing than their therapsid (protomammal) contemporaries. They went through the filter of the Permian extinction with two great competitive advantages; they were cold-blooded, and they could extract more oxygen out of the air than a mammal or mammal-like reptile. The deck was stacked against us mammals. We never really had a chance in this most consequential competition for not only survivability but for eventual dominance amid the crisis and chaos of mass extinction. The mammals of the Mesozoic eventually would rarely be larger than rats. Probably highly fearful as rats are as well, surrounded by dinosaurs.

THE AVIAN AIR-SAC SYSTEM

The last kind of lung found in terrestrial vertebrates is a variant on the septate lung. The best example of this kind of lung and its associated respiratory system is found in all birds. The lungs themselves in this system are small and somewhat rigid. Thus bird lungs do not greatly expand and contract as ours do on each breath. But the rib cage is very much involved in respiration, and especially those ribs closest to the pelvic region are very mobile in their connection to the bottom of the sternum, and this mobility is quite important in allowing respiration. But these are not the biggest differences. Very much unlike extant reptiles and mammals, these lungs have appendages added to them known as air sacs, and the resultant system of respiration is highly efficient. Here is why. We mammals (and all other nonavians as well) bring air into our dead-end lungs and then exhale it. Birds have a very different system.

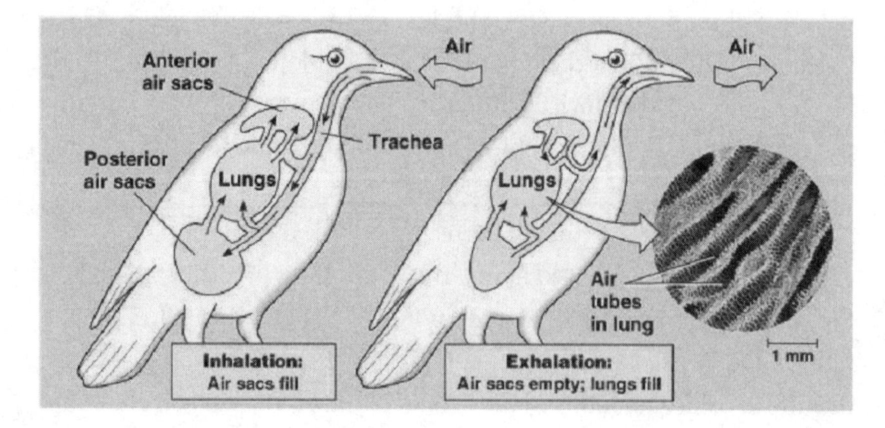

When a bird inspires air, the air goes first into the series of air sacs. It then passes into the lung tissue proper, but in so doing the air passes but one way over the lung, since it is not coming down a trachea but from the attached air sacs. Exhaled air then passes out of the lungs. The one-way flow of air across the lung membranes allows a countercurrent system to be set up: the air passes one direction, and blood in the blood vessels within the lungs passes in the opposite direction. This countercurrent exchange allows for more efficient oxygen extraction and carbon dioxide venting than are possible in dead-end lungs.

Anatomists have been dissecting and describing birds for centuries. It thus seems odd that an accurate understanding of bird air-sac anatomy did not occur until 2005. Two bird anatomists, Patrick O'Connor and Leon Claessens, injected substantial quantities of fast-jelling plastic into the respiratory systems of many different birds and then carefully dissected the corpses and described the anatomy of the filled cavities, the now plastic-filled air sacs.[4] To their surprise, they found that avian air sacs are much more voluminous and complicated than anyone had suspected. For the first time, the real relationship of air sac to bone in pneumatized bone—bones with large cavities in them—could be observed. In the same paper the two authors then compared the anatomy of pneumatized bird bones to pneumatized dinosaur bones. The similarity was remarkable, for there were the same shapes of holes in the same (or homologous) bones.

Those arguing that there was no air-sac system in dinosaurs have not denied that the dinosaur bones had holes in them. They said that the holes were there, all right, but that they were adaptations simply for lightening the bones. But there comes a point when arguing that the similarity is simply coincidence in shape collapses under the weight of too great a coincidence.

In the diagram on page 258, the various air sacs are shown with their communication to the lungs. It is clear that the volume of air sacs far exceeds the volume of the lungs themselves. The air sacs are not involved in removing oxygen; they are an adaptation that allows the countercurrent system to work. There is no question that the greater efficiency of this system compared to all other lungs in vertebrates is related to the two-cycle, countercurrent system produced by the air-sac lung anatomy in birds.

By 2005, the evidence that many dinosaurs had air sacs was overwhelming. Until then, one group of anatomists had vigorously argued that dinosaur lungs were no different than modern crocodile lungs, just large, and that the avian lung,[5] with its many auxiliary air sacs as well as a one-way airflow, did not appear until the Cretaceous, some 100 million years ago—and then was found only in birds! That view no longer became tenable. But in 2005 there was still no appreciation of the degree to which atmospheric oxygen levels had changed in the Early Mesozoic, or that such changes might have had any influence at all on the evolution of these various respiratory systems.

The air-sac system *is* better than the mammalian system. It has been estimated that a bird is 33 percent more efficient in extracting oxygen from air than a mammal at sea level. But at higher altitude this differential increases: a bird at five thousand feet in altitude may be 200 percent more efficient at extracting oxygen than a mammal. This gives the birds a huge advantage over mammals and reptiles living at altitude. And if such a system were present deep in the past, when oxygen even at sea level was lower than we find today at five thousand feet, surely such a design would have been advantageous, perhaps enormously so, to the group that had it in competing or preying on groups that did not.

We know that birds evolved from small bipedal dinosaurs that were of the same lineage as the earliest dinosaurs, a group called saurischians. The first bird skeletons come from the Jurassic (although there is now some controversy about just how "birdlike" the earliest species, such as the famous *Archaeopteryx*, really were, and we will return to this). But the air sacs attached to bird lungs are soft tissue, and would fossilize only under the most unusual circumstances of preservation. Thus we do not have direct evidence for when the air-sac system came about. But we do have indirect evidence, enough to have stimulated the air-sac-in-dinosaurs group to posit that all saurischian dinosaurs had the same air-sac system, as do modern birds. And like birds they were also warm-blooded. The evidence comes from holes in bones, places where these air sacs may have rested.

All credit for the first to make the audacious suggestion that dinosaurs had a bird launch system goes to Robert Bakker. It had been known since the late 1800s that some dinosaur bones had curious hollows in them, just as bird bones do. For decades this discovery was either forgotten or attributed to an adaptation for lightening the massive bones, for many of these bones with holes, later called pneumatic bones, came from the largest land animals of all time, the giant sauropods of the Jurassic and Cretaceous. The pneumatic bones were found mainly in vertebrae. Birds have similar pneumatic vertebrae, and while it can be said that some of the bird bones were light to enhance flying, it was also clear that some of the air sacs attached to bird lungs rested in hollows in bones. Thus, in birds, bone pneumaticity was an adaptation for stashing away the otherwise space-taking air sacs. The bodies of animals are filled with necessary organs, and putting the air sacs in hollowed-out bones make a lot of evolutionary sense. But Bakker made the leap and suggested that the pneumatic bones in his beloved fossil sauropods had evolved for a similar purpose, and were direct evidence that sauropods had and used the air-sac system.

Bakker's larger purpose was to try to add further evidence that dinosaurs were warm-blooded rather than make any claim about this being an adaptation to low oxygen. Birds, with their enormous energy and oxygen demands related to flying, were thought to have evolved the air-sac system as a way of satisfying the metabolic demands of their endothermy.

Following Bakker, other dinosaur workers took up the call, and the specific case of air sacs being present in sauropods was made by paleontologist and dinosaur specialist Matt Wedel in 2003, while similar arguments for bipedal species were made at about the same time by dinosaur specialist Greg Paul. In 2002 he suggested that the first of the so-called archosaurs, the primitive late Permian through early Triassic reptilian group (that we have called archosauromorphs), which would eventually give rise to crocodiles, dinosaurs, and birds, had air sacs. Examples of this group, which included the quadruped form Proterosuchus (described above as one of the earliest Triassic archosaurs), would have had a reptilian septate lung. Inspiration may have been aided by a primitive abdominal pump-diaphragm system (more primitive, perhaps, than this system as found today in modern crocodiles). What was unknown at the time, however, was that the crocodiles and their ilk back then had a far better respiratory system than was then agreed upon, thanks to their innovation of making air move one way through the respiratory system.

This discovery did not take place until 2010, and certainly affects our view of the relative evolutionary fitness of crocodiles, dinosaurs, and mammals. In fact, all of the Triassic reptiles seemed to have been better "breathers" than we mammals.

Successively, however, the evolution of the air-sac system may have fairly rapidly progressed in the lineage leading to dinosaurs at least. Alas (for them, anyway), the crocodiles ceased major innovation to their respiratory system with the newly evolved flow-through anatomy; they never experimented with pneumaticity in bones, and air sacs.

By the time the first true dinosaur is seen in the Middle Triassic, there may have been part of the air-sac system in place. The most primitive theropods from this time (the first dinosaurs) do not show bone pneumatization; the lung itself may have become inflexible and relatively smaller, both characteristics of extant bird lungs. With the Jurassic forms such as *Allosaurus*, the air-sac system may have been essentially complete but still much different from the bird system, modified as it has been for flying (for even the modern-day flightless birds came from fliers in the deep past) with large thoracic and abdominal air sacs.

By the time *Archaeopteryx* had evolved in the middle part of the Jurassic, there may have been a great diversity of respiratory types among the dinosaurs, some with pneumatized bones and some without. There also may have been a great deal of convergent evolution going on. For instance, the extensive pneumatization in the large sauropods studied with such care by Wedel may have arisen somewhat independently from the system found in the bipedal saurischians.

A final note about air sacs: While universal in saurischian dinosaurs, there is still no evidence of air sacs in the other giant dinosaur group, the ornithischian dinosaurs, including the well-known duckbills, iguanodons, and horned ceratopsian dinosaurs—not coincidentally (for three groups) all from the Cretaceous, not the Jurassic. The lack of an air-sac system in this group meshes well with their distribution in time. During the Jurassic times of very low oxygen they were minor elements of the fauna. It was not until the great oxygen rise of the Late Jurassic through the Cretaceous that this second great group of dinosaurs became common.

Perhaps the earliest dinosaurs were something like lions: sleeping twenty hours a day to conserve energy as dictated by the low oxygen, but when hunting, doing so actively, more actively than any of their competitors, which would have included the nondinosaur archosaurs (such as the early crocodiles), the cynodonts, and the first true mammals. All they needed to be was better than the rest. All evidence suggests that they were.

Metabolic complexes may have been far more diverse than our simple subdivision into endothermy and ectothermy. While modern birds, reptiles, and mammals are put into one of these two categories, there are, in fact, many kinds of organisms that can generate heat in their bodies without external heat sources. These include large flying insects, some fish, large snakes, and large lizards. Such animals are endotherms, but not in the mammalian or avian sense. There may have been many kinds of metabolism in the great variety of dinosaurs that existed.

Dinosaurs were not alone on the Jurassic stage, for our own ancestors were present, at very small size, as were other land animals and sea animals, including turtles on land and in the sea, as well as long-necked plesiosaurs and crocodiles. But the dinosaurs were certainly dominant on land. While at first it seems that there were many, many kinds of dinosaur body shapes, in fact there were really but three. All three shared a common characteristic with birds and mammals: a fully upright posture. The three kinds of dinosaurs were bipeds, short-necked quadrupeds, and long-necked quadrupeds. Each had a different time of origin and time of maximum abundance. Five distinct and successive assemblages of dinosaurian "morphotypes" (body plans) seem apparent to us. They are as follows:

1. *Late Triassic*. The earliest dinosaurs appeared in the last third of the Triassic but remained for their first 15 million years at low diversity. The majority of forms were bipedal, carnivorous saurischians. Toward the end of the period, quadrupedal saurischians (sauropods) evolved. Ornithischians diverged from the saurischians before the end of the Triassic but made up a very small percentage of dinosaur species and individuals. For much of the Triassic, dinosaur size is small, from one to three meters, and the earliest ornithischians (such as *Pisanosaurus*) were meter-long bipeds that had a new jaw system specialized for slicing plants. In the latest Triassic the first substantial radiation of dinosaurs occurs. It takes place among saurischians, with the evolution of both more and larger bipedal carnivores, and the first gigantism among early sauropods (such as *Plateosaurus* of the Late Triassic).
2. *Early to mid Jurassic*. Saurischian bipeds and long-necked quadrupeds dominated faunas. During this time, however, the ornithischians, while remaining small in size and few in number, diversify into the major stocks that will ultimately dominate dinosaur diversity in the Cretaceous. These stocks include the appearance of heavily armored forms (such as the thyreophorans). These are quadrupeds, and include the first stegosaurs of the middle part of the Jurassic. A second group is the unarmored neornischians (which include ornithopods—hypsilophodontids, iguanodons, and duckbills—and marginocephalians—the ceratopsians, which do not appear until the Cretaceous—and bone-headed pachycephalans). But it is the sauropods that are most evident in numbers. They split into two groups in the latest Triassic, the prosauropods and true

sauropods, and in the early and middle Jurassic the prosauropods were far more diverse than sauropods, but went extinct in middle Jurassic time, leading to a vast radiation of sauropods into the Late Jurassic.

The bipedal saurischians also showed diversity and success in the early and middle Jurassic. In latest Triassic time they had split into two groups (the ceratosaurs and tetanurans). The ceratosaurs dominated the early Jurassic, but by middle Jurassic time the tetanurans increased in number at the expense of the ceratosaurs. They too split in two, the two groups being the ceratosauroids and the coelophysids. The latter group eventually produced the most famous dinosaur of all: the late Cretaceous *Tyrannosaurus rex*, although its middle Jurassic members were considerably smaller. Their most important development in the Jurassic was the evolution of the stock that gave rise to birds.

3. *Late Jurassic*. This was the time of the giants. The largest sauropods come from late Jurassic rocks, and their dominance continues into the early part of the Cretaceous. Keeping pace with this large size were the saurischian carnivores, with giants such as *Allosaurus* typical. Thus the most notable aspect of this interval was the appearance of sizes far larger than in the early and middle Jurassic. And this was not only among the saurischians. During the late Jurassic the armored ornithischians also increased in size, most notably among the heavily armored stegosaurs. The diversification of ornithischians at this time with the appearance of stegosaurs, ankylosaurs, nodosaurs, camptosaurs, and hypsilophontids radically changes the complexion of the dinosaur assemblages.
4. *Early to middle Cretaceous*. While the dominants for the early part of this interval remained large sauropods, as the Cretaceous progressed a major shift occurred: ornithischians increased in diversity and abundance until they outnumbered saurischians. Sauropods become

increasingly rare as many sauropod genera go extinct at the end of the Jurassic.

5. *Late Cretaceous time*. Dinosaur diversity skyrocketed. Most of this diversification came through large numbers of new ornithischians: ceratopsians, hadrosaurs, and ankylosaurs among others. Only a small number of sauropods were present.

No evolutionary history can ever be pinned on one factor. Dinosaur morphology changed from predator-prey interactions, competition among themselves and others of their world, perhaps even climate change driven in large part by the incredible rises and falls of sea level during the Jurassic and Cretaceous—at one point a sea level rise so large that North America became two separate minicontinents, separated by a large if shallow north-south-running sea. Nevertheless, oxygen levels must have played a part.

The time of the first dinosaur grouping, the late Triassic assemblage, was a time of low oxygen levels, and this coupled with very high carbon dioxide levels—not asteroid impact—was the major cause of the Triassic-Jurassic mass extinction. The combination of low oxygen and high global temperatures was the killing mechanism. Yet studies of the number of land vertebrate taxa before and after the T-J mass extinction clearly show that saurischian dinosaurs survived this mass extinction event better than any group of vertebrates, and one important reason may have been because of their superior respiration system, because of their air-sac lungs, which gave them competitive superiority over other terrestrial animals with different lungs.

Ornithischian dinosaurs, on the other hand, did not possess as effective a respiratory system as did saurischians. However, they were competitively superior to herbivorous saurischians with regard to food acquisition, larger heads, stronger jaws, and better teeth. With the rise of oxygen to near present-day levels in the Cretaceous, ornithischians became the principal herbivores because of this superiority, leading to the extinction of many saurischian herbivores through competitive exclusion.

While the Jurassic to Cretaceous interval marked a relatively rapid and significant rise in atmospheric oxygen, other events were taking place. One of these was the breakup of the once-global continent of Pangaea into smaller continents. Another, and perhaps more significant for the distribution and taxonomic makeup of the later Mesozoic dinosaur faunas, was the radical change in flora. Dinosaurs evolved in a gymnosperm-dominated world—with conifers, but ferns, cycads, and ginkgos as well. But in the early part of the Cretaceous a new kind of plant appeared, a flowering plant.

With this new kind of reproduction and other adaptations, these plants, the angiosperms, underwent a rapid adaptive radiation. They outcompeted the earlier flora nearly everywhere on Earth, to the extent that by the end of the Cretaceous, some 65 million years ago, the angiosperms made up as much as 90 percent of vegetation. This transition in available food types would have affected the herbivores, and the kind of herbivores available as food would have directly affected carnivore body plans. Killing a late Jurassic sauropod would have been very different from killing a late Cretaceous hadrosaur.

Herbivory is dependent on the correct kind of teeth for the available plants. The sauropods may have lived on pine needles, their huge barrel bodies being, essentially, giant fermenting tanks for digestion of a relatively indigestible food source. The appearance of broad-leafed plants, the angiosperms, would have required different teeth and biting surfaces than those optimal for slicing pine needles off trees. Thus the transition from the sauropod-dominated faunas of the Jurassic to the ornithischian-dominated faunas of the Cretaceous was surely related in some part to the change in plant life. But respiration may have played a part as well, and perhaps if oxygen had not risen above 15 percent, the ornithischian takeover would not have taken place.

JURASSIC-TRIASSIC DINOSAUR LUNGS AND THE EVOLUTION OF BIRDS

Here it is proposed that the first dinosaurs were of a kind of animal never before seen or alive today: through upright posture and an evolving air-sac system they developed respiratory efficiency (the amount of oxygen extracted from air per unit time, or per unit energy expended in breathing) superior to any other then-extant animal. But these early forms may have lost endothermy, replacing it with a more passive homeothermy. That was their trick, using homeothermy to reduce oxygen consumption while at rest, and a superior lung system to allow extended movement without going into rapid anaerobic (and thus poisonous) states when active. We know that birds, a group of dinosaurs first appearing in the Jurassic, eventually had both endothermy and a very different kind of lung than in any extant reptile.

THE AVIAN DINOSAURS—AND DINOSAURIAN AVIANS

With perhaps the exception of the always-fascinating tyrannosaurids, no group of dinosaurs has received more attention in recent times than the basal birds. Vigorous debate centers on their body covering, but most important, on when flight first evolved, and why.

The first birds appeared about 150 million years ago, and the famous first bird remains *Archaeopteryx*. That is just before the start of the Jurassic. Oxygen had been rising for 50 million years at that time. Gigantism in dinosaurs was common. The immediate ancestors of the birds were fast, ground-running dinosaurs that may have used their forelimbs for a type of predation, a motion that was preadapted for a wing stroke in a flier, according to Berkeley paleontologist Kevin Padian. The fossil record suggests that the ancestors of the first bird were the bipedal carnivorous saurischians known as troodontids or perhaps the dromaeosaurids, forms that appear to have been already feathered.

Could *Archaeopteryx* fly? Most specialists now think so. But there is debate about when true flight took place. Could the late Jurassic "birds" really fly, at a time when their competition in the air would have been the diverse and successful pterodactyls? The fossil record does show that by the lower Cretaceous there is a bird fossil (*Eoluolavis*) that had evolved a "thumb wing," an adaptation that allows flight at slower speeds with greater maneuverability. Thus, within a few million

years after *Archaeopteryx*, fairly advanced flight was present. New discoveries from China have revealed an unexpectedly high diversity of birds already in place by the early part of the Cretaceous. Flight was an adaptation that stimulated a rapid evolution of new forms. We will return to the evolutionary history of birds in a subsequent chapter, as much of their story happened after the Jurassic.

Flight in birds is highly energetic. They use a great deal of energy to fly, and that, added to their relatively small size and endothermy, makes them great users of oxygen. So their air-sac system serves them well.

DINOSAUR REPRODUCTION AND OXYGEN LEVELS

One of the great discoveries of twentieth-century paleontology was the finding of dinosaur eggs.[6] In the latter half of that century, complex patterns found in associated fossil eggs hinted at behavioral complexity in dinosaur reproduction, or at least the egg-laying parts of it. In this century, ready access to new machines—small desktop CT scanners—has led to a third revolution in understanding dinosaur reproduction. Now eggs can be examined without mechanical damage to reveal delicate embryos within, and there is increasing understanding not only of the growth of these embryos, but of the construction of the eggs themselves—the how and why of a dinosaur egg.

Birds show little variation in at least one aspect of their reproduction. Extant birds, our best window to the dinosaurs, all lay eggs with a porous, calcareous shell. There is no live birth in birds, in contrast to extant reptiles, which have many lineages using live birth. There is also great variation in egg morphology between birds and some reptiles. While the eggshell in birds and reptiles consists of two layers, an inner organic membrane overlain by an outer crystalline layer, that amount of crystalline material varies greatly, from a thick calcium carbonate layer like that in birds to almost no crystalline material at all, so that the outer layer is a leathery and flexible membrane. Even the mineralogy of the crystalline layer varies, from calcite in birds, crocodiles, and lizards, to aragonite (a different crystal form of calcium carbonate) in turtles. Eggs are thus divided into two main types: hard or crystalline, and soft or parchment eggs. Some workers further subdivide the parchment eggs into flexible (used by some turtles and lizards) and soft (used by most snakes and lizards). Not surprisingly, the fossilization potential of these different hardness categories of eggs differs markedly. There are numerous fossil hard eggs known (many from dinosaurs), a few flexible eggs, and no undisputed soft eggs preserved.

Because of the great interest in dinosaurs there has been much speculation about their reproductive habits (the thought of two gigantic *Seismosaurus* mating rather boggles the imagination), and there are still many mysteries. One of the seminal discoveries about dinosaurs was that they laid large, calcareous eggs with calcite crystals making up the mineral layer, a find from the first expedition to the Gobi desert by the American Museum of Natural History expedition of the 1920s. Since then thousands of Cretaceous dinosaur eggs have been found, and the nesting patterns discovered and publicized by Jack Horner in Montana have opened a window into dinosaur reproduction. But are these Cretaceous finds characteristic of dinosaurs as a whole? This question remains unresolved and controversial. While most workers assume that all dinosaurs laid hard-shelled eggs, this is far from proven, and as we shall see below, there is indirect evidence that some early dinosaurs may have utilized parchment eggs or even live birth.

Almost all dinosaur eggs come from the Cretaceous. There is great variability in the nature of their crystal form and size, number, and the pattern of pores in them. But it is not the variety that is the most interesting scientific question. There are far fewer Jurassic dinosaur eggs, and almost none known from the Triassic. Why would that be? Has it to do with different kinds of preservation characteristics of the Cretaceous world and how things on land fossilized or did not? Or was it because of the far lower oxygen levels of the Triassic and Jurassic compared to the Cretaceous (and especially the late Cretaceous, where the vast majority of dinosaur eggs are recovered)?

There are several possibilities for this: perhaps there is indeed some preservation bias, with pre-Cretaceous eggs as common as those of the Cretaceous, but the lesser extent of Triassic and Jurassic dinosaur beds compared to the vast expanse of Cretaceous-aged beds has caused this difference. Thus the difference simply comes from sample size. Another (and different) possibility is that pre-Cretaceous eggs fossilized much less readily than those from the Cretaceous. This would certainly be the case if pre-Cretaceous eggs were leathery like those of extant reptiles, rather than calcified like birds. And if, like the marine ichthyosaurs, some dinosaurs utilized live birth rather than egg laying, there would certainly be fewer eggs to find. As in so many other aspects of the history of life, the level of atmospheric oxygen may have played a major role in dictating mode of reproduction.

Fossil eggs from Cretaceous deposits attributed to dinosaurs have a calcium carbonate covering like a chicken egg (but thicker), but unlike chicken eggs, which are smooth, the dinosaur eggs were usually ornamented with either longitudinal ridges or nodular ornamentation. Presumably ornamentation allowed the eggs to be buried after emerging from the female, with the ornament allowing airflow between the eggs and the burial material. The ability to bury eggs may have aided their fossil preservation potential, and perhaps helps explain why there are so many Cretaceous eggs, and so few other kinds. The heavy calcification would also help the eggs withstand the overpressure of burial in soil or sand.

Also now known is the complex behavior in nest making and orienting the eggs in burial mounds in the late Cretaceous, but not before. The Late Cretaceous bipedal dinosaur *Troödon* arranged its eggs tightly together in pairs, or vertically, while Jack Horner has shown complex burial patterns in late Cretaceous hadrosaurs from Montana.

The advantages of calcareous eggs is that they are strong, harder for predators to break into, and also aid in development: as the embryo grows inside the egg, some of the calcium carbonate is dissolved from the eggshell itself to be used in bone growth. They also might shield the egg from bacterial infections. But this comes at a price. Calcium carbonate, even an eggshell-thin layer of it, will not allow the passage of air or water into or out of the shell. But developing embryos need both water and oxygen. All calcareous eggs thus have pores so that oxygen-laden air can enter, but not so many pores that water quickly leaves by desiccation. To ensure sufficient water, the interior of the egg has a large amount of a compound known as albumin (familiar to us as the "white" of a chicken egg), which provides water to the embryo. This kind of egg is found in all birds and crocodiles.

The second kind of reptilian egg, the parchment egg, is found in turtles and most lizards. This kind of egg can take up water and actually expand in size with water uptake. But water permeability is a two-way street: parchment eggs can easily lose water as well. Burying these kinds of eggs in nests, the habit of many turtles and alligator-crocodiles, reduces water loss as well as hiding them from predators.

Burying an egg presents some danger: all developing embryos require oxygen, and thus the embryo requires an egg that can allow the passage of oxygen from the atmosphere into the egg. If the egg is buried too deeply or in impermeable sediment, the embryo will suffocate. And if the egg is laid at high altitude, it runs the risk of the same fate, even being smothered by parental care. So far, biologists have concentrated on temperature as the major variable affecting development rates in reptiles and birds. But the clues given by high-altitude lizards suggest that oxygen levels certainly play a part as well.

Modern lizards living at altitude often show viviparity, or live birth. They also hold the eggs in the birth canal for long periods of time. In both cases the explanation has been that they do this to maintain relatively high temperatures in an environment where there can be very cold temperatures that could slow development. But both of these adaptations would lessen or completely remove the time that the embryo is enclosed in a capsule that itself reduces the rate of oxygen acquisition. Calcareous eggs cannot be held in the mother because they do not allow oxygen to enter the egg until they emerge from the mother.

So we have a mystery. Reptiles show four different kinds of reproduction: live birth, parchment eggs that are held within the mother for extended periods of time, parchment eggs that are laid soon after formation within the mother, and calcareous eggs. And following birth there is also a series of possibilities: the eggs are buried or not, and when not buried, the eggs can be cared for by the parent or not. The advantages of each of these and the time they first appeared are still unknown.

And we have a second mystery. Most known dinosaur eggs are of Cretaceous, and as noted above, mainly late Cretaceous eggs, and are calcified; also, the appearance of burial behavior in dinosaurs is also characteristic of the Late Cretaceous. But what of the pre-Cretaceous dinosaurs? While there are eggs from sauropods and bipedal saurischians from the Late Jurassic—most spectacularly from deposits in Portugal, where the eggs contain the bones of embryos—earlier rocks are nearly barren of dinosaur eggs and/or nests. Only a few confirmed eggs are known from the Triassic.

Just when these various kinds of eggs first evolved thus remains a mystery. In 2005 it was proposed that calcareous eggs first appeared at the end of the Permian as an adaptation to avoid desiccation in the increasingly dry late Permian through Triassic global climates. Unfortunately, there is no fossil evidence to support this: there are no accepted Permian eggs in spite of the existence at that time of anapsids (the group that would give rise to turtles), diapsids (the group that would give rise to crocodiles and dinosaurs), and synapsids (the group that would give rise to us). Furthermore, only a small number of late Triassic eggs are known that may be from dinosaurs. But this poses a great dilemma: dinosaur eggs preserved commonly in Cretaceous sediments are not found in the same kinds of depositional environments of Permian and Triassic-aged strata. In all likelihood, if archosaurs had used hard eggs in the Permian or Triassic—if any group of reptiles had used hard eggs then—we would have already found them.

The absence of evidence is always a dangerous tool, but eventually numbers must be accepted. All evidence suggests that hard eggs were not commonly produced by pre-Cretaceous terrestrial egg-laying organisms. Even the 2012 discovery of hard dinosaur eggs in South Africa became but a single exception to the rule. It is hard to see how future collecting, no matter how intense, can now overcome this trend.

Only two shapes of dinosaur eggs are known: rounded and elongate. But seven different patterns of crystal arrangements making up the eggshells are now recognized. This diversity of egg wall morphology would be surprising if all dinosaurs had evolved from a single egg-laying ancestor. But it would be what we would predict if hard-shelled egg laying evolved *numerous* times by separate lineages of dinosaurs. If we add the additional (and different) eggshell morphologies found in extant reptiles and birds, there are a combined twelve separate eggshell microstructures that have evolved during the now-long history of reptile, nonavian dinosaurs, and the real avian dinosaurs: birds.

Perhaps each of these is an adaptation to a different kind of stress an egg belonging to a particular group or species is normally subjected to: a turtle egg in a deep burrow, for instance, has a very different series of challenges facing it than does a robin egg in a nest high in a tree. But another possibility is that the various calcareous eggs are evidence of an independent evolutionary history, in which hard eggs separately evolved in multiple lineages, including dinosaur lineages.

THE "IDEAL" OXYGEN LEVEL

One of the most interesting of new discoveries about the evolution of many of the modern stocks of land animals is that so many came from a fairly narrow time interval—during a time in the late Paleozoic when oxygen was higher than now. This is true for many groups of extant vertebrates, including the first members of families that would go on to be lizards, turtles, crocodiles, and mammals. But it is not just land vertebrates that suggest this trend: many of the terrestrial invertebrates, including the basal stocks of many insects, arachnids, and land snails, also began during the Carboniferous period of more than 300 million years ago. New experiments of the last five years are indicating that there is a "magic" level of oxygen for the fastest rate of embryonic development within both land vertebrate eggs and insect eggs, and that number is 27 percent.

Today we are at 21 percent atmospheric oxygen. But studies on alligators and insects shows that optimal development takes place at 27 percent. Eggs incubated in either higher or lower values of oxygen take longer to develop and hatch. In lower levels of oxygen—not coincidentally, perhaps, the 10–12 percent that occurred in the latest Triassic—many or most eggs never hatch at all, or do so only after such a long time that their probability of not being eaten by egg-eating predators becomes low indeed. Adding heat to the equation makes survivability even lower, because the eggs need holes to let oxygen in. But water escapes from these and causes a higher chance of death of the embryo. The worst combination would have been an atmospheric oxygen level of 10–12 percent in a world both hotter and drier than now. We know of such a time. It was the late Triassic. Those creatures laying eggs in the late Triassic period were in for trouble.

The problem is that reptiles were first evolved in a relatively high oxygen world: the Carboniferous, when oxygen was above 27 percent. These early reptiles pioneered the amniotic egg. But as oxygen levels dropped and temperature levels increased globally, the original reptilian eggs may have become death traps: not enough oxygen could diffuse in from outside the egg, while too much water was diffusing out. Seemingly a better response to heat and low oxygen (which is further magnified by the heat) would be live birth. The evolution of live birth thus may have come about in response to lowering global oxygen values in the late Permian. In spite of the enormous number of therapsid bones found in South Africa, Russia, and South America, there has never been an egg or nest found from these rocks. Therapsids may have already evolved live birth by this time, a trait carried on by their descendants, the true mammals that are first found at about the same time as the first dinosaurs appeared on the scene.

It may be that many lineages of dinosaurs evolved the calcareous egg in the Late Jurassic as a response to rising oxygen, and that the formation of calcareous eggs that are then buried was not viable in the late Permian through middle Jurassic environments of lower atmospheric oxygen.

The low oxygen–high heat conditions of the late Permian into the Triassic perhaps stimulated the evolution of live birth and of soft eggs that would have been effective at allowing oxygen movement into eggs and carbon dioxide out. On the other hand, the higher oxygen levels (and continued high temperatures) of the late Jurassic–Cretaceous interval stimulated the evolution of rigid dinosaur eggs and egg burial in complex nests.

Like characteristic metabolism, the contrasting patterns of live birth vs. egg laying is one of the most fundamentally important of all biological traits—and one that has received surprisingly scant attention by evolutionary biologists. Solving this problem by learning the time of origin and the distribution of one kind of birth strategy or the other should be a major research topic of the near future, but sadly may prove to be intractable because of the nonpreservation of parchment eggs.

CHAPTER XV

The Greenhouse Oceans: 200–65 MA

MOST discussion about the Mesozoic world (Triassic, Jurassic, and Cretaceous) concentrates on its land animals, especially the dinosaurs. But great changes were taking place in the marine world as well. The Mesozoic oceans progressively became more and more modern as the Mesozoic wore on in the shallow waters, but the mid-water to deepwater faunas remained very different from those of today. A transect going from shallow to deep water illustrates this, even near the very end of the Mesozoic era, in this case in the Late Cretaceous. Here is what such a dive might look like, a trip that can summarize a great deal about our current understanding of what we can call Mesozoic "greenhouse" oceans.[1]

The atmosphere that the greenhouse ocean sits under and interacts with importantly affects the chemistry and physical environment of any ocean.[2] The temperature of the atmosphere, the differences in temperatures from pole to equator, and the chemistry of seawater—including how much dissolved oxygen it contained—all dictated ocean conditions and the creatures the oceans contained. A crucial physical fact is that warm water holds less oxygen than cold water. During all of the Mesozoic, except the last 5 million years of the Cretaceous period, the global atmosphere from pole to equator was hot and humid. But the heat alone caused a lower overall oxygen content than we find in the ocean today. Coupling that with less oxygen in the air leads us to understand how different and less amicable to life the Mesozoic oceans would have been. The life that was there was, not surprisingly, evolved in many ways to deal with this low-oxygen world ocean.

While the Mesozoic world was different from now, in one way it might have seemed familiar. Just as the lower altitudes of our world's atmosphere are quite populated with a wide diversity and abundance of flying creatures, from insects to birds to bats, so too was the Mesozoic sky a place of movement and life. The air would have been filled with an assortment of flying organisms, including insects, but also with two very different groups than found today: the enormous pterosaurs (reptiles) as well as smaller reptilian pterodactyls, and many kinds of birds, with the latter composed of forms quite different from most birds of today, with and without teeth, and others with or without wings.

A wide lagoon of some sort would have fronted most Cretaceous oceans. Lagoons are formed when some kind of reef walls off an inner water body. Usually such lagoons are both hotter and of lower oxygen than the open ocean itself. The shallows of these lagoons would have been inhabited by clams and snails that were quite similar and in many cases of the same taxonomic group (such as genus) as those found in tropical lagoons and near shore environments of modern oceans.

Already present, for example, were burrowing clams, tusk shells, oysters, scallops, mussels, cowries, cone shells, tritons, conchs, whelks, sea urchins (both the globular surface dwellers, the "regular" urchins, and the burrowing or "irregular" urchins such as the sand dollars and "sea biscuits" of today). There would also have been spiny lobsters and crabs. All in all, the "modern" fauna was already well established in the shallows of the Late Cretaceous oceans, and in fact would be relatively little affected by the gigantic era-ending mass extinction that by the Late Cretaceous (from about 90 to 65 million years ago) was coming ever nearer in time.

In deeper water, the kinds of life would change, just as it does in our ocean, to forms adapted for the finer sediments found in deeper water rather than for species that live in the coarser sand environments of shallower water. There would have been many animals that were buried, including many clams still around today, as well as many other kinds of burrow dwellers. Hiding in the sediment was a major survival tactic, because by the Late Cretaceous many kinds of predators adapted either to break open or to drill into mollusk shells that were present. Also present in the shallow waters of the lagoon would have been hunks of hard limestone formed by reef-forming organisms, such as corals in our modern day. These patches are tiny reefs that then, as now, formed into horseshoe shapes, with the front or arch of the horseshoe growing into the prevailing wind.

Farther from shore would have been the large barrier reef, which would have grown right up to the sea surface. These enormous walls of limestone would have been hundreds to thousands of miles long, growing right at the edge of large islands or continents at the point where the continental shelf gives way to the continental slope and deep water. Both sides of the barrier reef rampart would have been home to many fish species, including the bony fish and cartilaginous sharks, skates, and rays.

This inner edge of the barrier reef—in fact its entire overall shape—would have been a direct look-alike of what many of today's barrier reefs, such as the Great Barrier Reef of Australia, look like. Yet a major difference is that while there are coral species living on reefs today, their main framework builders were not corals at all.

The three-dimensional wave-resistant structures that we call reefs have been a key community of life since the Ordovician period. All have been and continue to be composed of builders and binders; they are akin to brick houses made of coral "bricks," encrusting algae, flat coral, and carbonate-particle mortar. But perhaps a much better analogy is that they are like some ancient city, where centuries of buildings have been erected, existed for some time, and then have tumbled down or disintegrated but were only partially cleared away before new construction rose atop the older rubble. Over time, the great weight of ever-larger stone buildings often caused the very crust of the Earth beneath the ancient city to slowly but measurably subside.

Such is the nature of a coral reef: over centuries, the larger and blockier corals attach to an already-existing reef surface and build up, growing upward toward sunshine, in a true life-or-death race to grow faster than one's neighbors. Corals compete to avoid being grown over or shielded from life-giving sun and open water, as the sun is necessary for the millions of single-celled plants growing in each coral polyp, and open water gives the carnivorous coral polyps their own sustenance. The tiny plants allow the coral animals to build their gigantic skeletons, and in turn the plants, called dinoflagellates, receive nutrients and protection from predators. In this fashion, tiny coral larvae fall out of the plankton to settle onto any nonliving and hard substrate they can find, and then grow up toward the seas' surface. These microscopic larvae with luck can grow from one polyp into hundreds of thousands in a single gigantic colony and can live for centuries or more, with a vast calcareous skeleton weighing thousands of tons. Although there are single colonies now thousands of years in age or even older, huge colonies eventually die. After death the coral skeleton becomes fragmented and grown upon in turn.

The reefs of the Cretaceous greenhouse oceans were no different in this process and in the shapes eventually built, but their building materials were not coral reefs at all but clam reefs—created by quite large clams that looked nothing like any clam alive today. They were bizarrely shaped bivalve mollusks called rudists, and most looked like some kind of upright garbage can, complete with a lid that could open or close on the cylindrically shaped shell of the clam. Some approached the size of the modern-day *Tridacna*, the "giant clam" of today's tropics. But unlike *Tridacna*, which are solitary, the rudists grew side by side in gregarious fashion, much as modern-day mussels do, crowding to cover every square inch of substrate, even growing over one another.

Each large cylinder that is the bottom shell of a single rudist clam is jammed vertically next to others of its kind, all packed into a solid pavement of one- to two-foot-long, sometimes foot-wide cones, each topped with gorgeously colored flesh reaching up toward the light. Like corals, they had tiny symbionts, single-celled plants that need light for photosynthesis and in turn provide the clam with bountiful oxygen, as well as carbon dioxide and waste removal from its tissues. But unlike modern corals, which can take centuries to reach large size, the clams grew very quickly. Within a year after sinking down from the floating plankton onto shallow ocean bottoms (they probably needed light to survive because their flesh contained tiny plants), the small clams grew thick carbonate outer shells to mature size in a year or less. They were born, grew quickly, and more often than not soon died, as others of their kind descended on their hard shells and grew, smothering the immobile yet living real estate they squatted on. A coral skeleton would have needed a century to grow from one individual to a colony several feet high and wide, whereas the rudists could have done the same in five years at most.

Like all reefs, the rudist reefs grew right to the very surface. On their outer seaward side, water depth dropped off quickly. Outside of the reef lay the vast open oceans of the Mesozoic, and both above and on the bottoms of these oceans there existed other now-extinct creatures.

The surface of these oceans would have been patrolled by both large sharks and giant seagoing reptiles. These latter included long- and short-necked plesiosaurs, as well as the lizard-like mosasaurs. They probably lived much as modern-day seals do, diving for food but needing to surface for air. But they were far larger than any seal, larger than any other creature that needs to come out of the water on occasion to rest or breed.

The deeper bottoms of the greenhouse oceans were also different from those of most oceans. Only the present-day Black Sea is similar to the conditions of the deeper bottoms and even mid-water regions of the greenhouse oceans—warm environments with so little dissolved oxygen that even most fish cannot live there. The bottoms were made up of black mud, as are the bottoms of the Black Sea. The mud trapped great quantities of fine, particulate organic matter that is black in color. There was little oxygen in the seawater at these depths—so little, in fact, that normal decomposition of organic material cannot take place, or does so only at a rate far lower than that on an oxygenated sea bottom. An entirely different community of microbes lived within the first few inches of this muddy bottom sediment, one that lived on sulfur, and a by-product of their particular form of respiration are the compounds hydrogen sulfide and methane.

Only in a few places on the the Mesozoic ocean bottoms would there have been enough oxygen to support animals that require normal amounts of oxygen.[3] But in the greenhouse oceans, two different kinds of mollusks evolved specifically for the characteristic low-oxygen conditions. One, a bivalve mollusk, lived on the bottoms. The other, composed of a vast diversity of cephalopod mollusks, the ammonites, lived in the water column, but fed off the bottom.

The ammonites of the Cretaceous ocean we are profiling here belonged to a group that first appeared in the earliest Jurassic, and their sudden appearance in rocks of that age suggest that the devastating Triassic-Jurassic mass extinctinon, which took place almost 130 million years prior to the late Cretaceous, opened the door to new kinds of animals, including new designs of ammonites. Finding them is one of the delights of fossil hunting, and because we two coauthors have spent so much time doing research in strata with ammonites over the past two decades, this has been rather a constraint on our great friendship. Coauthor Ward will become totally mesmerized by the least trace of an ammonite fossil. Coauthor Kirschvink would just as soon drill a paleomagnetic core out of even a museum-quality specimen. And has.

The final group of ammonites, which began in the oldest Jurassic strata and continued to the very greenhouse ocean of this chapter, are of great importance not only to the history of life, but to the very science of geology and using fossils to tell time. There are many places in the world where marine strata of latest Triassic age are overlain by Jurassic strata. At such outcrops one can walk through time, and if the strata are continuous, the dramatic events of the late Triassic and early Jurassic are present for all to see. This interval of time and rock preserves evidence of the great Triassic mass extinction, one of the so-called big five mass extinctions, a dubious honor of species death. As you walk through upper Triassic beds you are first in strata packed with the fossils of the flat clam *Halobia*, and then you move into younger rocks with the even more abundant *Monotis*. But then the clams disappear, over only several meters of strata, leaving a long barren interval of rock and time—the last stage of the Triassic, an interval perhaps 3 million years in length known as the Rhaetian stage.

Finally, after this thickness virtually without fossils, a new group suddenly appears—ammonites. While there were ammonites in the upper Triassic rocks, they are never abundant. But most famously at the beach of Lyme Regis of England, as well as in southern Germany and at many other localities worldwide, the earliest Jurassic ammonites appear in huge numbers, and over only a few short meters of strata they diversify as well. These are not like the Triassic flat clams, where one species is all you get. These ammonites of the first part of the Jurassic are diverse and abundant, which tells us that the great drop of oxygen was finally over and a slow rise was in place. But the ammonites are not telling us that oxygen levels similar to today were suddenly in place. The ammonites appear because the surface of the early Jurassic seas began to have a modicum of oxygen, and the ammonites took full advantage. They did so because they may have been among the best animals on Earth for low oxygen and could and did seize ecological advantage in the greenhouse oceans of the Jurassic and Cretaceous.

Because of the overall similarity of the chambered shells in both nautiloids and ammonoids, we presume that they may have had somewhat similar modes of life. Nautiluses today live in highly oxygenated water over most of its range. But here and there they also live in hypoxic bottoms. This was a great curiosity, because conventional wisdom was always that the cephalopods in general need high-oxygen conditions, but not so the one remaining stock of externally shelled, chambered cephalopods, the *Nautilus*. These latter are very tough and resistant when taken out of the water. They can sit out for ten or fifteen minutes with no ill effects. And when they are in water they gain oxygen through one of the relatively largest and highest-powered pump gills ever evolved, streaming great volumes of oxygen over the gills, thus allowing for sufficient oxygen molecules even in low-oxygen water. If ever an animal was adapted for low oxygen, this is it. British zoologist Martin Wells, who measured oxygen consumption of various captive nautiluses in New Guinea, finally proved this. When a nautilus is confronted with low oxygen, it does two things. First, metabolism slows way down. Second, with its strong swimming ability it can travel vast distances in search of not only food but higher-oxygen water areas.

The mass appearance of ammonite fossils in lower Jurassic strata suggests that the ammonites were superbly designed to extract maximum oxygen from minimal dissolved volumes of the oh-so-precious gas. Jurassic-through-Cretaceous ammonite body plans thus may have evolved near the Triassic-Jurassic boundary in response to worldwide low oxygen. Their new body plan (compared to the ammonoids that came earlier) involved a much larger body chamber relative to the phragmocone. Because of this, they had to use thinner shells, and this required more complex sutures. The sutures also allowed faster growth by increasing the rate of chamber liquid removal for buoyancy change. Within the large body chamber was an animal that could retract far into the body chamber, and that had very long gills relative to its ancestors.

We do not know if ammonites had four gills (like the *Nautilus*) or two like modern-day squid and octopuses. The lack of streamlined shells of the majority of early Jurassic forms makes it clear that these animals were not fast swimmers. It is far more likely that they slowly floated or gently swam near the surface, using their air-filled shell like a zeppelin.

The ammonites of the Jurassic period changed only in detail right up until the Cretaceous, but then spectacular changes in shell design began to take place. Where there remained many of the original planispiral shell design (like a nautilus shell), other shell shapes came into being in the Cretaceous, and with this, let us return to our dive in a late Cretacous ocean, among the ammonites.

Regardless of shape, most ammonites searched the bottoms for crustaceans or other small food. More than a dozen different kinds of ammonites could exist in the same environments, each with a different shell shape. Some were tiny, no more than an inch in diameter, while others were up to six feet in diameter. Most in the Cretaceous seas had thick, intricately branching ribs or tubercles of some kind, defensive armament that is testament to the abundance and efficiency of the shell-breaking predators present in these greenhouse oceans, and in all probability the plesiosaurs and mosasaurs were their major predators.

The ammonites would have looked a bit like squids stuck in a nautilus shell. Today's *Nautilus* has ninety tentacles, while the ammonites would have had either eight or ten. Nautiluses are scavengers, while the squids of today and the ammonites of the Mesozoic were carnivores, needing living organisms for food.

The second mollusk of the greenhouse oceans were clams, not as bizarrely different in shape as the rudists, but certainly different from anything alive today. They were what we call flat clams, known as *Inoceramus*. Related to oysters, they came in a variety of species, all competing on the same muddy bottoms. None could burrow, but instead had to sit on the bottom. Some were veritable giants, with gently ribbed, almond-shaped shells that attained lengths of more than eight feet from their beaks to their broad apertures. Yet unlike any clam today, their shells were almost paper-thin relative to their size, and their gently ornamented upper shells were sometimes encrusted with a diversity of oysters, scallops, bryozoans, barnacles, and tube-worms. Usually, however, the *Inoceramus* clams lived on bottoms and in seawater that had too little oxygen for "normal" mollusks or other invertebrates to live. A great many of our colleagues have used geochemistry to better understand how different these clams were from perhaps any clam living. New work by Neil Landman of the American Museum of Natural History in conjunction with geochemist Kirk Cochran has wonderfully shown the strangeness of these Mesozoic communities.

Just looking at the sizes of the inoceramids compared to other clams tells us how strange they were. In the modern world, the largest clams, the *Tridacna* (giant clams) of the tropics, can be six feet from end to end, thus holding hundreds of pounds of flesh. But the next biggest clams, known as geoducks, are at most a foot in length, with no more than a pound or two of living tissue. Some oysters attain a foot in length as well, but not many. But the inoceramids fill that modern-day gap between the giant *Tridacna* and the far smaller geoducks. An enormous variety of the inoceramids existed from the

Permian to the end of the Cretaceous when they died out, and it was in the greenhouse oceans where they flourished. They contained microbes allowing these big clams to live off methane and other chemicals seeping out of the organic-rich, low-oxygen sea bottoms of the greenhouse oceans rather than filtering food out of seawater, as the modern clams do.

A final region of the greenhouse oceans was the mid-water region,[4] the nether regions of the oceans that are too deep for sunlight, but still hundreds to thousands of feet above the stagnant bottoms of the seas. This vast, mid-water environment in today's oceans is the largest single habitat on the planet, and is colonized by a variety of creatures adapted to a life in which they never encounter the surface, and its sun and atmosphere, nor do they ever come in contact with the sea bottom. Here, life depends on staying "in between," for to these creatures both the warm shallows and the deeper cold bottoms would be fatal, from predation to temperature and oxygen conditions, or both. Thus, adaptations for the attainment and then maintenance of neutral buoyancy are paramount for existence. In our oceans, the most common of the larger inhabitants in this region are the mid-water squid, animals that have evolved floating tentacles or sacs within their bodies concentrated with fat or other chemicals such as ammonia-rich solutions that renders the entire animal lighter than seawater.

Their prey was individually small but vast in quantity and was composed of a diverse and abundant assemblage of small swimming animals that combined are known as the deep scattering layer (DSL)—based on their discovery by some of the world's first sonar, used in the 1940s. This DSL is composed of untold numbers of small crustaceans and other arthropods such as amphipods and isopods, as well as a variety of other phyla. By day this enormous layer of life—extending from perhaps eight hundred to six hundred meters deep—extends for hundreds or even thousands of miles in all directions in the ocean regions far from shore. But as daylight fades, the entire layer slowly begins to swim upward toward shallower depths, and with full darkness the untold gigatons of animals making up the DSL arrive in the shallower, warmer, more nutrient-rich depths—depths, however, that would be fatal during daylight for the tiny, succulent arthropods making up the vast preponderance of the DSL fauna because of visual predators such as fish and squid.

We have good evidence showing that this fundamentally new kind of lifestyle in the sea first appeared in the Cretaceous. Before that there would have been no food resources worth pursuing in the mid-water regions, and hence there were no species making the extensive adaptations that a larger animal would need not only to float throughout its life, but be able to somehow migrate hundreds of meters upward each nightfall and then settle back down into greater depths each morning. But with the appearance of the mid-water arthropods, evolution quickly produced animals capable of feeding on them using new kinds of buoyancy devices, for the most fundamental adaptation was some kind of way of being weightless in the mid-water.

The carnivores that evolved to exploit this resource were mainly ammonites, but in shapes very different from the traditional and ancestral planispirals, which marked the species living just above the sea bottoms. The mid-water ammonites had shells that let them float for their entire lives; they had bizarre shells that would not have allowed any kind of rapid swimming speed. But once in the thick oceanic water mass or liquid stratum inhabited by the denizens of the deep scattering layer, food would have been so abundant that they only had to stay in the layer, rising when it rose, slowly sinking with it during the day, to have a food-rich and predator-free existence. They thus lived a slow, floating existence; in essence these strange creatures acted like hot air balloons, with a large flotation device above, attached to a small passengers' basket suspended below the flotation device.

The ammonites of the mid-water[5] needed to efficiently control their buoyancy. We know that the buoyancy system of the living nautiluses is quite crude and slow to operate. But it may be that ammonites in general used their complex septa with the beautiful sutures as part of a far better buoyancy apparatus, one that could pump water out or let it flood back with great rapidity into otherwise air-filled chambers as ballast. These then-new Late Cretaceous ammonites are informally called heteromorphic ammonites, because of their shells' departure from the original, traditionally coiled design of ammonites from their first appearance in the Devonian period until their final demise at the end of the Cretaceous. They were body plans never found before or since their approximately 60-million-year-long existence, and they lived right up to the day that the Chicxulub asteroid fried all ammonites out of existence.

Some of the heteromorphic ammonites looked like giant snail shells, but snail shells filled with chambers of air, with one last, longer shell portion hanging beneath and containing the soft parts of the ammonites, tentacles and all. Others looked like gigantic paper clips, and some were simply huge hooks. But the most common were long, straight cones. The pointed end of the cone was the first formed chamber, and by adulthood, these long, thin cones could reach up to six feet in length. They hung vertically in the water, their tentacle heads hanging straight down and beneath the chambered flotation portions of the shell. There were named *Baculites*, and in the Late Cretaceous they may have been the most common carnivores on the planet.

Vast schools of these *Baculites* filled the mid-water depths.[6] They are often illustrated in the many murals and paintings of the Cretaceous oceans, and invariably they are wrongfully depicted as long, arrow-like shapes living horizontally in the water, much as fish and squids do. But this would have been impossible. They were vertical in orientation, with their tiny, earliest shell portions up, and their heavy head and tentacles hanging downward. They were never able to swim sideways or even float in a sideways orientation. Everything was up and down to them. They probably were amazingly rapid, shooting upward using jet propulsion and then sinking slowly back downward. The predators of the *Baculites*, probably fish and sharks, would have been mystified again and again in attempting the normal predatory attack, in which the prey tries to escape by swimming ahead of the predator. But the attackers would have seen a rapid upward movement of the long vertical creature, like a puppet on a string, as they swam helplessly forward in the direction that all self-respecting prey were supposed to flee.

THE MARINE MESOZOIC REVOLUTION

During the later Mesozoic era, a revolutionary changing of the guard took place in the seas. Paleobiologist Gary Vermeij of the University of California at Davis called it, simply enough, the Mesozoic marine revolution.[7] It was nothing less than a world where the marine predators ran wild in terms of evolution.

Watching our friend and colleague Gary Vermeij, blind since early boyhood, "see" (as he describes the process) the intricate adaptations for the strengthening of post-Paleozoic mollusk shells is to watch a concert pianist's fingers: rapid, complex movements of fingers seemingly gone boneless as they "play" the many morphological keys of a snail shell, from the spines of the turret to the fat-lip-like callus on the snail's aperture; the gentle discovery of a calcareous filling of the otherwise dangerous umbilical region of the shell, to the rapid trilling of the tiny yet strength-giving teeth on the outside of the shell's apertural lip, also thickened. We guide him to museum drawers, and then he guides us to insights coming to him from a single sense—touch.

Touch has memory, but touch also visualizes, and it was with touch only that Vermeij's agile mind took his "vision" of the ever-increasing offensive, shell-breaking capabilities of the newly evolved, post-Permian predators, coupled to the coevolving and ever-better calcareous suits of armor of the invertebrate herbivores, and smaller predators into a generalization we now know as the Mesozoic marine revolution.

At first the concept was only that predation of the post-Permian mass extinction shifted toward shell breaking—toward new ways of gaining the rich meat to be found in formerly impregnable fortresses such as those of the Paleozoic clams, snails, echinoderms, and brachiopods. But the concept has expanded.

The adaptations of the prey were no less impressive. Clams that used to live only on the ocean bottom, or just beneath, evolved adaptations for deep burrowing. These new clams are called heterodont clams (because of the many "teeth" in their hinge line). They underwent major anatomical revisions by fusing part of their mantle into a pair of siphons (what we call the neck of a clam). These burrowing clams today remain the most diverse group of clams—a whole spectrum of species that are capable of rapid burrowing into sand, mud, or silt. There is but a single reason to do this: to escape predation. Sitting within sediment rather than atop it in no way increases feeding efficiency. But it immensely increases survivability. Others radically changing morphology to allow a burrowing (or wood-boring) lifestyle included snails, new kinds of polychaete worms, some kinds of fish, and totally new kinds of sea urchins.[8]

Another group of invertebrates showing radical innovations were the class of echinoderms called crinoids.[9] These immense flowerlike invertebrates (they are commonly called sea lilies) were typical of the Paleozoic in that they were attached: they could never move through life after settling from a planktonic larval stage; they attached to the bottom. Driving through the Midwest today provides stark evidence of their former abundance: every road cut is made of rocks that are almost nothing but the tiny round "ossicles" that make up the long stem of the attached crinoids. To produce such abundance, a vast, shallow, extremely clear and warm ocean was needed, whose bottom would have been obscured by the forest of crinoids. It is doubtful that the sun could penetrate to these shallow bottoms, which would have been fine with the crinoids; their food is microscopic plankton, and they live in the "slow lane," at least metabolically. But once attached, they could never move, and if detached by storm or predator, they would soon die.

There is nothing like wholesale death to stimulate new evolution. The Permian extinction virtually wiped crinoids off the planet, and in the new rules of the predator-rich Mesozoic, they were quick meals for any predator designed to extract some sort of food out of a crinoid—which would be difficult, as there is probably no other example of life where so much calcium carbonate skeleton protects so little flesh. But the attached crinoids gave way to a new group of crinoids without stalks. These exist to the present day, and are among the most beautiful of all life to be found in modern coral reefs. There are actually capable of swimming—slowly and with great stateliness, using their arms to gently wing themselves through the water.

The Mesozoic marine revolution was not only about predators and prey but also about ever-greater utilization of new habitats.[10] This included the evolution of morphologies allowing ever-deeper burrowing by clams and snails to escape predation, as well as the ever-increasing number of other invertebrates that were using the sediment also for food. These changes are evidenced by an increase in the diversity and abundance of trace fossils, similar to those that we described in the Cambrian explosion chapter. The net result was almost complete bioturbation of Mesozoic sediments.

It was not just the bottom and beneath the bottom of the Mesozoic oceans where radical change was occurring among animals. For the first time in the existence of animals, a wholesale exploitation of the water column from top to bottom was also taking place. Many of the newly evolved forms were not animals at all, but protozoans and even floating single-celled planktonic plants. Important new groups of microfossils are found in Mesozoic strata, including the evolution of a huge variety of the amoeba-like but skeletonized foraminifera, which lived both on the bottom as well as floating well above it. Other plankton was the siliceous radiolarian. But perhaps the greatest radical change within the plankton of the Mesozoic and onward was the evolution of a group of algae called coccolithophorids—whose tiny skeletons, when accumulated on a sea bottom and turned to stone, is the well-known substance chalk.

Coccoliths are tiny plants sporting a half dozen to a dozen microscopic calcium carbonate plates bonded to the upper part of their rather spherical bodies. After death these tiny plates would fall to the bottom and accumulate in untold numbers until such immense sedimentary units as the famous White Cliffs of Dover were formed. The entire northern tier of Europe is lined by such cliffs, from Britain to France, Poland, Belgium, Holland, all over Scandinavia, and through much of the old Soviet Union all the way down to the Black Sea. The coccoliths have played a large part in the planet's temperature. Coccoliths are white in color, and this whiteness reflects sunlight back into space, which in turn cools the planet.

Just as in the Cambrian explosion, in which animals were stimulated to produce new kinds of body plans based on respiratory systems, so too did animals of the Triassic seas show a multitude of new adaptations. As we have seen, the land fauna experimented with a variety of lung types. The same kind of exploration took place in the oceans. The bivalve mollusks were one group that evolved a new kind of body plan and even physiology in response to the nearly endless expanse of nutrient-rich but low-oxygen bottoms.

The very lack of oxygen on the bottoms made them, in one sense, wonderful places to live. Vast quantities of reduced carbon, in the form of dead planktonic and other organisms, fell to the seafloor and were buried there. On an oxygenated bottom this material would be soon consumed by filter- or deposit-feeding organisms and scavengers. But the low-oxygen conditions kept these organisms out, and not even the usual bacteria that decompose dead creatures on the sea bottom were around. As we have seen, this is one reason that oxygen levels plummeted in the Triassic. But the clams figured a way out of this. A few kinds, such as the *Inoceramus* clams described above, who lived on the seafloor of the bottoms that had at least some oxygen, fed not on the falling organics, but on methane coming up from the some fraction of the organic-rich sediment. Methanogens are a group of bacteria that thrive in low- or no-oxygen conditions, and even several centimeters down into the sediment on a sea bottom with some oxygen would have penetrated into an oxygen-free zone—thus allowing the existence of the methanogens. As they metabolized, they released methane as a by-product. The clams may have had other bacteria in their gills that could exploit the methane and other dissolved organic material, or they may simply have fed on the bacteria. A somewhat similar mechanism is found today in the deep-sea vent faunas, where giant tube worms and clams use these chemicals as food. But the difference is that the modern vent faunas are oxygenated. The animals down there do not even need gills. The clams of the Mesozoic were not so lucky.

Another and quite different body plan was evolved by the crustaceans in response to oceans characterized by a critical lack of oxygen: the crabs and lobsters. While the overall shrimp-like body form of crustaceans is found in Paleozoic rocks, crabs are a relatively new invention. A crab is simply a shrimp-like form in which the abdomen is tucked under the body. The fusion of the head and thorax into a heavily armored and calcified plate makes the crab a difficult nut to crack for its predators. And the placement of the abdomen under this armor plating is design genius. It is the abdominal regions that are most susceptible to breakage in any predatory attack, and by eliminating this chink in its armor the crab rose rapidly to marine prominence. Their large claws allow them to crack open mollusk shells among other prey—they are known as durophagous, or shell-breaking, predators. Prior to this, few predators were able to break into shelled organisms. Crabs and others figured out how.

Thus the accepted reason for the crab body plan, novel as it is, relates to defense (the tucking of the abdomen, the thickening and increased calcification of the head-thorax region) and offense, the evolution of a strong pair of jaws. But here is another: crab design came about in some part as a primary adaptation for increasing respiratory efficiency by putting the gills in an enclosed space under the cephalothorax (the head-thorax), and then evolving a pump to move water over the now-enclosed gills.

The crab gill design is a marvelous way to increase water passing over gills.

Crabs evolved from shrimp-like organisms, and in these ancestors we can see a progression toward the crab gill system. In shrimp, the gills are partially enclosed beneath the animal. While covered dorsally, the gills are attached to segments and are open to water underneath.

The Mesozoic greenhouse oceans changed over time. Yet two of their most characteristic features, the ammonites and inoceramid bivalves, might still be with us today but for a very bad day that happened some 65 million years ago, when the Chicxulub asteroid obliterated the Mesozoic biota out of its characteristic existence.

CHAPTER XVI

Death of the Dinosaurs: 65 MA

SOMETIMES it is the great science fiction writers who can best encapsulate the past. Here is one of the best descriptions of the most famous of all mass extinctions, the K-T event, and as stated in our introduction, we use this older term instead. We were delighted to discover this wonderful description from a book by the iconic writers William Gibson and Bruce Sterling, in their book *The Difference Engine*:

> Storms of cataclysm lashed the Cretaceous Earth, vast fires raged, and cometary grit sifted through the roiling atmosphere, to blight and kill the wilting foliage, till the mighty dinosaurs, adapted to a world now shattered, fell in massed extinction, and the leaping machineries of Evolution were loosed in a chaos, to re-populate the stricken Earth with strange new orders of being.

As we well know, those "strange new orders of being" included the many kinds of mammals populating the Earth today. Yet how is it that we came to know with such certainty that it was indeed an asteroid that did in the dinosaurs? This "fact" has held sway since around 1990, or ten years after the Alvarez group of Berkeley published their bombshell findings that have utterly changed our understanding not only of mass extinctions but also of geological processes in general.

The study of mass extinctions was intricately interwoven into the very fabric of the nascent developing field of geology in its most fruitful early period, from about 1800 to 1860. For decades there had been a dispute about the most basic of principles that allowed explanation of how geological material and formations, as well as the kind of animals and plants present on the Earth, came into being. The fight was between those promoting the principle of uniformitarianism, with the mantra that the present is the key to understanding the past on one side, and a group pushing a principle of catastrophism on the other. The latter was championed by the scholar who first recognized the reality of extinction, Baron Georges Cuvier of pre- and immediately postrevolutionary France, and by later disciples, the most important being Alcide d'Orbigny, whose lasting contribution to geology was the development and then modernization of geological time units. But for all the science they contributed, both Cuvier and d'Orbigny resorted to the supernatural to explain the startling evidence for mass extinction that they first discovered in their early studies of the fossil record. Both believed that a supreme being brought about occasional world-covering floods that wiped out most life and then repopulated the postflood lands and oceans.

There was ebb and flow in terms of the confidence that new generations of geologists and naturalists had in uniformitarianism and the contrasting catastrophism. Uniformitarianism eventually won out, as more and more sophisticated interpretation of rocks, their features, and their ages showed no evidence of a single world-covering flood, let alone the succession of them that would be needed to explain the ever-growing list of mass extinctions known to have occurred. From oldest to youngest, these were the Ordovician, Devonian, Permian, Triassic, and Cretaceous-Tertiary mass extinctions, now referred to as the big five. By the twentieth century, catastrophism was no longer accepted by anyone, save the occasional crackpot writers trying to cash in on the many humans hopeful that the past was more exciting than the grain-by-boring-grain accumulation of the thick stratigraphic record. But in one facet—the explanation of the mass extinctions—the uniformitarianists (including Charles Darwin) remained uneasy.

Geology concluded that the mass extinctions were very slow, stately events, and that given enough time, even observable changes in climate and even the level of the oceans could explain the extinction of so many species at each of the big five mass extinctions accepted by all geologists in the latter half of the twentieth century.

There were a few (but only a few) cries of dissent. One of the best was from Otto Schindewolf, professor of paleontology at the University of Tübingen in southern Germany. Schindewolf objected to slow and stately as the reasons for mass extinctions. Instead (but only after a career of long and careful study of the fossil record and its changes), he speculated that perhaps far more catastrophic and rapid events caused the mass extinctions, and suggested that the effects of a nearby star going supernova might have been sufficient to cause one or more of the known mass extinctions. He even gave a name to this throwback to the past—he called it neocatastrophism, and he meant it to be a very nonuniformitarianistic way of explaining the past.

Schindewolf's speculations fell on deaf ears in his profession. Slow climate change, slow sea level change were the "facts"—and presumed causes—of the great mass extinctions. For thirty years, from 1950 on, the field of geology slumbered, smug in its understanding that all was explainable by earthly (and slow) causes. That, then, was the state of play concerning mass extinctions from the time of Schindewolf's speculations in the 1950s until 1980. And then it all changed. On June 6, 1980, the thirty-sixth anniversary of the D-Day invasion of Europe, the Alvarez paper on the asteroid causing the Cretaceous-Tertiary mass extinction invaded the old, stately, yet already tottering edifice of uniformitarianism in general, and the accepted causes of mass extinctions in particular.[1] It was a shot that started a scientific war that in some sense continues to this day. The work of the Alvarez group answered that question.

IMPACTS AND MASS EXTINCTIONS

The presence of numerous impact craters on every planet or satellite of the solar system with a solid surface gives stark and impressive evidence of the frequency and importance of these events, at least early in the history of our solar system. It is probable that impact is a hazard in most, or perhaps all, extra-solar planetary systems as well. Impacts are probably the most frequent and important of all planetary catastrophes. They can completely alter the biological history of a planet by removing previously dominant groups of organisms and thus opening the way for either entirely new groups or previously minor groups. Thus the 1980 Alvarez work took on special significance on many fronts.

Two of the most important lines of evidence used to convince most workers in the field that the K-T extinction was indeed caused by large body impact were the discovery of both elevated iridium values within the boundary clays and abundant "shocked quartz" intermingled with the iridium. By 1997, high iridium concentrations had been detected at over fifty K-T boundary sites worldwide. Iridium is seen as an indicator of impact because it is quite rare on the Earth's surface, but at much higher than Earth concentrations in most asteroids and comets. Shocked quartz grains are seen as indicators of impact because creating the multiple lamellae on sand-sized quartz found at most K-T boundary sites can be produced only by high pressure events, such as would occur when a large asteroid hits rock containing quartz at high velocity. No "earthly" conditions naturally create such quartz grains with multiple shock lamellae.

In addition to iridium and shocked quartz grains, K-T boundary sites have also yielded evidence of fiery conflagrations that must have occurred soon after the impact.[2] Fine particles of soot were found in the same K-T boundary clays from many parts of the globe. This type of soot comes only from burning vegetation, and its quantity suggested that much of the Earth's surface was consumed by forest and brush fires.

Although originally controversial, the mineralogical, chemical, and paleontological data gathered during the 1980s persuaded most experts that a large (~10–15 km diameter) comet or asteroid impacted the Earth ~65 MA, and that, at the same time, more than half of the species then on Earth became extinct rather suddenly in the K-T. The discovery of a large impact crater of precisely the right age in the Yucatán region of Mexico (the Chicxulub crater) largely swept away remaining opposition to the impact hypothesis.

The ultimate killer, according to Alvarez et al., was a several-month period of darkness, or blackout as they called it, following the impact. The blackout was due to the great quantities of meteoric and Earth material thrown into the atmosphere after the blast, and lasted long enough to kill off much of the plant life then living on Earth, including the plankton. With the death of the plants, disaster and starvation rippled upward through the food chains.

Several groups have calculated models of lethality caused by such atmospheric change. Apparently a great deal of sulfur was tossed into the atmosphere as well. A small portion of this was reconverted into H_2SO_4, which fell back to Earth as acid rain; this may have been a killing mechanism, but was probably more important as an agent of cooling than direct killing through acidification. However, more deleterious to the biosphere may have been the reduction (by as much as 20 percent for eight to thirteen years) of solar energy transmission to the Earth's surface through absorption by atmospheric dust particles (aerosols). This would have been sufficient to produce a decade of freezing or near freezing temperatures on a world that at the time of impact had been largely tropical, and confirmed the earlier Alvarez statement that the blackout played a part in the mass extinction. The prolonged winter was brought about by vastly increasing aerosol content in the atmosphere over a short period of time.

After the impact there was also dust, lots of it.[3] The global climatic effects of atmospheric dust produced by the impact of a large (10 km diameter) asteroid or comet would include long-term (on the order of months) blackout producing light levels below that necessary for photosynthesis, accompanied by rapid cooling of land areas. But perhaps the most ominous prediction in this model is the formerly unappreciated effect that the giant volume of atmospheric dust generated by the impact has on the hydrological cycle. Globally averaged precipitation decreased by more than 90 percent for several months and was still only about half normal by the end of the year. In other words, it got cold, dark, and dry. This is an excellent recipe for mass extinction, especially for plants, as well as for the creatures feeding on plants.

Finally, it was realized that within hours after the impact, bits of rock would rain out of the sky at high speed, and coming through the atmosphere from the near outer space heights that they had been blasted up to, they came in hot enough to set the Earth's vegetation on fire. What may have been the largest forest fires in history, on all continents, ensued, and this alone may have killed off the dinosaurs on land.

PRECURSOR EXTINCTIONS

It is now known that as many as 75 percent of all species went extinct in the K-T mass extinction. On land, it was marked by the loss of dinosaurs on one side, and the appearance of mammals on the other. In the sea it was the disappearance of ammonites on the Cretaceous side of the event, and the appearance of marine biotas dominated by clams and snails on the Paleogene side. But as age dating improves, we are seeing that the "mass extinction" at the Cretaceous-Paleogene boundary is far more complicated than the original single-strike theory can account for. We now know of at least two "pre-K-T" pulses of mass extinction prior to the final blow, which remains fully supported. But the last few years of research show that once again the effects of flood basalt volcanism were part of the killing mechanisms.

There are very few dinosaur fossils, and thus using them to try to learn the rate at which they went extinct is nearly impossible. The fossil record is full of microfossils, and it has been the study of these that lend great support to the claim of sudden extinction caused by the asteroid or comet's fall. Yet we needed to know the fate of larger fossils both on land and sea, and the most studied of these latter have been the ammonite cephalopods, described in the last chapter.

The best place to study the extinction of the last ammonites that lived near the equator of the Late Cretaceous world is at the thick stratal outcrops lining the Bay of Biscay, a large region extending from southwestern France to northeastern Spain, and best of all is at the rocky coastline near the old Basque village named Zumaya.[4] There, hundreds of meters of stacked strata ranging in age from 72 to about 50 million years in age are laid out like the pages of an open book. The mass extinction boundary lies in a bay defined by the rocks, and there is even a change of lithology and color marking this unmistakable boundary.

The oldest of the rocks along the Zumaya coastline dates to around 71 million years ago. The strata are composed of individual stratal beds each about six to twelve inches thick, and each bed is a pair: a thicker limestone, and a thinner, more lime-free rock called a marl. Thousands and thousands of these pairs are stacked one upon another and long ago lithified into the rocky coastline. From the kind of rock and from the fossils, we know that the strata were deposited in fairly deep water, in the deepest part of the continental shelf, or even over its edge, at depths between two hundred and perhaps four hundred meters deep.

The strata are aligned perpendicular to most of the coastline and tilted at a steep angle to the north. Walking along the coastline from south to north takes one ever younger in time. But the very steep tilt of the rocks and the high range of daily tidal change make a visit to a sizable portion of these two outcrops low-tide dependent and very difficult walking.

At the start of the walk, well down the beach (and thus down in time as well) from the long stairway that is the only entrance to this rocky site, fossils are everywhere. Most are large clams, the inoceramids discussed in the last chapter, but there are many ammonites as well, and not a few sea urchins, looking like large stuffed hearts. There are no vertebrate bones, no shark's teeth—and certainly no dinosaurs—but these marine strata are the same age as nonmarine beds found around the world holding one of the great dinosaur assemblages known.

The inoceramids provoke the most wonder. They are up to two feet in diameter, looking like shallow but giant plates lying side by side with smaller versions of themselves—different species, in fact. For a hundred meters of stacked strata they are ubiquitous in each bed, and because of the dip of the beds, some of the bedding planes that can be examined are hundreds of square meters in area. Fossils are most easily found on the tops or bottoms of any stratum, rather than the sides, and the best hunting is always on the tops of large bedding planes. At Zumaya there are many of these to be explored and collected, and thus the fossil numbers are very large for any fossiliferous beds of any age. But then the large clams disappear, and they do so more than a hundred meters stratigraphically below the well-marked ammonite extinction horizon. The ammonites and echinoids continue in abundance right up to the point where they suddenly and dramatically disappear.

The work at the Bay of Biscay coastline, supplemented by work at other late Cretaceous deposits, tells us that the inoceramid bivalves died out gradually around 2 million years before the ammonites suddenly died out. In fact, using statistical methods developed by Charles Marshall of UC Berkeley, Marshall and coauthor Ward showed that at least twenty-two species of ammonites existed in this region right up to the layer that contains the most important evidence of impact: iridium, shocked quartz, glassy spherules (tektites, which are bits of rock blown upward by the titanic impact and then turning to tiny fragments of glass as they fell back to Earth at high speed).

The curious part about the inoceramids' extinction is not that they died out well before the ammonites, but that their extinction took place at different times in different places. For instance, the last inoceramids in Antarctic Cretaceous rocks are no younger than 72 million years in age, or about 7 million years prior to the ammonite extinction. We now know that these globally distributed bivalves experienced a wave of species death starting first in Antarctic regions and then gradually moving to the northern hemisphere. It was almost like a disease slowly moving northward and killing off the clams in gradual fashion. But this was no disease: it was cold and oxygen.

Near the end of the Cretaceous, an oxygenated kind of thermohaline circulation began to occur in the high southern latitudes, and over about 2 million years this cold, oxygenated bottom water spread into all the seas, moving from south to north. Its presence spelled the end of the clams that we affectionately called inos, and the disappearance of these clams was a signal event in the history of life, for they had been highly successful up to that point for more than 160 million years. But they were adapted to the other kind of ocean, the low-oxygen and warm bottom water variety. Cold and oxygen killed them off.

ONLY IMPACT?

We can now summarize current understanding of the primary event that appears to have caused the K-T mass extinction. There was but a single comet strike, coming soon (1–3 million years) after two rapid changes in global sea level, themselves sandwiched around a major change in ocean water chemistry.[5] The impact created the large (up to 300 km diameter) crater now named Chicxulub, located in the Yucatán Peninsula. Although there is still debate about the actual crater size, there is now no doubt that the structure is a crater. The impact target geology and geography may have maximized subsequent killing mechanisms. This is especially true because the presence of sulfur-rich evaporites in the target area, and sulfur within the incoming comet itself, may have contributed to subsequent lethality. The 65-million-year-old comet strike in an evaporate-rich carbonate platform, itself covered by a shallow sea at an equatorial latitude, seems to have created unbelievably dire consequences: worldwide change in atmospheric gas inventory, accompanied by temperature drop, acid rain (mainly from sulfur derived from evaporites at the impact site), and global wildfires are all proposed killing mechanisms. Most scientists (but not all) also agree that thick, coarsely clastic sedimentary deposits found at many places along the eastern coast of Mexico were formed by impact waves. The prolonged impact winter was thus the most important killing mechanism—and it was brought about by vastly increasing aerosols in the atmosphere over a short period of time.

Another recently published model describing atmospheric changes following impact suggests that greatly increased levels of atmospheric dust generated by the blast may have been lethal as well. The fine dust would be generated by impact into either an oceanic or a continental target area and would produce a long-term (on the order of months) blackout. This reduction in light levels (below that necessary for photosynthesis) would be accompanied by rapid cooling of land areas. This excess dust would also adversely affect the world's hydrological cycle. Advanced climate modeling has indicated that, following a large impact event, globally averaged precipitation decreases by more than 90 percent for several months and is still only about half normal by the end of the first year following the impact. The effect on the biota is now well established.[6]

BUT WHAT ABOUT THE DECCAN TRAPS FLOOD BASALTS?

The pages above make the case that the K-T mass extinction was largely a one-event mass extinction. The Earth was hit, and that hit led to environmental changes sufficient to kill off more than half of all species then on Earth. There remains only one nagging bit of unexplained information. The asteroid hit a planet that was already in the midst of one of the most extraordinary periods of flood basalt volcanism known from any time in Earth history. Called the Deccan Traps, this event caused untold tons of basalt to issue forth onto the Earth's surface, with an origin from deep within the Earth. Perhaps 84 million years ago, a gigantic mass of molten rock detached from near the mantle-core boundary to begin an upward voyage that would take around 20 million years. On its way up, this great mass of molten rock very likely caused the Earth to undergo episodes of true polar wander, events that happen when there is a mass imbalance such that internal balance dictated by laws governing the conservation of momentum of our spinning planet cause great landmass movements. These rapid movements might have destabilized some environments. For instance, much of western Canada and Alaska seems to have resided at the latitude of Mexico prior to 84 million years ago—but found itself far from Mexico as the Mesozoic ended.

Yet of all of the effects of flood basalts, the most consequential for life, as we have seen in multiple episodes reported earlier in this book, is the great outpouring of carbon dioxide and other greenhouse gases that accompany flood basalt volcanism. The Earth warmed quickly at the poles and other high-latitude regions, but less quickly at the equator. These conditions have led to what we call greenhouse extinctions. A big flood basalt causes the high latitudes to heat, pushing the oceans into stagnancy and then anoxia. Deep ocean waters filled with poisonous hydrogen sulfide rise to the surface. Things then died, as they did in the Devonian, Permian, and Late Triassic. Yet the dirty little secret is that we students of these mass extinctions have for too long swept this unfortunate evidence under the carpet. Who needs death by stagnation when there is more than enough death to be handed out by an impacting asteroid?

Science gets things right, eventually, if the question is interesting enough. And there are few more interesting questions than those pertaining to why dinosaurs (and so much else) died out 65 million years ago. Why were there no observable effects of the Deccan Traps,[7] when all of the other flood basalts did so much damage and caused so much obvious species extinction?

In fact, the Deccan did a lot of damage. Perhaps the best evidence of this, modesty aside, comes from our own work in Antarctica. In 2012, one of our students, Tom Tobin, showed that there indeed was warming of the oceans some hundred thousand years prior to the impact—and that species did die because of this.[8] As noted, global warming (which is the result of a flood basalt, ultimately) is larger in extent (temperature change) at high latitudes. The tropics are already about as warm as they can be. As we are seeing in our own world, it is the Arctic and Antarctic that bear the brunt of temperature change—and temperature-change-caused havoc and extinction.

So too with the K-T. Yes, a large asteroid hit us. But for hundreds of thousands of years prior to this, a suddenly warmer world was made stagnant by flood basalt. We can finish this chapter with a hoary boxing analogy. A knockout punch is by definition a single blow. Yet very few knockouts occur from the first blow, no matter how devastating. It is the many rounds of jabs and body blows that set the stage. The Deccan Traps softened the world. The asteroid finished the job.

CHAPTER XVII

The Long-Delayed Third Age of Mammals: 65–50 MA

THE earliest-known mammals were tiny, shrew-sized waifs named morganucodontids, living (probably fearfully) among the many larger predators of 210 million years ago in the latest Triassic—and then somehow surviving the great T-J mass extinction. Soon the morganucodontids were joined by other primitive but "true" mammals. All living mammals today, including us, descend from the one line that survived this extinction. This is what the world came to after its long dinosaur era came to a crashing, fiery end: a plague of rats. Or at least rat-sized survivors.[1]

Paleontologists have long believed that the ancestors of all mammals living today emerged on one of the northern continents as Pangaea slowly split apart throughout the Mesozoic era, and then only slowly migrated south, all the way to Antarctica and Australia, as land connections (or only narrow waterways) developed between the continents. This has been dubbed the Sherwin-Williams model of evolution, a reference to paint dripping over a globe from north to south by a long-lived US paint company. But this idea has to be tossed on the giant mound of discredited hypotheses, with new evidence coming from both fossils and genetics. It now looks like the wave of mammalian modernization went from south to north. Especially telling are the newly collected fossils of advanced mammals far older than any known in the north.

Geneticists have also joined in, once again replicating a familiar pattern of major new understandings coming from both DNA comparisons as well as evo-devo. There has been no end of surprises in the twenty-first century.[2]

Here are the three most important. First, the major mammalian "groups"—the eighteen living orders, as well as some suborders and even families still found today—actually diversified long before the extinction of the dinosaurs, which overturns the long-held idea that these groups did not evolve until after the K-T calamity. Fossils suggest that most modern groups appeared around 60 million years ago, after the dinosaurs were gone. Molecular data suggest they actually began diversifying about 100 million years ago.[3]

Second, most early mammalian evolution and subsequent divergence happened on southern rather than northern continents. Third, many groups thought to be very distant cousins are in fact next of kin. For example, paleontologists have always assumed that bats were in the same superorder as tree shrews, flying lemurs, and primates. But genetic data place bats with pigs, cows, cats, horses, and whales. Whales themselves are now known to have come from piglike ancestors, rather than from the same stock that gave rise to seals.

Much of mammalian success came from anatomical change, including the separation of the jaw and the ear bones, which allowed the skulls of later mammals to expand sideways and backward, a prerequisite for bigger brains. But the most important of all innovations was by the revolution of mammalian teeth. The upper and lower molars of morganucodontid jawbones interlocked, letting them slice their food into pieces.

Today's mammals are split into two major groups: the ancestral marsupials, which produce extremely small newborns and then keep them in a pouch—and their more diverse and abundant descendants, the placental mammals. New DNA studies now suggest that placental mammals began to diverge from marsupials as early as 175 million years ago.[4] Fossils have also chimed in, most spectacularly from China.[5] There, a complete new fossil of a protoplacental species found in Liaoning Province supports the DNA inference that placentals began evolving much earlier than previously thought. Named *Eomaia*, the fossil's age at 125 million years makes it easier for paleontologists to accept the genetic evidence that says the first protoplacentals began to evolve as much as 50 million years earlier yet, back in the Jurassic.[6]

The oldest group of living placental mammals include elephants, aardvarks, manatees, and hyraxes.[7] When the African continent split off from the former supercontinent of Pangaea, it carried these animals away to evolve on their own for tens of millions of years. The continental dispersion also split South America from Eurasia and North America for millions of years, and South America became home to sloths, armadillos, and anteaters. The northern continents have the youngest placental mammals on Earth, including seals, cows, horses, whales, hedgehogs, rodents, tree shrews, monkeys, and eventually humans.

Yet, if a great deal of mammalian diversification predated the K-T extinction, the most notable change—size increase—happened soon after the fall. Within 270,000 years mammals were diversifying and growing bigger, although the really large mammals did not appear until around 55 million years ago. Then a rapid increase in global temperature was coincident with a widespread growth of forests around the world, even near both poles, and this aspect of the history of plants may have helped stimulate a great increase in mammalian diversity.

THE TERRESTRIAL WORLD OF THE PALEOCENE

That there was a Paleocene at all is entirely due to the K-T mass extinction. That mass extinction was absolutely unequivocal in its cause and effect. And the world afterward was very, very different, on so many levels.

On land, the dinosaurs had ruled for so long that with their passing, a whole set of new ecological relationships among the survivors had to be rather quickly worked out. And with the sudden absence of so many land animals, the evolutionary faucet of new species formation opened with one of the greatest gushers of diversity that the world has ever seen. The mammals, obviously, were the big winners on land, but birds made a comeback as well, and for some time competed with land mammals for various resources.

So it was into this ocean regime that the rock from space dropped. The great climate effects reverberated through the ecosystems for thousands of years, and there was great climatic instability added to the already slightly cooled world, both on land and in the sea. The biotic changes were no less devastating. For one thing, the disappearance of the dinosaurs led to denser forests. Just as modern-day elephants go a long way toward maintaining open spaces in forests by their movement and destructive eating habits, so too the dinosaurs, of far greater size, must have really affected vegetation patterns. But with their sudden disappearance the forests thickened up; it was as if some exacting gardener suddenly walked off the job, letting long-tended and pruned trees run riot.

By the Late Paleocene, more than 7 million years after the catastrophic K-T extinction, global climate had stabilized. The planet had slowly warmed to produce globally warm temperatures. From oxygen isotope evidence we know that equatorial surface waters of the oceans were in excess of 20°C, reaching as much as 26°C in some places, and thus were quite similar to ocean temperatures in similar latitudes today. But the big difference to our world occurred at higher latitudes. In the Arctic and Antarctic the surface of the sea was between 10°C and 12°C, compared to the near-freezing temperatures of our time. Thus, the difference in heat from equator and pole was some 10°C to 15°C, which is about half of what it is today. Nevertheless, in spite of these temperature differences, the oceanic circulation patterns were fairly similar to those of today. Most important, oxygenated water masses that ultimately would end up on the bottom of the ocean formed at high-latitude sites, just as they do today.

After the K-T mass extinction of 65 million years ago, it took some millions of years for the surviving mammals to grow large enough to start affecting plant patterns. There have been many artistic images of tiny, rat-sized mammals crawling from bomb-shelter-like burrows in a world of stinking, rotting dinosaur corpses. For some months, those mammals that could eat carrion would have been in Nirvana. But soon enough there were but bones, and even these rotted away or were buried in fairly short order, forcing all of the mammals to strike out in a newly organizing series of food webs that were unprecedented. It was before the time of grass, so the herbivores of early Paleocene time were leaf or fruit eaters rather than grazers. Seemingly, there were few leaf eaters at all. Most of the teeth of the Paleocene mammals argue for a diet of insects, fruit, or soft shoots rather than tougher leaves;

others may have been root or tuber eaters. It was only in the latter half of the epoch that tooth morphology appropriate for eating leaves appeared in any number. But once opened, the evolutionary faucet fairly spewed out new kinds of mammals, ever-larger mammals among the new in a torrent of evolution. Then, only 9 million years after the great K-T mass extinction, once again the biotic world was affected by environmental crisis.

THE PALEOCENE EOCENE THERMAL (PETM)

By the early Cenozoic era, the Earth had suffered through at least nine mass extinctions that we know of: the first was the great oxygenation event and the snowball it triggered, the second more than one billion years later was during the Cryogenian, then, in order, the late Ediacaran, late Cambrian, late Ordovician, late Devonian, late Permian, late Triassic, and late Cretaceous mass extinctions. The causes were amazingly varied: from sudden oxygen to too little of it; from the appearance of predators to the onset of anoxia coupled with hydrogen sulfide emissions to asteroid impact. But at the end of the Paleocene epoch, only 9 million years after the dinosaurs died out, there was to be a new assassin: methane, which precipitated one of the most rapid rises in global temperatures known. It is called the PETM: the Paleocene-Eocene thermal event.

This event was first discovered by oceanographers[8] who were not at all looking for any sort of temperature event of late Paleocene age. They were trying to get new data on the K-T mass extinction from deep-sea cores drilled by the US Ocean Drilling Program (ODP). But to drill down into the Cretaceous, the drills first had to pass through Eocene and then Paleocene sediment. Cores were taken from those depths while the drills went ever deeper toward their real quarry.

When these younger cores were eventually examined and measured for the carbon and oxygen isotope found in the shells of tiny, single-celled protists known as benthic Foraminifera, the registered temperatures, as well as the ration of carbon 12 to 13, looked like they had to be in error: they showed that when a series of cores were compared, those with strata pulled up from ancient, deeper-water parts of the ocean showed warmer paleotemperatures than those from shallower paleolocalities. Even in the frigid Antarctic today, water cools with depth, and back in the surely much warmer Paleocene, deeper water should be obviously colder than shallower. But the numbers here said just the opposite. Warm, deep waters and cool, shallow waters. Over a relatively short period of time the deep ocean had anomalously warmed.

Near the Paleocene-Eocene boundary there is a striking increase in global volcanic ash.[9] Like dust, this fine material makes its way to the seafloor from the atmosphere, but is put up there by volcanic eruption, not atmospheric storms. This increase could only be due to a sudden increase in global volcanic activity, about 58 to 56 million years ago. Further work in many places around the globe confirmed these findings as being global phenomena, not anomalous events limited to one ocean basin.

The late Paleocene tropics remained about the same (hot) temperature, but the Arctic and Antarctic regions warmed markedly. In the Paleocene the difference in seawater temperature between equator and pole was a hefty 17°C (it is an even heftier 22°C now). By early Eocene times, however, the difference had shrunk to only 6°C. And as the high latitudes warmed, the heat exchange between the two regions slowed, reducing both the number and ferocity of storms. The world went calm and got very hot, just as it did so many times before. This was yet another greenhouse mass extinction.

The carbon isotope record across the Paleocene-Eocene boundary in two cores also yielded a surprise. They showed a short-lived negative excursion—the kind of record that occurs when the amount of plant life is reduced—a hallmark of mass extinction. Other paleontologists began looking at the survival record of bottom dwellers from the region, looking specifically at the common benthic, or bottom-dwelling, Foraminifera—and found evidence of a catastrophic mass extinction on the bottom. Was it simply sudden warming of the deep that wiped out the cold-adapted species in short order? These results were published in the early 1990s, and soon after a Japanese paleontologist named K. Kaiho published studies inferring that the fate of the benthic forms was decided not by rising temperature in the great depths of the sea, but by falling oxygen levels on the bottom. This made a lot of intuitive sense, for warm water can often become eutrophic and oxygen poor.

A deep bottom warming and a lowering of bottom oxygen, even a warming of the surface waters. What was the ultimate cause? The K-T asteroid impact event caused sufficient havoc in shallow waters to kill off almost all of the surface and upper water column plankton, but left the deep relatively unscathed but for the loss of nutrients from above. Warming the deepest part of the ocean could conceivably happen if the large parts of the sea bottom quickly warmed, but this would require an entirely new kind of deep ocean volcanism. The sea bottom does have areas of high heat flow, but these are confined to the relatively narrow mid-ocean mountain chains where seafloor spreading—the ocean bottom growth phase of plate tectonics—takes place. Even much faster plate movement by increased rate of volcanism along these mid-ocean rift systems would not do the trick. It was correctly surmised that the entire warm bottom had come from the warm, tropical surface waters where evaporation would make the surface waters saltier and denser. This warm and saline water was then transported along the sea bottom, even as far as the cold, high-latitude sites of Paleocene age.

Some aspect of ocean currents and the normal export of cold, oxygenated surface water down onto the deep-sea bottoms was not working in the Paleocene ocean. The deep, thermohaline circulation system—the main way that the ocean stays mixed—was just the opposite of how such currents work in our current ocean. The first victims were the tiny organisms requiring oxygen, the benthic forams of the deep sea. Many of these species died out, and did so relatively quickly in an event that lasted about four hundred thousand years. Still, to count as a mass extinction at all, it would have to be shown that it was not just the ocean that was affected, but land fauna as well. So the search was on for events on land.

Wholesale changes among oceanic organisms because of this greenhouse event occurred on the land as well.[10] The newly discovered extinction in the deep sea stimulated paleontologists to look anew (and collect anew as well) at the fossil record of Paleocene land animals, to see if there was an extinction on land at the end of the Paleocene epoch. It did not take long to see that a great turnover *had* occurred among the mammals. Accurate dating soon showed that the extinctions on land and sea took place simultaneously.

In terms of the fossil record on land, the event itself seemed to mark nothing less than the start of our modern-day mammalian fauna. While there were numerous kinds of mammals by the latter part of the Paleocene (thirty distinct families are recognized from the collected fossils), many of these were small, and some belonged to groups no longer present, including survivors of small and rodent-like forms, many kinds of marsupials, some raccoon-like ungulates (a strange paradox, having the new entirely herbivorous ungulates taking on a meat-eating role in the Paleocene). There were also true insectivores and the first primates (like the insectivores, still at small size). But by late Paleocene time there were larger forms as well, and some of these were truly bizarre.

Dog- to bison-sized forms called pantodonts were leaf eaters that branched out into living a semiaquatic lifestyle like hippos, or living in trees, as well as having larger forms moving about on all fours on the forest floors. In general they were stout of body, with short legs, and one cannot help but surmise that, at least compared to modern herbivores, they were very clumsy and inelegant walkers. Yet large as they were, by the end of the Paleocene they were joined by even larger herbivores, the giant Dinocerata, which looked like huge rhinos even to the strange sets of knobs and horns on their skulls.

In the piles of strata marking the transition from Paleocene to Eocene, a reduction of species occurs, and over time—not instantly—new kinds of bones appear. Many come from kinds more familiar to us. The first even- and odd-toed ungulates appeared; more modern carnivores related to current groups soon evolved to eat the new herbivores, and all had to take into account an event that changed the very climate of the world. The lesson from past mass extinctions is that new occurrences would not have evolved as they did unless substantial extinction had opened the door to the possibility of new morphologies. This too happened at the end of the Paleocene.

Our colleague Francesca McInerney has given us a wonderful summary based on her work in the North American West that can help us describe the PETM. First, she noted that this event is highly relevant to us humans, as the amount of carbon released into the atmosphere, about 12,000 to 15,000 gigatons, is roughly equivalent to what we humans are releasing over time by our industries and energy use. The temperature change caused elevated greenhouse gases during the PETM made the world 5 to 9 centigrade warmer than it is now. The actual event lasted on the order of 10,000 years. Plants before and after were different from those during the event, when all the gymnosperms, the pines and their kind, disappeared. The plants that were present in her field area, as discovered by paleobotanist Scott Wing of the Smithsonian, were mainly plants that until the PETM lived in lower latitudes and thus at higher temperatures. After the event the old plants came back, as did the insects that were present prior to the 10,000 years of literal hell on Earth. But not so the mammals. This event caused a wholesale change in the North American mammalian fauna.

A final note. Had there been large ice sheets such as we have today they would have rapidly melted. That causes sea level to rise. In our view this is the single most dangerous aspect of human-caused warming: we are melting Antarctic and Greenland ice that will over the coming centuries inundate huge areas of current human farmland. The highest known rate of sea level rise is currently on the south China coast, one of the most heavily populated areas on Earth with sea-level rice farms.

GRASSLANDS AND MAMMALS OF THE COOLING CENOZOIC WORLD

From the Eocene to the start of the 23.5–5.3-million-year-ago Miocene epoch, the world slowly began to cool. At first, during the Eocene, this was almost imperceptible, and in fact there was a global tropical forest with crocodiles living inside the present-day Arctic Circle. But in the Oligocene this cooling accelerated, creating a different kind of major climate, and changing what had been a near uniform global climate to one with extreme seasonality. At the same time, giant continental ice sheets began to form on Antarctica, and probably Greenland as well. These swelling ice sheets caused a rapid and dramatic fall in sea level. At higher latitudes, forests gradually gave way in many places to grassland meadows and savannas. But other changes were taking place as well, changes in the atmosphere that would prove to have enormous consequences to the history of life.

Plants need carbon dioxide. Yet the history of carbon dioxide through the billions of years on Earth has been one of short-term rises and falls that in fact are only minor variations in a much longer-term trend—the long-term reduction in this gas. With this long decline, our planet is gradually cooled, especially over the last 40 million years. Yet it is far more than the change in temperature that affected the evolution of plants during the Cenozoic era. Perhaps even more important has been an evolutionary formation of a more efficient form of photosynthesis, called C4 photosynthesis, which in many plants supplanted the more archaic mechanism, named C3 (the 3 and 4 in these terms is derived from different chemical changes taking place as sunlight and carbon dioxide are combined to form living plant cells and tissue). C4 photosynthesis, in fact, has shown an extraordinarily rapid rise in importance in terms of the number of plants using one over the other.

Plants that use the C3 pathway leave a different carbon isotope signature than those that use C4. Not only do the plants show this signature, which can be measured when any tissue from the plant is analyzed using a mass spectrometer specialized in looking at living tissue, but any animal eating those plants will leave a trace of it as well. Thus we know from the fossil record whether given herbivorous species ate C3 or C4 plants (or even a combination of the two).

We have two lines of evidence demonstrating when C4 plants first arose. The first is the molecular clock. By comparing the genomes of C4 to C3 plants, geneticists deduced that the differences were large enough that the C4 mechanism could not have arisen less than 25 million years ago (or any earlier than 32 million years ago as well). However, the fossil record yields a quite different answer to the question of when the C4 photosynthetic method first appeared, for the first fossils of C4 plants are only 12 to 13 million years of age.

The evolution to the C4 pathway was not a breakthrough that was then passed down to ever-greater numbers of plant species. In fact, it may have separately evolved more than forty times in the past, by that many separate lineages of plants. The eventual C4 plants are fire and desiccation-resistant plants adapted to heat and dry climates.

The most important C4 plants are grasses because of the dominance of the grass diet to so many kinds of herbivores, large grazing mammals as well as many kinds of birds, including the ubiquitous geese now found on most urban lawns near bodies of water. The reduction in carbon dioxide, especially over the last 20 million years, greatly abetted the expansion of C4 grasslands.[11] Most grasses cannot live on forest floors, where the cooler, shadier conditions do not favor their growth.

Deforestation, however, creates a more open habitat, and therefore one that is far better for grasses. While the main idea has long been that the long-term drop in carbon dioxide sparked the evolution to dominance of C4 grasses, an alternative and newer idea is that a change in forest cover of the planet was as important as or perhaps even more important than a drop in carbon dioxide levels. But what would have caused radical reduction in forestation? The answer seems to be forest fires.

A dramatically underappreciated aspect of a planet with plants is the effect of forest fires. Fire, of course, is affected by oxygen levels. In times of higher oxygen, especially during the Carboniferous period of around 320 to 300 million years ago, forest fires may have been ongoing. A view from space during this interval would have shown an atmosphere darkly smudged and thick with smoke, so that there would have been a world-covering global haze that would have made a clear sunny day a rarity. But such smoke covering much of the continents itself would have had a highly significant effect on global temperatures, because much of the smoke from a forest fire can be light in color when viewed from above. The global haze and smoke would have reflected more sunlight back into space than would otherwise have happened, thus changing the albedo (the degree of reflectivity of the sun's rays hitting the planet).

All of this would have created a chain of events radically changing not only global climate but also the entire history of life from that point onward. The rise of oxygen concentration and its prolonged high for more than 30 percent of all the Carboniferous period would have caused more forest fires. As noted above, this caused global temperature to drop, setting off a chain of events ending in one of the most prolonged polar glaciations in all of Earth's history. Although it was not global in extent like the snowballs, it was nearly as long as some of them. That time of ice may have lasted more than 50 million years, a time interval coincident with some of the most important of all events in Earth history, including the conquest of land by animals, the evolution of new and advanced (for the time) land plants that were capable of colonizing upland regions of the continents previously uninhabitable by plants, and the first appearance of some of the most important of all vertebrate groups—including the earliest reptiles, and soon after the ancestors of the mammals. But there is another aspect of fire that would've affected the history of plant life, and therefore the history of life in general.

New studies on Amazon basin fires have demonstrated that wildfires can greatly influence climate, and not just in the tropics. David Beerling, in his book *The Emerald Planet*, noted that during April 1988, smoke from fires may have inhibited cloud formation over parts of North America—which in turn affected rainfall patterns. This interval of time, in fact, was one of severe drought—and resulted in one of the driest months of the twentieth century. This spring drought followed some of the most extensive wildfires ever, two of them present in North America during July of 1988, a year when gigantic areas around Yellowstone National Park extensively burned. Beerling invokes a new way of understanding the spread of C4 grasslands—very positive feedback system may have been put in place.[12]

Positive feedbacks are those that increase environmental change within one particular direction. In our world today, the warming atmosphere causes ever more of the Arctic ice pack to melt, so that there is an ever-smaller percentage of highly reflective white ice in the northern hemisphere. The white, ice-covered oceans reflect sunlight back into space, but when the ice melts and is replaced by dark-colored, open water, the oceans absorb much more heat—and the seas warm. As the seas warm, more ice is melted and the cycle continues. The positive feedback is that the warming causes more warming.

David Beerling suggested that there is a positive feedback in forest fires causing ever more forest fires. The fires change the climate, causing more drought, which makes ever greater areas susceptible to burning, causing a greater extent of fire damage. And so the cycle goes—burning causes more burning.

We enter a time when global temperatures are rapidly rising. The eventual effects this will have on the planet is not entirely unknown. Less predicable is the effect a new, warmed, high-sea-level world will have on human industry, population, and civilizations.

CHAPTER XVIII

The Age of Birds: 50–2.5 MA

THE history of life as often first taught to us as children is broken down thusly: fish began in what we call the Age of Fish; some crawled ashore to start the Age of Amphibians, which then began what was once called the Age of Reptiles or sometimes the Age of Dinosaurs. Things finished off with an Age of Mammals. It is not hard to see why this has become the common knowledge: humans like to pigeonhole things, and a succession of "ages" is pigeonholing at its best. But among the many, many problems with this account is one of many other truths: there are no pigeons at all in this succession. Let us change that here and consider what we might call an Age of Birds.[1]

The evolution of birds is a major topic of research.[2] It has been a controversial area of research as well, with two major schools of "belief"—one, that birds evolved from a nondinosaur diapsid, something akin to one of the many reptilian-like forms that gave rise to the dinosaurs themselves, or two, that dinosaurs were the direct ancestors of birds. This school even invokes the methodology of cladistics to reinforce the claim that what we call birds are in fact dinosaurs, just highly modified.[3]

A host of fossils have shown that not only did many smaller bipedal, carnivorous dinosaurs resemble birds in the way they laid their eggs but that these eggs also looked like the eggs of birds. Even more striking has been the new discoveries that many dinosaurs both before and after the first appearance of *Archaeopteryx* even showed evidence of winglike arms, with feathers, suggesting a second attempt by dinosaurs to gain the ability to fly. The question was whether or not this famous fossil was even a dinosaur.[4]

The dispute goes back to about 1996, when paleobiologist Alan Feduccia investigated the then newly discovered fossil of what he interpreted to be an intriguing bird that lived about 135 million years ago, just after *Archaeopteryx*. The bird, *Liaoningornis*, did not look like a dinosaur bird at all.[5] It had massive flight muscles attached to a breastbone similar to modern birds. Yet it was it was found alongside fossils of ancient birds not unlike *Archaeopteryx*. How could such advanced evolution have taken place so quickly? Instead, Feduccia concluded, birds may have been already very widespread by the time that *Archaeopteryx* first appeared in the time interval roughly from 140 to 135 million years ago, and that by that time they were already occupying a variety of habitats. While more "advanced" than *Archaeopteryx*, they were still very primitive by modern bird standards. So where are they? Feduccia believes that most of them died out with the dinosaurs, about 65 million years ago, and that the ancestors of all today's birds evolved later, between 65 and 53 million years ago, independently of the dinosaurs. This is the so-called big bang theory of birds.[6] Feduccia and his colleagues view any similarity between birds and dinosaurs as simply due to convergent evolution, where natural selection independently produces similar morphologies.

This school of thought has the modern birds appearing late—either coincident with the 65-million-year-ago K-T extinction—or some tens of millions of years later. This is certainly not mainstream understanding about bird evolution anymore.[7] Within the last decade a large number and variety of birds have been found in Cretaceous rocks ranging from 130 to 115 million years ago, most from China. Some of these fossils show that a great diversity of birds with long, bony tails preceded the evolution of birds with the familiar short, bony tail.[8] However the dinosaur-to-birds theory was further supported by the discovery of two species of feathered dinosaurs in China, dating from between 145 million and 125 million years ago, followed by younger, early Cretaceous birds.

In fact, a great deal of scientific interest has gone into feather research: why did they evolve in the first place (in terms of function), and how did the wing feathers necessary to allow flight come about in the first place? Much of this research involves the concept of exaptation—where a particular adaptation is coopted to do something else. We all know the value of feathers in down vests and sleeping bags. Clearly feathers are good for insulating and staying warm, but the feathers used for warmth are far different from those used and necessary for a bird to fly. Feathers are rarely preserved, and like so much else in paleontology, working out their origin, first appearance, and use involved a fossil record of scant help. Yet as so often in the last several decades, fossils from China have come to the rescue. In this case, exquisite dinosaur fossils that do preserve feathers,[9] and sometimes (and not just from China) even soft parts.[10] Yet as is also so often, no evidence dealing with bird evolution ever gains acceptance without clamor of dissent and noisy opposition.[11] The evolution of flight (not just gliding), a major innovation that was successfully undertaken by arthropods, reptiles, dinosaurs (the bird version), and mammals, has been and remains a fertile topic of study.[12]

At present, over 120 avian species are known from the Mesozoic, from all continents except mainland Africa.[13] Despite this new information, controversy surrounds several aspects of avian evolution, including the timing of the origin and diversification of modern birds (Neornithes).[14]

The birds found in the oldest time intervals of the Cretaceous (which is divided into an Early Cretaceous, which began at about 145 million years ago and ended around 100 million years ago, and was succeeded from 100 to 65 million years ago by the Late Cretaceous). Birds of the Early Cretaceous must have rapidly evolved into a wide range of shapes and sizes. Some were crow sized, with strong beaks, such as *Confuciusornis*, a form that also possessed enormous claws in its wings. Others from this time, such as *Sapeornis*, had very long and narrow wings like those of a seagull. There were also smaller birds, such as the sparrow-sized *Eoenantiornis* and *Iberomesornis*. Yet for all their improvements in flying, these early Cretaceous birds still had toothed jaws similar to those of *Archaeopteryx*. But the variety of skulls, wings, and feet indicate that these Early Cretaceous birds had already specialized into a variety of different lifestyles, including seed feeders, fish eaters, insect feeders, sap eaters, and meat eaters. Their wings and rib cages suggest that soon after *Archaeopteryx*, birds evolved flying abilities not very different from modern birds.

For all the improvements of the Early Cretaceous birds, one remaining archaic feature of these early birds was their teeth. All modern birds have horny beaks, of a spectrum of morphologies that are adaptations for the many kinds of feeding that modern birds undertake. But when did the toothless birds first appear? This remains a contentious question—perhaps just answered in the cold wastes of the Antarctic Peninsula.

Modern toothless birds evolved from the toothed ancestors in the Cretaceous. But this was not so much a replacement as it was an addition, because the earlier primitive birds, toothed and long tailed, continued to thrive and diversify alongside the winged reptiles of the Cretaceous, including the dominant and largest fliers of the last half of the Cretaceous, the pterosaurs. The toothed relics continued to the end of the Cretaceous, but underwent final extinction in the K-T event—at least according to the most recent summation of bird fossils found in the most complete of all latest Cretaceous strata—those in the western interior of the United States, the Hell Creek Formation, home of *Triceratops, T. rex*, and a host of still-primitive birds.

The surviving lineages of birds were the comparatively primitive Paleognathae. These include the large flightless birds such as ostriches, Rheas, cassowaries—and the true giants we have just missed seeing, the enormous moa of New Zealand and elephant birds of Madagascar, wiped out by humans in the last thousand years. Some common birds of today, the aquatic ducks, the terrestrial fowl, and the best fliers of today, the Neoaves, have their roots in the Paleognathae.

The Hell Creek Formation and equivalent rocks in North America have now yielded a total of seventeen species, including seven of the most ancient species of archaic birds of all, including the diving, toothed birds belonging to the hesperornithes group, named for a four-foot-long, stubby diving bird named *Hesperornis*. The recovered assemblage includes both smaller forms and some of the largest fliers known from the Jurassic or Cretaceous, and this tells us quite emphatically that a great deal of avian diversification had happened by the end of the dinosaurs' reign.

In fact, the "avifauna" from these rocks seem to be highly slanted toward marine birds, which is no surprise because of the nearby inland sea that carved North America into two large subcontinents during the Late Cretaceous. None of these groups are known to survive into the Paleogene, and their presence in the Hell Creek Formation, which included the last 2–3 million years of the Maastrichtian age of the Late Cretaceous, tells us that a mass extinction of archaic birds coinciding with the Chicxulub asteroid impact in fact did take place.[15] But here is where controversy still exists: while most of the birds found in the North American beds represent "advanced" birds from a morphological sense, none can be categorically placed in the all-important group called the Neornithes. This avifauna is the most diverse known from the Late Cretaceous, although in diversity and disparity (the number of kinds of body plans) it is lower than in modern birds. But this group of fossils is key in helping us understand to what extent the K-T mass extinction affected birds.

If any group of vertebrates could survive the effects of an impact extinction, it surely would be the birds. The burning of most of the world's forests in the first few days after the collision of the giant rock from space with the Earth; the subsequent acid rain, followed by six months of darkness, and thus starvation and the extirpation of surely every terrestrial ecosystem as well as all of those in the marine and freshwater realms, save for the deep-sea communities—the effects of this impact were immense. Yet even the deep-water ecosystems would eventually have suffered grievously with the slowdown or cessation of the main source of food to the deep sea, the sinking bodies of shallower water plankton and dead animals. On land, the size of animals dictated their survivability, as the larger animals had no chance. But birds are not large animals.

With the ability to disperse rapidly and fly quickly toward less-affected land, it should be expected that birds as a group should show a lower extinction rate than nonflying animals—and nonflying birds. Unfortunately, birds rarely fossilize because they have fragile, hollow bones. Thus bird fossils are rare to begin with. Yet from great diligence in collecting, there is now enough information to make at least educated guesses about the fate of birds across the Mesozoic-Cenozoic transition, a transition burned into life's history.

By the Late Cretaceous, birds had already been on Earth longer than the time since the large Chicxulub impact turned a dinosaur-rich world into one with only avian dinosaurs left.

THE GREAT BIRD DIVISION

There is another source of information about the timing of modern bird diversification: the use of DNA. In the first decade of the twenty-first century a number of separate studies[16] proposed new "evolutionary trees" of birds, based on the DNA of extant species (evolving from presumed survivors from the times of archaic birds). This new tree contains several surprises. For example, the closest relatives of common freshwater diving birds known as grebes are—flamingos! Hummingbirds are a specialized form of nighthawk, while falcons are more closely related to songbirds than to other hawks and eagles. And surprising as these new conclusions may be, there were even greater jaw-droppers coming from this study.

For example, the new tree puts an order of flying birds, known as tinamous, on a branch of the tree also shared by the flightless ostriches, emus, and kiwis. The importance of this is that it indicates that flightlessness evolved at least twice in this lineage, or else that the tinamous re-evolved flight from a flightless ancestor. And still more: the new tree demonstrated that the closest relatives of perching birds, or passerines (which are by far the largest and most successful order of birds), are parrots. Yet with all of this new information, the age of the most fundamental split of surviving birds, the fork in the evolutionary road that resulted in the Neognaths and the presumably more primitive Paleognaths remains obscure.

Modern birds are classified in Neornithes, which are only lately known to have evolved into some basic lineages by the end of the Cretaceous, based on the discovery of a bird fossils now named *Vegavis*, known from Vega Island. The Neornithes are split into the Paleognaths (tinamous, ostriches, emus, and kiwis) and Neognaths (all the rest of the birds). The date for the split of the Neognaths into the familiar birds of today is also poorly understood. The best evidence suggests that a basic split in the Neornithes occurred before the K-T extinction. But how long before, if at all? As noted above, there remains a very convinced group of specialists (such as Alan Feduccia) believing that the modern birds evolved only after the K-T extinction event, as well as those having doubts about whether the radiation of the Neognaths occurred before or after the extinction of the other dinosaurs.

Thus new results from Vega Island in Antarctica are crucial. Vega Island is a small island north of James Ross Island that had previously yielded one of the most significant finds in bird evolution. This discovery offered the first evidence that modern birds existed alongside nonavian dinosaurs at the end of the Cretaceous.

A last question has tantalized paleontologists for years. In the mid-Cenozoic, birds tried to become giant carnivorous dinosaurs once again. The most famous of these were the "terror birds," with a number shown in the illustration here. Clearly there must have been severe competition with the then-rising modern carnivores, ancestral to all the major terrestrial carnivorous mammals of today (dogs, cats, bears, weasel groups).

The evolution of the large, flightless birds (ratite) still found today—such as the iconic ostrich, cassowary, Rhea, and others—has always seemed like a return to the bipedal dinosaur body plan. But because these giant birds cannot jump from island to island, or cross vast continents in the kinds of migrations so prevalent in the flying birds of the present and past, it has long been assumed that each of the major flightless groups evolved independently through the formation of isolated species. As most of these kinds of birds are found in the modern-day southern continents, which were in the Mesozoic all combined in one large landmass, the implication is that the ostriches of Africa, rheas of South America, and cassowaries of Australia were products of the breakup of the ancient landmass Gondwanaland. But a major surprise of the new DNA work is that these flightless birds actually evolved into their groups not *after* they lost the ability to fly, but *before*.[17]

Because Africa and Madagascar were among the first large hunks of landmasses splitting apart from the Gondwanaland supercontinent, it was predicted that the early isolation of Africa and Madagascar would have allowed evolutionary forces to create the oldest of the ratites, the ostrich of Africa and the and even larger elephant bird of ancient Madagascar. But because of the closeness of Madagascar to Africa, ostriches and elephant birds should have been closely related to each other but quite distinct from the other flightless birds, including those of South America and New Zealand, where the ancient moa (now extinct) and still-living kiwi evolved in their own isolation—at least supposedly. Yet when DNA became available surprises were present.

The DNA work showed that the Madagascar elephant birds were more closely related to the New Zealand birds than to the nearby African ostriches. This unexpected result strongly supports the conclusion that these groups became evolutionarily distinct before they lost the ability to fly.

The living ratites are—and were not—the only large, dinosaur-like birds of the past. The largest land birds, now extinct, showed an evolutionary return to the body plan of the bipedal Mesozoic carnivorous dinosaurs. Known scientifically as phorusrhacids, the terror birds evolved in South America about 60 million years ago, and lasted until about 2 million years ago, the time the great ice sheets were spreading across the globe in the first Pleistocene ice advances. Some, at least, made it into North America as well, and for most of the Cenozoic these were the top carnivores of South America. There are nothing like the terror birds living today, which might be a blessing.

New research in 2010, using CT scanning technology, has given us a new understanding of how these behemoths lived and died. The scans revealed that the gigantic beaks of these monster predators were hollow, which was a surprise. This must have made the beak weak and vulnerable to breaking when moved from side to side. Instead they may have used the beak like an ax, and also used their powerful, talon-equipped legs for help in killing prey.

Like most flightless birds, terror birds had stubby little wings but long, powerful legs that ended in large, taloned feet. The muscular legs produced great ground speed; it has been estimated that some species of terror bird could reach speeds of nearly seventy miles per hour over flat terrain, and on the vast South American pampas, there was plenty of room to run. In this they were probably comparable to a cheetah. The combination of running, a monstrous beak, and the deadly talons on powerful legs surely made the terror birds very effective predators.

They had very big heads and the largest brains of any bird. This leads to some uncomfortable realizations. Recent work on the intelligence of African gray parrots has led neuroscientists and psychologists alike to realize that we have vastly underestimated the level of bird intelligence. While primatologists try to link various primates to higher cognitive function, birds in general—and perhaps terror birds in particular—may have been among the most intelligent species to ever walk the Earth.

CHAPTER XIX

Humanity and the Tenth Extinction: 2.5 MA to Present

SOME decades ago, several books came out suggesting that the world might be entering into a new mass extinction (including two books by coauthor Ward, *The End of Evolution*, and an update of that book retitled *Rivers in Time*).[1] One of these books, Richard Leakey's *The Sixth Extinction*,[2] was overtly referring to the big five mass extinctions we have profiled in this book, those with more than 50 percent species loss: the events of the end of the Ordovician, Devonian, Permian, Triassic, and Cretaceous. Here we advocate that there have actually been *ten major* mass extinctions so large that they merit differentiating from the more minor events, such as the PETM of a previous chapter, and several smaller events in the Jurassic and Cretaceous periods. These top ten are as follows:

1. *The Great Oxidation Extinction*. This may have been the most catastrophic of all extinctions, if judged on the percentage of species and individuals killed off. Oxygen would have been a deadly poison to virtually all microbes at the time. Combined with the nearly contemporaneous first snowball Earth, this might have been the worst of them all, and also the first out of the gate. Imagine that you walk outside and there is no longer breathable air. Air, all right, but different. So it was to those aquatic organisms that constituted life on Earth. The seas were filled with poison gas: oxygen.
2. *The Cryogenian Extinctions*. The combined snowball Earth episodes of the late Proterozoic. Thick dirty ice covers oceans and land. Photosynthesis slows and mainly stops. A rich, diverse assemblage of life on land and in the sea (far, far richer in the sea) dies out. Not just diversity, but biomass tumbles.

3. *The late Vendian-Ediacaran Extinction*. This included stromatolites, microbial mats and especially Ediacarans at the Proterozoic-Paleozoic Boundary. That supposed garden of Ediacara, invaded by voracious—and more important—active, moving animals, eating everything in their path and ravishing the slow-moving, microbially draped oceans and land.
4. *The late Cambrian SPICE Extinction*. Extinction of most trilobites, many "weird wonders" of the Burgess Shale, and so much else. Most important, there was a wholesale change of trilobites away from having primitive segmentation and eyes, incapable of enrolling for defense, little defensive ornament—probably due to an increase of predators as much as anything else. The first really large, mobile, and armored carnivores, the nautiloid cephalopods, were involved in this extinction, as were chemical changes.
5. *The Ordovician Mass Extinction*. Wholesale extinction of tropical species. Caused by cold or perhaps by sea level change.
6. *The Devonian Mass Extinction*. Benthic and water column animals in the sea—the first greenhouse extinction?
7. *The Permian Mass Extinction*. Land and sea greenhouse extinction.
8. *The Triassic Mass Extinction*. Land and sea greenhouse extinction.
9. *The Cretaceous-Paleogene Extinction*. Combined greenhouse and impact extinction.
10. *The Late Pleistocene-Holocene Mass Extinction*. From 2.5 million years ago to today—climate change and human activities.

It is the last on this list that should worry us. The others, especially the greenhouse extinctions, should terrify us, but they do not, because they were—and would be—too slow moving. The slow death . . . and not for our species. We are pretty extinction-proof. We would be alive, yes, but happy? On an empty planet? Surrounded by our domestic animal and plants, whose jumping genes will make their own perverse and unpredictable Cambrian explosion in the long run.

GETTING TO THE TENTH EXTINCTION

In 2010, a traveling exhibit from Ethiopia[3] brought one of the most famous of all fossils to the United States: Lucy, the early hominid.[4] At about three and a half feet tall, with remains that total only 40 percent of her original skeleton, in fact there is not a lot to Lucy. But she has told us a great deal.

Sexual dimorphism is the term used to describe the two different morphologies of males and females of a species. It is certainly not limited to hominids, and it is not the case that the larger of the dimorphs is always the male. In many animals, for instance, including a variety of cephalopods (excepting *Nautilus*, interestingly enough), the female morph is the larger. Apparently it takes more organ mass to produce eggs than sperm. In hominids, however, from chimps to us humans, the male is the larger. The dimorphism in humans is statistically significant, and appears to range from females being about 90 to 92 percent the height of males, depending on race. In Lucy's kind, however, it was quite a different story.

Lucy is far from the only fossil skeleton of her kind. Her species, *Australopithecus afarensis*, is now far better known compared to our understanding (or lack thereof) when a team led by Don Johanson found her in 1974. One of the more recent finds is of a male skeleton of her kind that is complete enough to allow a good estimate of his height in life. He is called Big Man. He was five feet tall to Lucy's three and a half. Her chin would have come just above his navel if standing face-to-face—except it had been face to lower chest.

If Lucy and Big Boy are representative of their genders in *A. afarensis*, it means that females were only 70 percent as large as their men. There had to be consequences to this—behavioral as well as cultural. In 2012, when anthropologist Patricia Kramer of the University of Washington did a detailed study[5] on the relative walking speeds of males and females, based on their leg lengths, she discovered that Big Man's optimal walking speed would have been 2.9 mph, but Lucy's would have been a rather slower 2.3 mph. Keeping up with males would have been taxing for females—and living in a world filled with predators, being constantly in a state of anaerobic respiration would not be a very good survival tactic. Kramer thus suggested that like chimpanzees, male and female hominids spent much of the day apart, ranging separately as they foraged and hunted for food.

Other new fossil finds from Africa are also turning over some long-held views. Lucy and her kind are invariably reconstructed in dioramas or illustrations as walking upright through the Late Pliocene world of north and eastern Africa—a place with a mosaic of grassland and small patches of open forest. But for the first time ever the shoulder blades of a female of Lucy's species—but coming from a time interval about a hundred thousand years before Lucy—show features that suggest she and her kind were tree climbers as well as adapted for walking on the ground. The question of whether these distant ancestors of ours also spent significant time in trees has been hotly debated,[6] largely because until this new find, there was no way to see the morphological adaptations necessary for a tree climber. The new view seems to be that australopithecines may not have come down from the trees as early as currently believed.

While hominids are new arrivals on Earth, our group, the primates, dates well back into the Cretaceous, and we have an ancestor, *Purgatorious*, that survived the K-T mass extinction itself—which is lucky for us. Some of the earliest primates belonged to the lemur branch. By 45 million years ago, more advanced primates—the first true anthropoids, which today include monkeys, apes, and humans—appear in the fossil record of Asia. The oldest of these was found in China and is now named *Eosimias*.

About 34 million years ago, surely smarter, definitely bigger, and perhaps more aggressive monkeys evolved. One of these, named *Catopithecus*, has a skull the size of a small monkey's, a relatively flat face, and is the first primate to sport the same arrangement of teeth humans have—two incisors, one canine, two premolars, and three molars. We now have a good idea of our own evolutionary tree, right up to where and when "humans" can be said to first appear—the African genesis of the australopithecines.

Paleoanthropologists have done a remarkable job of deciphering the where and when of the speciation event that produced our species. The human family, called the Hominidae, seems to begin as much as 5–6 million years ago with the appearance of Lucy and her kind, the *Australopithecus afarensis* described above. Since then, our family has had as many as nine species, although there is ongoing debate about this number, which seems to change years as both new discoveries and new interpretations of past bones make their way into print. But the most important descendant of the early pre-Pleistocene hominids is the first member of our genus, *Homo*, a species named *Homo habilis* (handyman) for its ability to use tools, which is about 2.5 million years old. This creature gave rise to *Homo erectus* about 1.5 million years ago, and *H. erectus* either gave rise to our species, *Homo sapiens*, directly about 200,000 years ago, or through an evolutionary intermediate known as *Homo heidelbergensis*. Our species has been further subdivided into a number of separate varieties. Some workers consider the Neanderthals to be a variety, while others interpret it as separate species, *Homo neanderthalensis*. A great deal of new work on recovered and decoded Neanderthal DNA[7] is one of the most intriguing aspects of human paleobiology, with the latest evidence suggesting that the human and Neanderthal lineages diverged before the emergence of contemporary humans and our current DNA. They did not come from us, nor did we come from them. We both evolved from a common extinct ancestor different from both species.[8]

Each formation of new human species occurred when a small group of hominids somehow became separated from a larger population for many generations. In the 1960s and 1970s there was a view that modern humans came about from what has been called a candelabra pattern of evolution—that all over the planet separate stocks of archaic hominids such as *Homo erectus* all evolved into *Homo sapiens* at different times and places. This notion now seems laughable.

The fossil record tells us that the so far oldest member of our species—variably called a modern to distinguish it from more archaic forms of *Homo sapiens*—lived 195,000 years ago in what is now Ethiopia. It is unknown and not terribly important whether this fossil represents the oldest tribe of us or was from a group that wandered in from the true origin place and was fortuitously fossilized in Ethiopia. But very soon after, this band set out walking to the farthest southern regions of the African continent, and then to the north as well, finding a way out of Africa through Eurasia—and in so doing they spread out across the globe,[9] effectively isolating themselves from others of our species, and thus adapting to the very different environmental concision in which these wanderers found themselves. Quite different adaptations, morphological to physiological, were necessary for survival in the sun-starved, ice-covered north than on the plains of Africa, as well as all areas in between. As our numbers grew, so too did our variation—and our various evolutionary changes. But all of this was within the same species.

THE LAST ICE AGE AND LIFE

Climatologists have long theorized that climate change observed over the past two and a half million years—the alteration between long periods of very cold climate with growing ice sheets and dropping sea level alternating with shorter times of warmth—were the result of the orbital changes described above as having been first articulated by Milutin Milanković. Until the ice cores became available, with their unprecedented resolution in discerning climate through recent time, the changes were thought to have been slow. But with that resolution a newer view became apparent.

The ice core records and other sources of climate information such as deep-sea paleontological and isotopic records indicate that over the past eight hundred thousand years the interglacial periods—the warmer times between the much cooler glacial intervals—have lasted on average about eleven thousand years. That's almost half the Earth's precessional cycle, orbital changes occurring every twenty-two thousand years. The current interglacial has already lasted more than eleven thousand years and some records suggest that we have been in the warm period for as long as fourteen thousand years. Does this mean that the glaciers are advancing at this moment? The answer to that question is a decided no, for several reasons. First of all, precession is not the only orbital aspect that affects climate. Records show that between 450,000 to 350,000 years ago there was an interglacial stage that lasted much longer than eleven thousand years. This interglacial was coincident with a time when orbital eccentricity was at a minimum. Just such a pattern of minimal orbital eccentricity is under way at this time, suggesting that the present interglacial could continue for thousands or perhaps a few tens of thousands of years into the future—or it could end at any time.

The Pleistocene epoch signaled a significant kind of climate change beginning about 2.5 million years ago. The large cool grasslands and tundra of the high latitudes during the last pre-ice-age epoch of the Cenozoic era gave way to a new kind of land cover—ice. Year by year a slow excess of snow and ice caused the formations of glaciers, which slowly crawled southward. Eventually continental glaciers began to coalesce and merge with mountain glaciers, uniting in unholy matrimony to grip the land in glacial ice and glacial winter.

By no means was the entire planet gripped in ice, as seems to be popularly imagined. There were still tropics and coral reefs and warm sunny climes pleasant the year around. But probably no place on Earth was unaffected in at least some minor way; the global climate changed, causing shifts in wind and rain patterns. Even those places far from the ice were climatically changed, perhaps colder or even warmer, often quite dryer. Gigantic cold deserts and semideserts expanded in front of the advancing ice sheets, while regions normally dry, such as the Sahara desert of northern Africa, experienced increased rainfall. Conversely, the great rainforests covering the Amazon basin and equatorial Africa, regions of relative climatic stability for tens of millions of years prior to the onset of the ice age, experienced a pronounced cooling and drying such that large tracts of jungle retreated into pockets of forest surrounded by wider regions of dryer savannas.

THE SPREAD OF HUMANITY

Many of these rapid climate changes occurred while humanity was colonizing the globe. By about thirty-five thousand years ago, it appears that the final evolutionary tweaks had occurred, making us as we are now. We can call these new humans the moderns, and they conquered the world bit by bit. They arrived in each new region slowly yet inexorably. It didn't happen in a century. It didn't parallel the taming of North America by Europeans, when several centuries saw the transformation of a giant native-vegetation-covered continent to a giant agriculture- and concrete-covered continent. It was instead a slow conquest, with millennia falling away like leaves as the moderns slowly spread over the globe. Even the island continent of Australia had become the habitat of *Homo sapiens* thirty-five thousand years ago. Northern Asia, however, was still undiscovered. And beyond Asia, an even bigger territory, North and South America, had still not experienced the first human footfall.

The first people to arrive in the vast tract of what is now Siberia were Paleolithic big game hunters. They arrived as long as thirty thousand years ago, with a tradition already in place for existing in this harsh climate. Eastern Siberian stone tools show some differences from the European traditions of the time, and were clearly influenced by the flake cultures of Southeast Asia. Yet the major technology, the construction of large spearpoints, was formulated for killing large animals.

The arrival of the first humans in Siberia was set against a time of slight warming, and this warmer period, following a cooler time, may have encouraged the spread of humans into an otherwise hostile region. Yet soon after their arrival in Siberia the Earth began to cool again, and by twenty-five thousand years ago a major glacial event was well under way.

In western Europe and North America the great continental ice sheets were inexorably spreading downward to cover vast regions with ice a mile thick. In Siberia, however, there was so little moisture that the ice was unable to form. Into this vast treeless frozen territory, humans expanded ever eastward. Because there was so little wood, the hides and antlers of their prey became important resources, and the very bones of the largest quarry—the mastodons and mammoths—were used for housing. These people became—by necessity—big game hunters, and their principal prey may have been the mammoth and mastodon.

As humanity crossed Asia and settled in Beringia, in a succession of small waves between perhaps thirty and twelve thousand years ago, the continental ice sheets covering large portions of North America expanded to maximum size in a long series of cooling episodes, followed by rapid warming. As the ice increased in volume, the level of the sea began to fall, causing huge land areas long underwater to become dry land—land that would in some areas serve as migratory paths between formerly isolated islands and large landmasses. But when the ice finally began to melt, the level of the sea began to rise as well. As late as fourteen thousand years ago, the continental glaciers covering most of Canada and large portions of what is now the United States were still slowly melting under gradually rising temperatures.

Soon thereafter, however, a new event accelerated the melting process. When enough ice had melted so that the glaciers no longer extended out to sea from the coast, the calving of icebergs from the eastern and western coastlines of what are now Canada and the northern portions of the United States could no longer occur. Each spring during the period of glacial maxima (about eighteen thousand to fourteen thousand years ago) great fleets of icebergs were launched into the coastal oceans, and this in turn kept the waters cool and created very cold winds that cooled the lands as well. Yet with the cessation of iceberg formation, warmer onshore winds arose, and the ice began to melt everywhere on the continents in earnest.

The melting fronts of the glaciers must have been extraordinary and extraordinarily harsh places—incessant strong winds characterized the retreating glacial walls. So strong was the wind that it created great piles of sand and silt carried by the wind, sediment called loess. The winds also carried in seeds, so that the drifting soils in front the glaciers were soon colonized by pioneering plants. First came the ferns and then more complex plants. Willow, juniper, poplar, and a variety of shrubs were the first stable communities to transform this ancient glacial regime; and soon thereafter successive communities of plants arrived, depending on location. In the more temperate west, low forests dominated by spruce were the norm; in the middle colder parts of the continent permafrost and tundra were the norm. Yet everywhere the glaciers were in retreat, and as they migrated, or more correctly, melted north, they were followed by a front of advancing tundra, which soon was followed by vast spruce forests.

Spruce communities of large areas in North America were as much open woodland as dense forest, with copses of trees interspersed with grass and shrubs. By no means was it similar to the great, thick Douglas fir communities found in the few remaining old-growth forests of the Northwest, places where dense underbrush and fallen rotting logs make passage by large game, or humans, exceedingly difficult.

South of the ice in North America, throughout the ice age, a variety of habitats existed. There was forest tundra, grassland, and deserts, and plants sufficient to sustain enormous herds of giant mammals. With the end of ice and cold over so much of the world, human populations began to increase markedly.

By ten thousand years ago humans had successfully colonized each of the continents except Antarctica, and adaptations to the many locales led to what we now call the various human races. While it was long thought that such obvious features as skin color were purely adaptations to varying amounts of sun, more recent work suggests that much of what we call "racial" characteristics might simply be adaptations brought about by sexual selection, rather than increases to fitness in various environments. But many other adaptations, most invisible to morphologists, were happening as well.

Africa is revered for its abundance of large mammals. Nowhere else on Earth can such diversity of larger herbivores and carnivores be found. Yet this animal paradise, instead of being the exception, was the rule: all of the world's temperate and tropical grazing regions were quite recently of African flavor. But like the forces that wiped out the elephants in the Karoo, an extraordinary event has depleted the Earth's biodiversity of large mammals over the past fifty thousand years.

Although the disappearance of larger animals poses a tremendous challenge to those studying extinction, a significant lesson from the past is that the extinction of larger animals has a far more important effect on the structure of ecosystems than does the extinction of smaller animals. The extinction at the end of the Cretaceous was significant not because so many small mammals died out, but because the very large dinosaurs did.

It was the removal of the very large land-dwelling dinosaurs that reconfigured terrestrial environments. In similar fashion, the removal of the majority of larger mammals species across most of the world over the last fifty thousand years is an event whose significance is only now becoming apparent, and one that should have lasting effects for additional millions of years into the future.

One time period in particular was the Late Pleistocene epoch of about fifteen thousand to twelve thousand years ago, when a significant proportion of large mammals in North America went extinct. At least thirty-five genera (and thus at least this number of species) became extinct. Six of these lived on elsewhere (such as the horse, which died out in North and South America but lived on in the Old World). The vast majority, however, died out. In fact, most belonged to a wide spectrum of taxonomic groups, being distributed across twenty-one families and seven orders. The only unifying characteristic of this rather diverse lot is that most (but certainly not all) were large animals.

The most well-known and iconic were the elephant-like animals, the Probiscideans, including mastodons and gomphotheres as well as mammoths, which were closely related to the two types of still-living Old World elephants. Of these, the most widely distributed in North America was the American mastodon, which was widespread from coast to coast across the unglaciated parts of the continent. It was most abundant in the forests and woodlands of the eastern part of the continent, where they browsed off trees and shrubs, especially spruce trees. The gomphotheres, a bizarre group quite unlike anything now alive, are questionably recorded from deposits in Florida, but otherwise were widely distributed in South rather than North America. The last group, the elephants, was represented in North America by the mammoths, comprised of two species, the Columbian mammoth and woolly mammoth.

The other group of large herbivores iconic of the ice age in North America was the giant ground sloths and their close relatives the armadillos. Seven genera comprising this group went extinct in North America, leaving behind only the common armadillo of the American Southwest. The largest of this group were the ground-living sloths, ranging in size from the size of a black bear to the size of a mammoth. An intermediate-sized form is commonly found in the tar pits of present-day Los Angeles, while the last, the Shasta ground sloth, also the most well known, was the size of a large bear or a small elephant. Also spectacular was the North American glyptodont, a heavily armored creature ten feet in length with a heavily armored, turtle-like shell. Also going extinct was an armadillo, a genus survived by the common nine-banded armadillo.

Both even- and odd-toed ungulate animals died out as well. Among the odd-toed forms, the horse, comprising as many as ten separate species, went extinct, as did two species of tapir. Losses were greater among the even-toed ungulates. Thirteen genera belonging to five families went extinct in North America alone in the Pleistocene extinction, including two genera of peccaries (wild pigs), a camel, and two llamas, the mountain deer, the elk-moose, three types of pronghorns, the Saiga, the shrub-ox, and Harlan's muskox.

With so many herbivores going extinct, it is no surprise that many carnivores also underwent extinction. These included the American cheetah, a large cat known as the scimitar cat, the saber-toothed tiger, the giant short-faced bear, the Florida cave bear, two types of skunks, and a dog. Some smaller animals round out the list, including three genera of rodents and the giant beaver. But these were exceptions. Most animals dying out were large in size.

The extinction in North America coincided with a drastic change in plant community makeup. Vast regions of the northern hemisphere changed from plant assemblages dominated by highly nutritional willow, aspen, and birch trees, to far less nutritious spruce and alder groves. Even in those areas dominated by spruce (itself a relatively poorly nutritious tree), a diverse assemblage of more nutritious plants were still available. However, as the number of nutritious plants began to decrease due to the climate changes, the herbivorous mammals would have increasingly foraged on the still-remaining, more nutritional plant types, thus exacerbating their demise, and perhaps thus leading to the reduction in size of many species of mammals depending on vegetation for food. As the Pleistocene ended, the more open, high-diversity spruce forests and nutritional grass assemblages were rapidly replaced by denser forests of lower diversity and lower nutritional value. In the eastern parts of North America the spruce stands changed to large, slow-growing hardwoods such as oak, hickory, and southern pine, while in the Pacific Northwest great forests of Douglas fir began to cover the landscape. These types of forests have a far lower carrying capacity for larger mammals than the Pleistocene vegetation that preceded them.

It was not just North America that suffered such severe losses.[10] North and South America had been isolated from one another, and hence their faunas underwent quite separate evolutionary histories, until the Isthmus of Panama formed, some 2.5 million years ago. Many large and peculiar mammals evolved in South America, including the enormous, armadillo-like glyptodons, as well as the giant sloths (both of which later migrated northward and become common in North America), giant pigs, llamas, huge rodents, and some strange marsupials. When the land connection formed, free interchange between the two continents began.

As in North America, large-animal extinction occurred among South American mammals soon after the end of the ice age. Forty-six genera went extinct between fifteen thousand and ten thousand years ago. In terms of the percentage of fauna affected, the mass extinction of large animals occurring in South America was even more devastating than those in North America.

Australia suffered even greater losses, but at an earlier time than either North or South America. Since the age of dinosaurs the

Australian continent had been an isolated landmass, surrounded by ocean. It was thus cut off from the mainstream of Cenozoic-era mammals. The Australian mammals followed their own evolutionary path, resulting in a great variety of marsupials, many of large size.

The mass extinction striking the Australian fauna during the last fifty thousand years killed off forty-five species of marsupials belonging to thirteen genera. Only four of the original forty-nine large (greater than twenty pounds in weight) marsupial species present on the continent a hundred thousand years ago survived. No new arrivals from other continents bolstered the disappearing Australian fauna. The victims included large koala bears, several species of hippo-sized herbivores called *Diprotodon*, several giant kangaroos, several giant wombats, and a group of deer-like marsupials. Carnivores (also all marsupials) were lost as well, including a large lionlike creature and a doglike carnivore. In more recent times, a third predator, a catlike creature found on offshore islands, has also disappeared. Large reptiles also disappeared, including a giant monitor lizard, a giant land tortoise, a giant snake, as well as several species of large flightless birds, among others. The larger creatures that did survive were those capable of speed or those that had nocturnal habits, as noted by our great friend Tim Flannery of Australia,

The wave of extinctions affecting the faunas of Australia, North America, and South America coincides both with the first appearance of humanity in all three regions and with substantial climate change. Reliable evidence now shows that humans reached Australia between thirty-five and fifty thousand years ago. Most of the larger Australian mammals were extinct by about thirty to twenty thousand years ago.

A different pattern emerges in the areas where humankind has had a long history, such as Africa, Asia, and Europe. In Africa, modest mammalian extinctions occurred 2.5 million years ago, but later losses, compared to other regions, were far less severe. The mammals of Northern Africa, in particular, were devastated by the climatic changes that gave rise to the Sahara desert. In eastern Africa, little extinction occurred, but in southern Africa, significant climate changes occurring about twelve thousand to nine thousand years ago were coincident with the extinction of six species of large mammals. In Europe and Asia there were also fewer extinctions than in the Americas or Australia; the major victims were the giant mammoths, mastodons, and woolly rhinos.

The extinction can thus be summarized as follows:

Large terrestrial animals were the primary creatures going extinct: smaller animals and virtually all sea animals were spared.

Large mammals survived best in Africa. The loss of large mammalian genera in North America was 73 percent; in South America, 79 percent; in Australia, 86 percent; but in Africa, only 14 percent died out during the last hundred thousand years.

The extinctions were sudden in each major land group, but occurred at different times on different continents. Powerful carbon dating techniques allow very high time resolution. These types of techniques have shown that some species of large mammals may have gone completely extinct in periods of three hundred years or less.

The extinctions were not the results of invasions by new groups of animals (other than mankind). It has long been thought that many extinctions take place when new, more highly evolved or adapted creatures suddenly arrive in new environments. Such was not the case in the ice age extinctions, for in no case can the arrival of some new fauna be linked to extinctions among the forms already living in the given region.

These various lines of evidence suggested to many scientists that humanity provoked this mass extinction. Others argue just as vigorously that the cause of the megamammal extinction was change in resource patterns in vegetation that occurred during the intense climate changes accompanying the end of the Pleistocene glaciation. Most discussion about this extinction deals exclusively with this argument over cause, with the two major camps being called overkill (human hunting) and climate change.

Whatever its cause, a major reorganization of terrestrial ecosystems occurred on every continent save Africa. Today, Africa is losing its megamammals as the large herds of game become restricted to game parks and reserves, where they become easy prey to poaching within their newly restricted habitats.

The end of the megafauna is not a clearly defined line. But then we are looking at it from the present, and it is just a moment away. Intervals of time lasting ten thousand years are insignificant and probably beyond the resolution of our technology, when viewed from times tens to hundreds of millions of years away. The end of the age of megamammals looks protracted from our current vantage point, but will look increasingly sudden as it disappears into the past, one of the odd aspects of time.

The megamammals still left on Earth now make up the bulk of endangered species, and many more large mammalian species are now at risk. If the first phase of the modern mass extinction was the loss of megamammals, its current phase seems concentrated on plants, birds, and insects, as the planet's ancient forests are turned into fields and cities.

CHAPTER XX

The Knowable Futures of Earth Life

THE future is a never-reachable time, the fast-moving bait to race-track greyhounds. If there is any lesson from life's history, it is that chance has been one of the two major players at the game of life, with evolution the other, and chance makes any attempt at prognosticating events and trends in the *future* history of life a very chancy proposition. But the planetary scientist and brilliant writer Don Brownlee of the University of Washington has responded to this seemingly impenetrable obfuscation of the future. Brownlee claims that there is a "knowable" future, and that seemingly paradoxically events become more knowable the further into the future they are. On this topic, Brownlee was talking about physical and predictable changes in the properties of our planet and our sun. One example of a knowable future that can be quite accurately predicted is the future history of our sun, which we know will become a red giant star with a diameter larger than the orbit of Earth and probably Mars (and thus certainly consuming the Earth and probably Mars as well) in 7.5 billion years, give or take a quarter billion.

The study of biological evolution on Earth has increased scientists' understanding of the distant past, and this too offers clues to the future. One characteristic is that evolutionary history has been importantly affected not only by the interplay of life (competition and predation) but also by the course of the physical evolution of Earth, its atmosphere, and its oceans. While many events will remain dictated by chance, such as the rate and future history of asteroid impact with the Earth, we *can* make highly refined estimates about predictable changes in global temperatures, atmospheric and oceanic chemistries, and large-scale geophysical events that will necessarily take place over Earth's remaining lifetime.

The concept of a habitable planet is based on planetary nurture, with life being the ultimate result of planetary formation and change.

We have already looked at the most important elemental renewal systems that recycle important nutrients and maintain near-constant global temperatures, and changes in (or total cessation of) these, like the rate at which the sun expands, are knowable. For life, the most important of these fluxes are the movement and transformation of the elements carbon, nitrogen, sulfur, phosphorus, and various trace elements. The energetic underpinnings of the various systems largely come from two sources: the sun and heat generated from the breakdown of radioactive material beneath Earth's surface. Of these, and because of its importance to life as the source of energy through photosynthesis, the sun is the more important of the two.

The sun is a powerful nuclear reactor, but its stability is a matter of debate. As the sun evolves, the number of particles in its core decrease as hydrogen atoms are fused into helium atoms, but seemingly paradoxically, as the number of atoms in the core of the sun decrease, its energy output (as light and heat) slowly but inexorably increases.

All stars like the sun share this same characteristic. The sun has increased in brightness by about 30 percent in the last 4.5 billion years of its life. The rise in brightness increases the intensity of the sunlight that illuminates its planets. A continuation of this change will cause the loss of oceans and create hellish conditions, similar to those that exist on Venus. (The oceans do not "boil" away, as seen in some garish depictions of the Earth's future, but one by one, oceans' water molecules are stripped of their hydrogen, which ascends high into the atmosphere. The oxygen stays behind.)

For all its history, Earth has been within the "temperate zone" of the solar system. That is, Earth has been in the "right" range of distance from the sun to have surface temperatures that allow oceans and animals to exist without freezing or frying. This *habitable zone* (actual geography in space) extends from a well-known limit just inside Earth's orbit to a less understood outer limit near Mars or possibly beyond. The habitable zone moves outward as the sun becomes brighter, and in the future the zone will pass Earth and leave it behind. Earth will in essence become the Venus of today. The inner edge of the habitable zone is only about 9.3 million miles (15 million km) away, and it will effectively reach Earth in half a billion or a billion years from now (or possibly less). After this time, the sun will be too bright for organisms to survive on Earth.

The steadily rising amount of energy hitting Earth from the sun over the past 4.567 billion years *should* have ended life on Earth long ago, as it did on Venus (assuming that Venus ever had life), except for one of the most important of all of the planetary life support systems, the planetary thermostat described in the first chapter. For more than 3 billion years (and perhaps 4 billion years) this system has kept the global average temperature of Earth between the freezing and boiling points of water (except for the occasional snowball Earth event), thus allowing the most important requirement for life—liquid water—to continually exist on the surface of the planet for that immense amount of time. Just as important, life, which evolved within tight temperature limits, has been able to maintain essentially similar physiologies and internal chemical reactions that are temperature dependent. Rising temperature because of the sun and an increasing reduction in atmospheric carbon dioxide are the two processes that in combination will have the greatest effect on future biotic evolution.

The rises and falls of CO_2 are now fairly well documented for the last 500 million years—the time of animals. Oxygen, a requirement of all animals, is obviously important too. We have already highlighted levels of these two gases from past to present. But like the knowledge about the rate of the enlarging and ever more energetic sun, the future trajectory of both carbon dioxide and oxygen are also knowable and thus predictable.

The long-term prediction for carbon dioxide is that it will continue in the same trend it has shown over at least the last billion years—a slow but inexorable decrease. The lowering levels are because of both life and plate tectonics: as more and more CO_2 is used to make the skeletons of organisms, especially in the oceans, CO_2 is consumed. If these skeletons stay in the oceans, the skeletally confined CO_2 (now in calcium carbonate) will recycle. But plate tectonics makes the continents ever larger, and an increasing amount of limestone, which is the grave of atmospheric CO_2, becomes locked to the continents as sedimentary deposits.

One would think that the long-term trend of lowering CO_2 would be a plunge into inescapable snowball Earth conditions. But it is not cooling from a lowering of the concentration of CO_2 in the atmosphere that will be a hallmark of the aging Earth. It will be heating. The increasing heat from the sun will utterly dwarf the cooling effects of diminishing carbon dioxide and its greenhouse gas effects. When the average global temperature rises to perhaps 120 to 140°F (50 to 60°C), Earth will begin to lose its oceans to space.

But long before the oceans are lost in 2 to 3 billion years, life will have died out on Earth's surface because photosynthetic organisms, from microbes to higher plants, will no longer be able to survive in the low-CO_2 atmosphere. This dwindling carbon resource will then cause a further reduction of planetary habitability, because the CO_2 drop will trigger a drop in atmospheric oxygen to a level too low to support animal life.

This process is already observable. When vascular plants first colonized Earth's surface some 475 million years ago, they did so in an atmosphere rich in carbon dioxide. There was no need for conserving carbon in physiological processes. Even today, many plant species require a minimum of 150 ppm of CO_2, and James F. Kasting pointed out in a 1997 article that there is a second large group of plants, including many of the grassy species so common in the mid-latitudes of the planet, that use a quite different form of photosynthesis and can exist at lower CO_2 concentrations, sometimes as low as 10 ppm—the C4 plants described in an earlier chapter. These plants will last far longer than their more CO_2-addicted cousins and will considerably extend the life of the biosphere even in a world in which CO_2 levels have fallen far, far below present-day values.

We can safely predict that the future evolution of plant life will be toward plants that can live at lower CO_2 levels than that of their stock ancestral C3 plants. Also, because global temperatures will be rising, keeping water within a plant will be an increasing problem. Plants will have two conflicting needs—ever larger holes in their exterior to let the small amount of carbon dioxide in the atmosphere get into the interior, where photosynthesis can take place, at the same time trying to reduce the loss of water molecules through these same pores. At a minimum, one can expect a future flora of tough, waxy plants that would completely close down all portals to the outside world when there is no sunlight for photosynthesis.

With new plants with tougher exteriors, leaves—at least in their present form—might be expected to disappear. The same will happen to grass; the loss of water from plants with relatively high surface-area-to-volume ratios will doom grass blades and thin leaves alike. All of this, of course, will require a marked change in animal life.

As early as 500 million years from now, or perhaps as much as 1 billion years or so into the future, the level of CO_2 in the atmosphere will reach a point at which familiar plant life will no longer be able to exist. The changeover at first will be in no way dramatic. All over the world, the plants will slowly die. But the planet will not immediately become brown. For as one suite of plants dies, their places will be taken immediately by another cohort of plant life that may look nearly identical to those dying. Deep inside the tissues of these two groups of plants, however, fundamental processes of photosynthesis will be radically different. After this changeover, life on Earth will continue in ways probably not too dissimilar from that which came before—at least for a time.

There is also the possibility that plants will continue to evolve other photosynthetic pathways to compensate for lower CO_2 levels. In this case, some sort of plant life may survive at minimal CO_2 levels. Eventually, however, even these last holdouts will die out. All models suggest that CO_2 will continue to drop in volume, ultimately arriving at the critical level of 10 ppm.

The most important questions about any future evolution concern the future of biodiversity—the number of species on Earth. Two questions that arise are: Will there be more species than now? And if so, for how long? But as is so often the case, to begin to answer these questions one needs to look into the past.

More than just the flora on land will be traumatized by the lower CO_2 levels. Larger marine plants and perhaps plankton as well will be affected similarly. Marine communities thus will be strongly affected, because the base of most marine communities is phytoplankton, a single-celled plant that floats in the seas. A reduction in CO_2 will directly affect these as well as the land plants. Yet the disappearance of land plants will also cause a drastic reduction in the biomass of marine plankton, even without accounting for CO_2 effects on plant volumes in the seas.

Marine phytoplankton is severely nutrient limited in most ocean settings. The influx of nitrates, iron, and phosphates into the oceans each season causes phytoplankton to bloom. But the source of this phosphate and nitrate is rotting terrestrial vegetation, brought into the oceans through river runoff from the land. As land plants diminish in volume, so too will the volume of nutrients be diminished. The seas will be starved for nutrients, and the volume of plankton will decline catastrophically. This decline will never be reversed, for even if land plants rebound at low levels, as outlined above, they will never again reach the enormous mass of material that is present in a world (such as that of today) where CO_2 starvation does not exist.

On land and sea the base of the food chains as they are constructed today will disappear. The loss of plants will suddenly cause global productivity—a measure of the amount of life on the planet—to plummet. But there will still be life: great masses of bacteria, such as cyanobacteria will continue to live, because these hardy single-celled organisms can live at lower CO_2 levels that are below those necessary to keep multicellular plants alive, and they also do not require oxygen, something multicellular plants do.

The disappearance of plants will drastically affect landforms and the nature of the planet's surface. As roots disappear and surface layers become less stable, the very nature of rivers will change. The large, meandering rivers of the modern era date back, at most, to the Silurian period of some 400 million years ago, when land plants first colonized the surface of the planet, for it takes root stability to maintain the banks of meandering rivers. When plants die out or are not present because of slope, soil, or other inopportune environmental conditions, a different kind of river exists—braided rivers or streams, the kinds of flows found on desert alluvial fans or in front of glaciers, two types of environments not conducive to rooted plant life. This was the nature of rivers before the advent of land plants, and it will again be the way that rivers flow when CO_2 drops to the plant die-off threshold.

The loss of soils will be no less dramatic. As soils are blown away, they leave behind bare rock surfaces. As this condition begins to occur over the surface of the planet, it will change the albedo—the reflectivity of Earth. Far more light will reflect back into space, thereby affecting Earth's temperature balance. The atmosphere and its heat transfer and precipitation patterns will be radically changed. Blowing wind will begin to carry the grains of sand created by the action of heat, cold, and running water on the bare rock surfaces. While chemical weathering will lessen as a result of the loss of soil, this mechanical weathering will build up an enormous volume of blowing sand. The surface of the planet will become a giant series of dune fields.

Although this event could signal the final extinction of all plant life on land (and perhaps in the sea as well), it is more likely that a long period of time (perhaps in the hundreds of millions of years) will ensue in which CO_2 levels hover at the level causing plant death. As the levels drop to lethal limits, plants will die off, reducing weathering and allowing CO_2 to again accumulate in the atmosphere, once again allowing any small surviving seeds or rootstocks to germinate and, at least for some millennia, to flourish at least at low population numbers. As plant life again spreads across land surfaces, weathering rates will again increase and thus increase the rate of carbon dioxide uptake out of the atmosphere.

Animal life is dependent on an oxygen atmosphere. There are almost no animals capable of living in zero- or even low-oxygen conditions (although in 2010 a tiny invertebrate able to live with anoxia was discovered deep in the Mediterranean Sea). David C. Catling of the University of Washington has suggested that by about 15 million years after the death of plants, less than 1 percent of the atmosphere will be oxygen in contrast to the 21 percent volume that Earth's atmosphere contains today.

FUTURE EVOLUTION OF HUMANS

Life is one of the main agents of both its evolution and its extinction. Coauthor Ward's Medea hypothesis was based on the conclusion that life has been more enemy to itself than friend; that the various ecosystems and their species do not become ever better adapted and successful the longer they last. As we have seen, in fact, the actual killing done by the major mass extinctions was caused by various toxins produced by microbial life. Thus it seems appropriate to us to end this book with some comments about one of the most Medean of all species ever evolved: our own. What will the future of evolution be for our own species?

The science fiction trope concerning our own future is one of an even larger head, containing a *much* larger brain, high foreheads, and higher intellect. But bigger brains are probably not in humanity's future. The fossil record shows that the days of rapid brain increase, at least based on skull sizes over the past several thousand generations, seem to be over, and those conditions causing the rise in brain size (theorized to have been largely climatic in origin) are not likely to be repeated. But if not giant brains, what *might* evolution hold for the human species? Another intriguing question is whether the human species has undergone *any* significant evolution since its formation some two hundred thousand years ago.

The surprising revelation based on genetic study is that not only has the human genome undergone some major reshuffling since the species' formation, some two hundred thousand years ago, but it appears that the rates of human evolution, if anything, have been increasing over the past thirty millennia. A study by Henry C. Harpending and John Hawks suggests that over the past five thousand years alone, humans have evolved as much as a hundred times more quickly than at any time since the split of the earliest hominid from the ancestors of modern chimpanzees some 6 million years ago. Moreover, rather than seeing a reduction of evolution of those characteristics that in combination are used to distinguish human races, until very recently the human races in various parts of the world have become *more*, not less, distinct. Only in the past century, through the revolution in human travel and the more open behavioral attitudes of most humans to those of other races, has this pattern slowed. There are two main reasons: agriculture and cities. Food and crowding.

Humans thus seem to be first-class evolvers, or at least they were until very recently. With that known, it is possible to speculate about what the future might hold for the human species in terms of further evolutionary change—assuming that the species gets its few million years that appears to be the average longevity of any mammal species. Because much of the observed evolutionary changes of the past five thousand years involved adaptation to particular environments, it is fair to ask how the future world, with the expectation of larger populations than now, and with larger cities and agricultural fields among the other offshoots of technology, might affect the species' evolutionary outcome—or will it be affected at all? There are many questions: Will humans become larger or smaller, gain or lose intelligence, be it intellectual or emotional? Will humans become more or less tolerant of oncoming environmental problems, such as a dearth of freshwater, an abundance of ultraviolet radiation, and an increase in global temperatures? Will humans produce a new species, or is the species now evolutionarily sterile? Might the future evolution of humanity be not within the species' genes, but through the addition of silicon expression and memory augmentation to human brains through neural connections with inorganic machines? Is humanity but the builder of the next dominant intelligence on Earth—the machines?

THE END OF HISTORY

For those concerned that the "End is nigh!"—and even those who worry that life on this planet is at least under the shadow of a new mass extinction or is already in one—there should be solace. We seem to be at a high point of species numbers in all of the (at least) 3.4 billion-year-old history of life. Our view is that it is impossible to prove what percentage of life is now going extinct—the metric of deciding if a mass extinction is major (greater than 50 percent), minor (between 10 to 50 percent), or not an extinction at all—if the denominator is not known. Clearly there are more than 1.6 million species on Earth. If it is determined that a new mass extinction is taking place, there is some slight solace in the fact that after every past mass extinction, biodiversity bounced back to even greater levels.

The latter was the argument of the great Frank Drake, in a debate with one of us years ago about whether or not Earthlike planets are rare. Author of the eponymous Drake equation, a way to try to estimate the number of other intelligent species in the galaxy, he took the view that a giant mass extinction, such as the Permian extinction, was actually a good thing for any planet. But the price to pay . . . It took 5–10 million years following the Permian extinction for biodiversity to finally arrive at even the pre-extinction levels. The world went back to the Proterozoic in terms of biodiversity and even the kinds of life—a situation we have elsewhere only somewhat humorously described as the empire strikes back—in this case, the Precambrian empire of anoxic and toxic microbes.

A final prediction of Ward's Medea hypothesis is that it should pertain to every planet with life, and that there is only one way out of this suicidal box that life creates simply through existing: intelligence. The intelligence to see the future. One such future is that our species expands its habitat first to Mars, then the asteroid belts, and finally to other stars. Another future is that the carbon dioxide we are pumping into the atmosphere causes all the ice on Earth to melt, raising sea levels, slowing the thermohaline circulation patterns, bringing stagnation followed by anoxia to the ocean bottoms, and then into ever-shallower waters, at the same time liberating toxic levels of hydrogen sulfide to percolate out of every single ocean. In that future, only animals with very good gas masks will survive.

History is an early warning system.

LAST WORD

Nothing lasts. That goes for planets to organisms to scientific careers. While funerals are among the saddest events that we humans can participate in, at least they are definitive moments marking change:

from living to dead. But perhaps even sadder is the life near its end, such as a human with a fatal malady given a highly definitive death sentence. Such is the case for the chambered nautilus, an animal featured in this book as the best model for the extinct ammonites, and an animal that dodged the bullet of the major mass extinctions, if not in its present guise, at least as a major taxonomic order. Nautiloids first appeared in the Cambrian explosion of 500 million years ago. They are still with us but in dwindling numbers, and are now on the brink of extinction in various Pacific countries because of the demand for their shell: the past mass extinctions did not kill off organisms because of their perceived beauty. The human-induced mass extinction acts otherwise.

But even before the nautiluses became the objects of commerce, such that a half million of their shells were shipped to the United States alone in the five years between 2005 and 2010, they had received their own death sentence. The nautilus body plan evolved to work in warm, shallow water. It uses an osmotic pump that empties its chambers of the liquid each is completely filled with at its formation; they evolved to grow their shells in shallow, calcium-rich seawater. But along came the Mesozoic marine revolution that we profiled earlier in this book. Nautiluses were previously impregnable in their hard outer shells until new kinds of fish evolved during the Cretaceous and later that could easily break the durable outer shell of the nautilus. Life became impossible in the shallow waters. The shallows became a death sentence.

Life is about change. The nautiluses dealt with these new evolutioary and ecological stresses over the millions of years by slowly and steadily living in ever deeper water. Our new results show that in the last five million years, they were living at average depths of 200 to 300 meters, but their design is ill-equipped for these depths. They grow slower: it used to take a year to reach full size, while now it takes ten to fifteen. Now they live as a deep sea animal, few in number, in a dark, low-resource environment that is difficult at best. And the predators are following them down. They can go no deeper, as their shells have a depth limit below which they implode, causing instant death. There is no further place to hide.

The fate of the nautilus is a metaphor for all animal life. Sooner or later evolution, competition, and the natural changing of our Earth and sun as they age will make any body plan obsolete. For us land animals, it is not the predators that will do us in, but an enlarging sun and too little carbon dioxide. There will be no place on Earth to survive. The only hope for our species, if we wish to do what the nautilus has done—or better yet, do what the cyanobacteria have done and last two to three billion years—is to leave. The last chapter here has been about the history of life—*on Earth*. But there can be a whole new book. A whole library of new books, in fact.

Perhaps life did start on Mars, our kind of life. The choice was to leave Mars or die. Survival is literally in our genes.

Notes

INTRODUCTION

1. J. Loewen, *Lies My Teacher Told Me: Everything Your American History Textbook Got Wrong* (New York: Touchstone Press, 2008).
2. J. Baldwin, *Notes of a Native Son* (Boston: Beacon Press, 1955).
3. N. Cousins, *Saturday Review*, April 15, 1978.
4. P. Ward, "Impact from the Deep." *Scientific American* (October 2006). The actual first use of the term "greenhouse mass extinction" is difficult to ferret out, although I used it overtly in a *Discover* magazine article in the 1990s.
5. G. Santayana, *The Life of Reason, Five Volumes in One* (1905).
6. Fortey's book was (and remains) a masterpiece, not just for the "facts" but also for the stories of science in the presentation of what in lesser hands is often delivered as more dry history. And yet it is now dated (how could it not be, including much new work by Richard himself). We have used his book as a bit of a stalking horse and straw man, and hope we are forgiven. Just for starters, we take some exception to the title: assuming life is already 4 billion years old on Earth might have seemed like a good bet in the mid-1990s when the book was written, but that might not be the case now. Perhaps he is still right on this, but our own arguments are to come. The reference of the book is: R. Fortey, *Life: A Natural History of the First Four Billion Years of Life on Earth* (New York: Random House, 1997).
7. A great read on this entire aspect of the philosophy of those building the then-nascent field of geology (and its subfield of paleontology) is in M. J. Rudwick, *The Meaning of Fossils: Episodes in the History of Palaeontology* (London: Science History Publications, 1972). This book, early on hard to find, was later republished in more accessible presses. Rudwick's take on the late 1700s into the early 1800s, when the foment over geological time and processes was intersecting with early ideas about the stratigraphic ranges of fossils as well as early ideas about evolution, is seminal and remains a must-read for anyone interested in time and natural history.
8. We tell our undergraduate classes that Charles Darwin was a geologist before all. His understanding of the fossil record, as well as the many kinds of fossils he saw whenever he got off that tiny ship the *Beagle* (which was whenever possible, as Darwin was severely affected by seasickness), was crucial in preparing his mind for the observations that would lead to his celebrated hypotheses about evolution. A good read about all of this training is A. Desmond, *Darwin* (New York: Warner Books, 1992).
9. M. Rudwick, *Georges Cuvier, Fossil Bones, and Geological Catastrophes: New Translations and Interpretations of the Primary Texts* (University of Chicago Press, 1997).
10. There are many works on the number of species through time, and we will look at this in detail in the pages to come. One of the most recent is by John Alroy and

a host of other authors, "Phanerozoic Trends in Global Diversity of Marine Invertebrates," *Science* 321 (2008): 97.

11. N. Lane, *The Vital Question: Why Is Life the Way It Is?* (London: Profile Books, 2015); *Life Ascending: The Ten Great Inventions of Evolution* (London: Profile Books, 2009); *Power, Sex, Suicide: Mitochondria and the Meaning of Life* (Oxford: Oxford University Press, 2005); *Oxygen: The Molecule That Made the World* (Oxford: Oxford University Press, 2002).

CHAPTER I: TELLING TIME

1. A good guide to stratigraphic usage comes from the International Subcommission on Stratigraphy. This is a very formal group that angsts over every term and name. They are online: one useful chapter is stratigraphy.org-upload-bak-defs.htm.
2. Variety of age dating, brief synopsis of: Uranium, potassium argon, uranium lead, strontium isotope dating, and magnetostratigraphy are all used. To come to grips with all of this we recommend the works of Martin Rudwick, all available in both libraries and online bookstores. These include, most recently, M. Rudwick, *Earth's Deep History: How It Was Discovered and Why It Matters* (Chicago: University of Chicago Press, 2014).
3. The first system was indeed rock type: each kind of lithology, such as volcanic rock, metamorphic rock, and particularly the kinds of sedimentary rock (such as sandstone, chalk, shale) were thought to be distinctive and characteristic of a specific time. Hence the Cretaceous period was first named for the kind of rock found commonly in Europe, white chalk. Later it was found that the same rock types could be made at any time. See M. Rudwick, *The Meaning of Fossils: Episodes in the History of Paleontology* (London: Science History Publications, 1972).
4. The use of fossils for telling time, and William "Strata" Smith's role in the revolution of understanding and defining the geological time scale can be found in many books. A very useful one is from our late friend Bill Berry of UC Berkeley, a paleontologist and scientist sorely missed: W. B. N. Berry, *Growth of a Prehistoric Time Scale* (Boston: Blackwell Scientific Publications, 1987): 202.
5. J. Burchfield, "The Age of the Earth and the Invention of Geological Time," D. J. Blundell and A. C.Scott, eds., *Lyell: the Past is the Key to the Present* (London,Geological Society of London, 1998), 137–43.
6. By the late 1800s a great deal of fame came to be associated with being the author of a geological period. One such grab for glory was by Lapworth. See the ever-readable M. Rudwick, *The Great Devonian Controversy: The Shaping of Scientific Knowledge Among Gentlemanly Specialists* (Chicago: University of Chicago Press, 1985).
7. K. A. Plumb, "New Precambrian Time Scale," *Episode* 14, no. 2 (1991): 134–40.
8. A. H. Knoll, et al., "A New Period for the Geologic Time Scale," *Science* 305, no. 5684 (2004): 621–22.

CHAPTER II: BECOMING AN EARTHLIKE PLANET: 4.6–4.5 GA

1. Earthlike planets and estimations of number of ELPs: The many definitions of what is "Earthlike" vary tremendously. The number of such planets, or at least our estimates, vary as well. A good scientific reference is here: E. A. Petigura, A. W. Howard, G. W. Marcy, "Prevalence of Earth-Size Planets Orbiting Sun-Like Stars," *Proceedings of the National Academy of Sciences of the United States of America* 110, no. 48 (2013). doi:10.1073-pnas.1319909110, and the NASA publicity view, www.nasa.gov-mission_pages-kepler-news-kepler20130103.html.
2. NASA's sense of it can also be found at science1.nasa.gov-science-news-science-at-nasa-2003-02oct_goldilocks-are Earth reference, discussion. Also interesting and up to date is S. Dick, "Extraterrestrials and Objective Knowledge," in A. Tough, *When SETI Succeeds: The Impact of High-Information Contact* (Foundation for the Future, 2000): 47–48.
3. While not the scientific papers that began the revolution, this later article by Geoff Marcy is a good entry into the subject: G. Marcy et al. "Observed Properties of Exoplanets: Masses, Orbits and Metallicities," *Progress of Theoretical Physics Supplement* no. 158 (2005): 24–42.
4. D. McKay et al., "Search for Past Life on Mars: Possible Relic Biogenic Activity in Martian Meteorite AL84001," *Science* 273, no. 5277 (1996): 924–30.
5. P. Ward, *Life as We Do Not Know It: The NASA Search for and Synthesis of Alien Life* (New York: Viking, 2005); P. Ward and S. Benner, "Alternative Chemistry of Life," in W. Sullivan and J. Baross, eds. *Planets and Life: The Emerging Science of Astrobiology* (Cambridge: Cambridge University Press, 2008): 537–44.
6. W. K. Hartmann and D. R. Davis, "Satellite-Sized Planetesimals and Lunar Origin," *Icarus* 24, no. 4 (1975): 504–14; R. Canup and E. Asphaug, "Origin of the Moon in a Giant Impact Near the End of the Earth's Formation," *Nature* 412, no. 6848 (2001): 708–12; A. N. Halliday, "Terrestrial Accretion Rates and the Origin of the Moon," *Earth and Planetary Science Letters* 176, no. 1 (2000): 17–30; D. Stöffler and G. Ryder, "Stratigraphy and Isotope Ages of Lunar Geological Units: Chronological Standards for the Inner Solar System," *Space Science Reviews* 96 (2001): 9–54.
7. A. T. Basilevsky and J. W. Head, "The Surface of Venus," *Reports on Progress in Physics* 66, no. 10 (2003): 1699–1734; J. F. Kasting, "Runaway and Moist Greenhouse Atmospheres and the Evolution of Earth and Venus," *Icarus* 74, no. 3 (1988): 472–94.
8. D. H. Grinspoon and M. A. Bullock, "Searching for Evidence of Past Oceans on Venus," *Bulletin of the American Astronomical Society* 39 (2007): 540.
9. A good general reference for the age of the Earth is G. B. Dalrymple, *The Age of the Earth* (Redwood City: Stanford University Press, 1994), while his more technical take is "The Age of the Earth in the Twentieth Century: A Problem (Mostly) Solved," *Special Publications, Geological Society of London* 190 (2001): 205–21.
10. This concern that the heavy bombardment would have adversely affected life and its early history was first illuminated by Kevin Maher and David Stevenson of Caltech in 1988, in a short letter to *Nature*. "Impact Frustration of the Origin

of Life," *Nature* 331, no. 6157 (1988): 612–14. Many have followed on, including Kevin Zahnle and Norm Sleep. An early reference is K. Zahnle et al., "Cratering Rates in the Outer Solar System," *Icarus* 163 (2003): 263–89; F. Tera et al., "Isotopic Evidence for a Terminal Lunar Cataclysm," *Earth and Planetary Science Letters* 22, no. 1 (1974): 1–21. The origin of the bombardment has recently been reexamined, concerning possible migration of outer planets several hundred million years after the major phase of accretion: W. F. Bottke et al., "An Archaean Heavy Bombardment from a Destabilized Extension of the Asteroid Belt," *Nature* 485 (2012): 78–81; G. Ryder et al., "Heavy Bombardment on the Earth at ~3.85 Ga: The Search for Petrographic and Geochemical Evidence," in *Origin of the Earth and Moon*, R. M. Canup and K. Righter, eds. (Tucson: University of Arizona Press, 2000): 475–92.

11. A great deal has been written about the origin of the Earth's atmosphere. A good website concentrating on the role of life in this process is www.amnh.org-learn-pd-earth-pdf-evolution_earth_atmosphere.pdf.

    A reference article can be found from K. Zahnle et al., "Earth's Earliest Atmospheres," *Cold Spring Harbor Perspectives in Biology* 2, no. 10 (2010).

12. The evaporation of the early oceans by impact from "Texas-sized asteroids" always got an uneasy chuckle from undergraduate classes when we brought it up during the George W. Bush presidency. As time separates us from those days, the concept now seems a bit different and purely scientific. A good PDF (but with a strange name) gets into the physics of it all in an understandable way: www.breadandbutterscience.com-CATIS.pdf.

13. Figuring out how much carbon dioxide was in the early Earth atmosphere is difficult. There are no real direct methods. References include: J. Walker, "Carbon Dioxide on the Early Earth," *Origins of Life and Evolution of the Biosphere* 16, no. 2 (1985): 117-27. For Phanerozoic era (the time of "visible life"), here are two seminal papers: D. H. Rothman, "Atmospheric Carbon Dioxide Levels for the Last 500 Million Years," *Proceedings of the National Academy of Sciences* 99, no. 7 (2001): 4167–71, and D. Royer et al., "CO_2 as a Primary Driver of Phanerozoic Climate," *GSA Today* 14, no. 3 (2004): 4–15. For much of the rest of this chapter there is no better primer than the wonderful college text by L. Kump et al., *The Earth System*, 3rd ed. (Upper Saddle River, NJ: Prentice Hall, 2009). This amazing if pricey textbook is a great doorway into what is called Earth system science. The discussions of the carbon cycle as well as other elemental systems leading to habitability come from this text.

14. Ward has dealt with this topic in book-length treatment (P. Ward, *Out of Thin Air*. Washington, D.C.: Joseph Henry Press, 2006). The various Robert Berner references include R. A. Berner, "Models for Carbon and Sulfur Cycles and Atmospheric Oxygen: Application to Paleozoic Geologic History," *American Journal of Science* 287, no. 3 (1987): 177–90. Also highly relevant are: L. R. Kump, "Terrestrial Feedback in Atmospheric Oxygen Regulation by Fire and Phosphorus," *Nature* 335 (1988): 152–54; L. R. Kump, "Alternative Modeling Approaches to the Geochemical Cycles of Carbon, Sulfur, and Strontium Isotopes," *American Journal of Science* 289 (1989): 390–410; L. R. Kump, "Chemical Stability of the Atmosphere and Ocean," *Global and Planetary*

Change 75, no. 1–2 (1989): 123–36; L. R. Kump and R. M. Garrels, "Modeling Atmospheric O_2 in the Global Sedimentary Redox Cycle," *American Journal of Science* 286 (1986): 336–60.

15. W. F. Ruddiman and J. E. Kutzbach, "Plateau Uplift and Climate Change," *Scientific American* 264, no. 3 (1991): 66–74, and M. Kuhle, "The Pleistocene Glaciation of Tibet and the Onset of Ice Ages—An Autocycle Hypothesis," *GeoJournal* 17 (4) (1998): 581–95; M. Kuhle, "Tibet and High Asia: Results of the Sino-German Joint Expeditions (I)," *GeoJournal* 17, no. 4 (1988).
16. The life and work of Robert Berner: R. A. Berner, "A New Look at the Long-Term Carbon Cycle," *GSA Today* 9, no. 11 (1999): 1–6; R. A. Berner, "Modeling Atmospheric Oxygen over Phanerozoic Time," *Geochimica et Cosmochimica Acta* 65 (2001): 685–94; R. A. Berner, *The Phanerozoic Carbon Cycle* (Oxford: Oxford University Press, 2004), 150.; R. A. Berner, "The Carbon and Sulfur Cycles and Atmospheric Oxygen from Middle Permian to Middle Triassic," *Geochimica et Cosmochimica Acta* 69, no. 13 (2005): 3211–17; R. A. Berner, "GEOCARBSULF: A Combined Model for Phanerozoic Atmospheric Oxygen and Carbon Dioxide," *Geochimica et Cosmochimica Acta* 70 (2006): 5653–5664; R. A. Berner and Z. Kothavala, "GEOCARB III: A Revised Model of Atmospheric Carbon Dioxide over Phanerozoic Time," *American Journal of Science* 301, no. 2 (2001): 182–204.

CHAPTER III: LIFE, DEATH, AND THE NEWLY DISCOVERED PLACE IN BETWEEN

1. Perhaps the best way to understand the Mark Roth work is his TED talk: www.ted.com-talks-mark_roth_suspended_animation.
2. T. Junod, "The Mad Scientist Bringing Back the Dead. . . . Really," Esquire.com, December 2, 2008.
3. E. Blackstone et al., "H_2S Induces a Suspended Animation–Like State in Mice," *Science* 308, no. 5721 (2005): 518.
4. D. Smith et al., "Intercontinental Dispersal of Bacteria and Archaea by Transpacific Winds," *Applied and Environmental Microbiology* 79, no. 4 (2013): 1134–39.
5. K. Maher and D. Stevenson, "Impact Frustration of the Origin of Life," *Nature* 331 (1988): 612–14.
6. E. Schrödinger, *What Is Life*? (Cambridge: Cambridge University Press, 1944), 90.
7. P. Davies, *The Fifth Miracle: The Search for the Origin and Meaning of Life.* (New York: Penguin Press, 1998), 260.
8. P. Ward, *Life as We Do Not Know It* (New York: Viking Books, 2005).
9. W. Bains, "The Parts List of Life," *Nature Biotechnology* 19 (2001): 401–2; W. Bains, "Many Chemistries Could Be Used to Build Living Systems," *Astrobiology*, 4, no. 2 (2004): 137–67; and N. R. Pace, "The Universal Nature of Biochemistry," *Proceedings of the National Academy of Sciences of the Unites States of America* 98, no. 3 (2001): 805–808; S. A. Benner et al., "Setting the Stage: The History, Chemistry, and Geobiology Behind RNA," *Cold Spring Harbor Perspectives in Biology* 4, no. 1 (2012): 7–19; M. P. Robertson and G. F. Joyce, "The Origins of the RNA World," *Cold Spring Harbor Perspectives in*

Biology 4, no. 5 (2012); C. Anastasi et al., "RNA: Prebiotic Product, or Biotic Invention?" *Chemistry and Biodiversity* 4, no. 4 (2007): 721–39; T. S. Young and P. G. Schultz, "Beyond the Canonical 20 Amino Acids: Expanding the Genetic Lexicon," *The Journal of Biological Chemistry* 285, no. 15 (2010): 11039–44.

10. F. Dyson, *Origins of Life*, 2nd ed. (Cambridge: Cambridge University Press, 1999), 100
11. Nick Lane is an iconoclast with rather unerring judgment. For a good take on energy complexity, see N. Lane, "Bioenergetic Constraints on the Evolution of Complex Life," in P. J. Keeling and E. V. Koonin, eds., *The Origin and Evolution of Eukaryotes. Cold Spring Harbor Perspectives in Biology* (2013).
12. J. Banavar and A. Maritan. "Life on Earth: The Role of Proteins," J. Barrow and S. Conway Morris, *Fitness of the Cosmos for Life* (Cambridge: Cambridge University Press, 2007), 225–55.
13. E. Schneider and D. Sagan, *Into the Cool: Energy Flow, Thermodynamics, and Life* (Chicago, IL: University of Chicago Press, 2005).

CHAPTER IV: FORMING LIFE: 4.2(?)–3.5 GA

1. Dr. D. R. Williams, Viking Mission to Mars, NASA, December 18, 2006.
2. www.space.com-18803-viking.
3. ntrs.nasa.gov-archive-nasa-casi.ntrs.nasa.gov-19740026174.pdf. Also see R. Navarro-Gonzáles et al., "Reanalysis of the Viking Results Suggests Perchlorate and Organics at Midlatitudes on Mars," *Journal of Geophysical Research* 115 (2010).
4. P. Rincon, "Oldest Evidence of Photosynthesis," BBC.com, December 17, 2003 and S. J. Mojzsis et al., "Evidence for Life on Earth Before 3,800 Million Years Ago," *Nature* 384 (1996): 55–59; M. Schidlowski, "A 3,800-Million-Year-Old Record of Life from Carbon in Sedimentary Rocks," *Nature* 333 (1988): 313–18; M. Schidlowski et al., "Carbon Isotope Geochemistry of the 3.7×10^9 Yr Old Isua Sediments, West Greenland: Implications for the Archaean Carbon and Oxygen Cycles," *Geochimica et Cosmochimica Acta* 43 (1979): 189–99.
5. K. Maher and D. Stevenson. "Impact Frustration of the Origin of Life," *Nature* 331 (1988): 612–14
6. R. Dalton. "Fresh Study Questions Oldest Traces of Life in Akilia Rock," *Nature* 429 (2004): 688. This work is continuing; see Papineau et al., "Ancient Graphite in the Eoarchean Quartz-Pyroxene Rocks from Akilia in Southern West Greenland I: Petrographic and Spectroscopic Characterization," *Geochimica et Cosmochimica Acta* 74, no. 20 (2010): 5862–83.
7. J. W. Schopf, "Microfossils of the Early Archean Apex Chert: New Evidence of the Antiquity of Life," *Science* 260, no. 5108 (1993): 640–46.
8. M. D. Brasier et al., "Questioning the Evidence for Earth's Oldest Fossils," *Nature* 416 (2002): 76–81.
9. D. Wacey et al., "Microfossils of Sulphur-Metabolizing Cells in 3.4-Billion-Year-Old Rocks of Western Australia," *Nature Geoscience* 4 (2011): 698–702.
10. M. D. Brasier, *Secret Chambers: The Inside Story of Cells and Complex Life* (New York: Oxford University Press, 2012), 298.

11. "Ancient Earth May Have Smelled Like Rotten Eggs," *Talk of the Nation*, National Public Radio, May 3, 2013.
12. www.nasa.gov-mission_pages-msl-#.U4Izyxa9yxo.
13. www.abc.net.au-science-articles-2011-08-22-3299027.htm.
14. J. Haldane, *What Is Life*? (New York: Boni and Gaer, 1947), 53.
15. L. Orgel, *The Origins of Life: Molecules and Natural Selection* (Hoboken, NJ: John Wiley and Sons, 1973).
16. J. A. Baross and J. W. Deming, "Growth at High Temperatures: Isolation and Taxonomy, Physiology, and Ecology," in *The Microbiology of Deep-sea Hydrothermal Vents*, D. M. Karl, ed. (Boca Raton: CRC Press, 1995), 169–217, and E. Stueken et al., "Did Life Originate in a Global Chemical Reactor?" *Geobiology* 11, no.2 (2013); K. O. Stetter, "Extremophiles and Their Adaptation to Hot Environments," *FEBS Letters* 452, nos. 1–2 (1999): 22–25. K. O. Stetter, "Hyperthermophilic Microorganisms," in *Astrobiology: The Quest for the Conditions of Life*, G. Horneck and C. Baumstark-Khan, eds. (Berlin: Springer, 2002), 169–84.
17. Y. Shen and R. Buick, "The Antiquity of Microbial Sulfate Reduction," *Earth Science Reviews* 64 (2004): 243–272.
18. S. A. Benner, "Understanding Nucleic Acids Using Synthetic Chemistry," *Accounts of Chemical Research* 37, no. 10 (2004): 784–97; S. A. Benner, "Phosphates, DNA, and the Search for Nonterrean life: A Second Generation Model for Genetic Molecules," *Bioorganic Chemistry* 30, no. 1 (2002): 62–80.
19. G. Wächtershäuser, "Origin of Life: Life as We Don't Know It," *Science*, 289, no. 5483 (2000): 1307–08; G. Wächtershäuser, "Evolution of the First Metabolic Cycles," *Proceedings of the National Academy of Sciences* 87, no. 1 (1990): 200–204; G. Wächtershäuser, "On the Chemistry and Evolution of the Pioneer Organism," *Chemistry & Biodiversity* 4, no. 4 (2007): 584–602.
20. N. Lane, *Life Ascending: The Ten Great Inventions of Evolution* (New York: W. W. Norton & Company, 2009).
21. W. Martin and M. J. Russell, "On the Origin of Biochemistry at an Alkaline Hydrothermal Vent," *Philosophical Transactions of the Royal Society B-Biological Sciences* 362, no. 1486 (2007): 1887–925.
22. C. R. Woese, "Bacterial Evolution," *Microbiological Reviews* 51, no. 2 (1987): 221–71; C. R. Woese, "Interpreting the Universal Phylogenetic Tree," *Proceedings of the National Academy of Sciences* 97 (2000): 8392–96.
23. S. A. Benner and D. Hutter, "Phosphates, DNA, and the Search for Nonterrean Life: A Second Generation Model for Genetic Molecules," *Bioorganic Chemistry* 30 (2002): 62–80; S. Benner et al., "Is There a Common Chemical Model for Life in the Universe?" *Current Opinion in Chemical Biology* 8, no. 6 (2004): 672–89.
24. A. Lazcano, "What Is Life? A Brief Historical Overview," *Chemistry and Biodiversity* 5, no. 4 (2007): 1–15.
25. B. P. Weiss et al., "A Low Temperature Transfer of ALH84001 from Mars to Earth," *Science* 290, no. 5492, (2000): 791–95. J. L. Kirschvink and B. P. Weiss, "Mars, Panspermia, and the Origin of Life: Where Did It All Begin?" *Palaeontologia Electronica* 4, no. 2 (2001): 8–15. J. L. Kirschvink et al., "Boron, Ribose, and a Martian Origin for Terrestrial Life," *Geochimica et Cosmochimica Acta* 70, no. 18 (2006): A320.

26. C. McKay, "An Origin of Life on Mars," *Cold Spring Harbor Perspectives in Biology* 2, no. 4 (2010). J. Kirschvink et al., "Mars, Panspermia, and the Origin of Life: Where Did It All Begin?" *Palaeolontogia Electronica* 4, no. 2 (2002): 8–15.
27. D. Deamer, *First Life: Discovering the Connections Between Stars, Cells, and How Life Began* (Oakland: University of California Press, 2012), 286. But also see the great new work from our friend Nick Lane: N. Lane and W. F. Martin, "The Origin of Membrane Bioenergetics," *Cell* 151, no. 7 (2012): 1406–16.
28. www.nobelprize.org-mediaplayer-index.php?id=1218.

CHAPTER V: FROM ORIGIN TO OXYGENATION: 3.5–2.0 GA

1. J. Raymond and D. Segre, "The Effect of Oxygen on Biochemical Networks and the Evolution of Complex Life," *Science* 311 (2006): 1764–67.
2. J. F. Kasting and S. Ono "Palaeoclimates: The first Two Billion Years," *Philosophical Transactions of the Royal Society B-Biological Sciences* 361 (2006): 917–29
3. P. Cloud, "Paleoecological Significance of Banded-Iron Formation," *Economic Geology* 68 (1973): 1135–43.
4. M. C. Liang et al., "Production of Hydrogen Peroxide in the Atmosphere of a Snowball Earth and the Origin of Oxygenic Photosynthesis," *Proceedings of the National Academy of Sciences* 103 (2006): 18896–99.
5. J. E. Johnson et al., "Manganese-Oxidizing Photosynthesis Before the Rise of Cyanobacteria," *Proceedings of the National Academy of Sciences* 110, no. 28 (2013): 11238–43; J. E. Johnson et al., "O_2 Constraints from Paleoproterozoic Detrital Pyrite and Uraninite," *Geological Society of America Bulletin* (2014), doi: 10.1130-B30949.1.
6. J. E. Johnson et al., "O2 Constraints from Paleoproterozoic Detrital Pyrite and Uraninite," *Geological Society of America Bulletin*, published online ahead of print on February 27, 2014, doi: 10.1130/B30949.1.
7. R. E. Kopp et al., "Was the Paleoproterozoic Snowball Earth a Biologically Triggered Climate Disaster?" *Proceedings of the National Academy of Sciences* 102 (2005): 11131–36.
8. J. E. Johnson et al., "Manganese-Oxidizing Photosynthesis Before the Rise of Cyanobacteria."
9. Ibid.
10. R. E. Kopp and J. L. Kirschvink, "The Identification and Biogeochemical Interpretation of Fossil Magnetotactic Bacteria," *Earth-Science Reviews* 86 (2008): 42–61.
11. Ibid.
12. D. A. Evans et al., "Low-Latitude Glaciation in the Paleoproterozoic," *Nature* 386 (1997): 262–66.
13. J. L. Kirschvink et al. "Paleoproterozoic Snowball Earth: Extreme Climatic and Geochemical Global Change and Its Biological Consequences," *Proceedings of the National Academy of Sciences* 97 (2000): 1400–1405.
14. J. L. Kirschvink and R. E. Kopp, "Paleoproterozic Ice Houses and the Evolution of Oxygen-Mediating Enzymes: The Case for a Late Origin of Photosystem-II," *Philosophical Transactions of the Royal Society of London, Series B* 363, no. 1504 (2008): 2755–65.

15. D. A. D. Evans et al., "Paleomagnetism of a Lateritic Paleoweathering Horizon and Overlying Paleoproterozoic Red Beds from South Africa: Implications for the Kaapvaal Apparent Polar Wander Path and a Confirmation of Atmospheric Oxygen Enrichment," *Journal of Geophysical Research* 107, no. 2326.

CHAPTER VI: THE LONG ROAD TO ANIMALS: 2.0–1.0 GA

1. H. D. Holland "Early Proterozoic Atmospheric Change," in S. Bengtson, ed., *Early Life on Earth* (New York Columbia University Press, 1994), 237–44.
2. D. T. Johnston et al., "Anoxygenic Photosynthesis Modulated Proterozoic Oxygen and Sustained Earth's Middle Age," *Proceedings of the National Academy of Sciences* 106, no. 40 (2009), 16925–29.
3. A. El Albani et al., "Large Colonial Organisms with Coordinated Growth in Oxygenated Environments 2.1 Gyr Ago," *Nature* 466, no. 7302 (2002): 100–104.2; www.sciencedaily.com-releases-2010-06-100630171711.htm.
4. D. E. Canfield et al., "Oxygen Dynamics in the Aftermath of the Great Oxidation of Earth's Atmosphere," *Proceedings of the National Academy of Sciences* 110, no. 422 (2013).
5. A. H. Knoll, *Life on a Young Planet: The First Three Billion Years of Evolution on Earth* (Princeton: Princeton University Press, 2003).

CHAPTER VII: THE CRYOGENIAN AND THE EVOLUTION OF ANIMALS: 850–635 MA

1. R. C. Sprigg, "Early Cambrian 'Jellyfishes' of Ediacara, South Australia and Mount John, Kimberly District, Western Australia," *Transactions of the Royal Society of South Australia* 73 (1947): 72–99.
2. M. F. Glaessner, "Precambrian Animals," *Scientific American* 204, no. 3 (1961): 72–78.
3. Jim Gehling is one of the giants of Australian science, but more, he has collaborated with a veritable who's who of international science in his career-long work on the Ediacarans. The new exhibit he organized is worth a trip to Adelaide alone. See J. G. Gehling et al., in D. E. G. Briggs, ed., *Evolving Form and Function: Fossils and Development* (Yale Peabody Museum, 2005), 45–56; J. G. Gehling et al., "The First Named Ediacaran Body Fossil, Aspidella terranovica," *Palaeontology* 43, no. 3 (2000): 429; J. G. Gehling, "Microbial Mats in Terminal Proterozoic Siliciclastics; Ediacaran Death Masks," *Palaios* 14, no. 1(1999): 40–57.
4. P. F. Hoffman et al., "A Neoproterozoic Snowball Earth," *Science* 281, no. 5381 (1998): 1342–46; F. A. Macdonald et al., "Calibrating the Cryogenian," *Science*, 327, no. 5970 (2010): 1241–43.
5. F. A. Macdonald et al., "Calibrating the Cryogenian," *Science* 327, no. 5970 (2010): 1241–43.
6. B. Shen et al., "The Avalon Explosion: Evolution of Ediacara Morphospace," *Science* 319 no. 5859 (2008): 81–84; G. M. Narbonne, "The Ediacara Biota: A Terminal Neoproterozoic Experiment in the Evolution of Life," *Geological Society of America* 8, no. 2 (1998): 1–6; S. Xiao and M. Laflamme, "On the Eve of Animal Radiation: Phylogeny, Ecology and Evolution of the Ediacara Biota," *Trends in Ecology and Evolution* 24, no. 1 (2009): 31–40.

7. R. Sprigg, "On the 1946 Discovery of the Precambrian Ediacaran Fossil Fauna in South Australia," *Earth Sciences History* 7 (1988): 46–51.
8. S. Turner and P. Vickers-Rich, "Sprigg, Martin F. Glaessner, Mary Wade and the Ediacaran Fauna," Abstract for IGCP 493 conference, Prato Workshop, Monash University Centre, August 30–31, 2004.
9. A. Seilacher, "Vendobionta and Psammocorallia: Lost Constructions of Precambrian Evolution," *Journal of the Geological Society, London* 149, no. 4 (1992): 607–13; A. Seilacher et al., "Ediacaran Biota: The Dawn of Animal Life in the Shadow of Giant Protists," *Paleontological Research* 7, no. 1 (2003): 43–54. Dolph Seilacher was one of a kind. He and his wife, Edith, were world travelers. He was a champion of science, and one of the warmest scientists we have known. For a full list of his work, see Derek Briggs, ed., *Evolving Form and Function: A Special Publication of the Peabody Museum of Natural History* (New Haven, CT: Yale University 2005).
10. Martin Glaesner, a longtime faculty member at the University of Adelaide, lives on in that institution with his well-stocked Glaesner Room in Mawson Hall, where many of the fossil he collected and his copious notes can be found.
11. South Australian Museum, Ediacaran fossils. www.samuseumn.sa.gov.au/explore/museum-galleries/ediacaran-fossils.
12. B. Waggoner, "Interpreting the Earliest Metazoan Fossils: What Can We Learn?" *Integrative and Comparative Biology* 38, no. 6 (1998): 975–82; D. E. Canfield et al., "Late-Neoproterozoic Deep-Ocean Oxygenation and the Rise of Animal Life," *Science* 315, no. 5808 (2007): 92–95; B. Shen et al., "The Avalon Explosion: Evolution of Ediacara Morphospace," *Science* 319, no. 5859 (2008): 81–84.
13. B. MacGabhann, "There Is No Such Thing as the 'Ediacaran Biota,'" *Geoscience Frontiers* 5, no. 1 (2014): 53–62.
14. N. J. Butterfield, "*Bangiomorpha pubescens* n. gen., n. sp.: Implications for the Evolution of Sex, Multicellularity, and the Mesoproterozoic-Neoproterozoic Radiation of Eukaryotes," *Paleobiology* 26, no. 3 (2000): 386–404.
15. M. Brasier et al., "Ediacaran Sponge Spicule Clusters from Mongolia and the Origins of the Cambrian Fauna," *Geology* 25 (1997): 303–06.
16. J. Y. Chen et al., "Small Bilaterian Fossils from 40 to 55 Million Years before the Cambrian," *Science* 305, no. 5681 (2004): 218–22; A. H. Knoll et al. "Eukaryotic Organisms in Proterozoic Oceans," *Philosophical Transactions of the Royal Society* 361, no. 1470 (2006): 1023–38; B. Waggoner, "Interpreting the Earliest Metazoan Fossils: What Can We Learn?" *Integrative and Comparative Biology* 38, no. 6 (1998): 975–82.
17. A. Seilacher and F. Pflüger, "From Biomats to Benthic Agriculture: A Biohistoric Revolution," in W. E. Krumbein et al., eds., *Biostabilization of Sediments*. (Bibliotheks- und Informationssystem der Carl von Ossietzky Universität Odenburg, 1994), 97–105; A. Ivantsov, "Feeding Traces of the Ediacaran Animals," Abstract, 33rd International Geological Congress August 6–14, 2008, Oslo, Norway; S. Dornbos et al., "Evidence for Seafloor Microbial Mats and Associated Metazoan Lifestyles in Lower Cambrian Phosphorites of Southwest China," *Lethaia* 37, no. 2 (2004): 127–37.
18. The data from Svalbard are from A. C. Maloof et al., "Combined Paleomagnetic, Isotopic, and Stratigraphic Evidence for True Polar Wander from the

Neoproterozoic Akademikerbreen Group, Svalbard, Norway," *Geological Society of America Bulletin*, 118, nos. 9–10 (2006): 1099–124; the matching data from Central Australia are from N. L. Swanson-Hysell et al., "Constraints on Neoproterozoic Paleogeography and Paleozoic Orogenesis from Paleomagnetic Records of the Bitter Springs Formation, Amadeus Basin, Central Australia," *American Journal of Science* 312, no. 8 (2012): 817–84.

19. R. N. Mitchell, "True Polar Wander and Supercontinent Cycles: Implications for Lithospheric Elasticity and the Triaxial Earth," *American Journal of Science* 314, no. 5 (2014): 966–78.
20. J. Kirschvink, R. Ripperdan, D. Evans, "Evidence for Large Scale Reorganization of Early Cambrian Continental Masses by Inertial Interchange True Polar Wander," *Science* 277, no. 5325 (1997): 541–45.

CHAPTER VIII: THE CAMBRIAN EXPLOSION: 600–500 MA

1. Sadly, this great book is no longer required of college students. At the University of Washington we have tried to reverse that, requiring students enrolled in the course A New History of Life to read C. Darwin *On the Origin of Species by Natural Selection* (London: 1859).
2. A wonderful guide to the Cambrian as well as Darwin and the Burgess Shale is in the indispensable book by S. J. Gould, *Wonderful Life: The Burgess Shale and the Nature of History* (New York: W. W. Norton & Company, 1989). Steve, a friend to us both, was the greatest lecturer either of us has ever heard. His was a voice that had to be heard in person. His power as a lecturer came from his enormous intellect and mastery of both the science of evolution and Darwin, the English master. How that voice of reason, eloquence, and science is missed. If Huxley was Darwin's bulldog, Gould was his pit bull.
3. K. J. McNamara, "Dating the Origin of Animals," *Science* 274, no. 5295 (1996): 1993–97.
4. A. H. Knoll and S. B. Carroll, "Early Animal Evolution: Emerging Views from Comparative Biology and Geology," *Science* 284, no. 5423 (1999): 2129–371.
5. K. J. Peterson and N. J. Butterfield, "Origin of the Eumetazoa: Testing Ecological Predictions of Molecular Clocks Against the Proterozoic Fossil Record," *Proceedings of the National Academy of Sciences* 102, no. 27 (2005): 9547–52.
6. M. A. Fedonkin et al., *The Rise of Animals: Evolution and Diversification of the Kingdom Animalia* (Baltimore: Johns Hopkins University Press, 2007), 213–16.
7. It is hard to argue with the view that the Cambrian explosion is one of the—if not the—preeminent events in paleontology. However, those studying how life first arose view animals as latecomers of not much importance: that getting to life was the hard part, and animals were then foreordained once life arose. We remain split on this. There are many good papers of recent vintage dealing with this relative importance. Among them are: G. E Budd and J. Jensen, "A Critical Reappraisal of the Fossil Record of the Bilaterian Phyla," *Biological Reviews* 75, no. 2 (2000): 253–95; and S. J. Gould, *Wonderful Life*.

8. Oxygen levels in the Cambrian remain controversial. We continue to trust Bob Berner's work using his GEOCARBSULF models: R. A. Berner, "GEOCARBSULF: A Combined Model for Phanerozoic Atmospheric Oxygen and Carbon Dioxide," *Geochimica et Cosmochimica Acta* 70 (2006): 5653–64.
9. N. J. Butterfield, "Exceptional Fossil Preservation and the Cambrian Explosion," *Integrative and Comparative Biology* 43, no. 1 (2003): 166–77; S. C. Morris, "The Burgess Shale (Middle Cambrian) Fauna," *Annual Review of Ecology and Systematics* 10, no. 1 (1979): 327–49.
10. D. Briggs et al., *The Fossils of the Burgess Shale* (Washington, D.C.: Smithsonian Institution Press, 1994).
11. H. B. Whittington, Geological Survey of Canada, *The Burgess Shale* (New Haven: Yale University Press, 1985), 306–308.
12. J. W. Valentine, *On the Origin of Phyla* (Chicago: University of Chicago Press, 2004). See also J. W. Valentine and D. Erwin, *The Cambrian Explosion: The Construction of Animal Biodiversity* (Roberts and Co. Publishing, 2013). 413; J. W. Valentine, "Why No New Phyla after the Cambrian? Genome and Ecospace Hypotheses Revisited," abstract, *Palaios* 10, no. 2 (1995): 190–91. See also S. Bengtson, "Origins and Early Evolution of Predation" (free full text), in M. Kowalewski and P. H. Kelley, *The Fossil Record of Predation. The Paleontological Society Papers 8* (Paleontological Society, 2002): 289–317.
13. P. Ward, *Out of Thin Air* (Joseph Henry Press, 2006).
14. S. Carroll, *Endless Forms Most Beautiful: The New Science of Evo Devo and the Making of the Animal Kingdom* (New York: W. W. Norton & Company, 2004).
15. H. X. Guang et al., *The Cambrian Fossils of Chengjiang, China: The Flowering of Early Animal Life*. (Oxford: Blackwell Publishing, 2004).
16. This nasty fight between two very literate writers would have ended up on the dueling ground, with one or both dead, were it still the 1800s. Gould was nothing but appreciative and polite to Simon. The reverse did not hold. A good description can be found at www.stephenjaygould.org-library-naturalhistory_cambrian.html.
17. M. Brasier et al., "Decision on the Precambrian-Cambrian Boundary Stratotype," *Episodes* 17, nos. 1–2 (1994): 95–100.
18. W. Compston et al., "Zircon U-Pb Ages for the Early Cambrian Time Scale," *Journal of the Geological Society of London* 149 (1992): 171–84.
19. A. C. Maloof et al., "Constraints on Early Cambrian Carbon Cycling from the Duration of the Nemakit-Daldynian-Tommotian Boundary Delta C-13 Shift, Morocco," *Geology* 38, no. 7 (2010): 623–26.
20. M. Magaritz et al., "Carbon-Isotope Events Across the Precambrian-Cambrian Boundary on the Siberian Platform," *Nature* 320 (1986): 258–59.

CHAPTER IX: THE ORDOVICIAN-DEVONIAN EXPANSION OF ANIMALS: 500–360 MA

1. There is no better source to refer to about ancient reefs than our friend George Stanley of the University of Montana. A good place to start is his magnificent book: G. Stanley, *The History and Sedimentology of Ancient Reef Systems* (Springer Publishing, 2001). Another good source is E. Flügel in W. Kiessling,

E. Flügel, and J. Golonka, eds., *Phanerozoic Reef Patterns* 72 (SEPM Special Publications, 2002), 391–463.

2. Archaeocyathids are one of the most curious of all fossils. In the twentieth century they were thought to belong to no known phylum. Now they are put into the sponges. But they have a curious structure of a "cone in a cone"—as if one hollow ice-cream cone is stacked into a second. They are prominent in being the very first reef-forming organisms that we know of, as they formed three-dimensional wave-resistant structures built by organisms—our definition of a reef. F. Debrenne and J. Vacelet, "Archaeocyatha: Is the Sponge Model Consistent with Their Structural Organization?" *Palaeontographica Americana* 54 (1984): 358–69.
3. T. Servais et al., "The Ordovician Biodiversification: Revolution in the Oceanic Trophic Chain," *Lethaia* 41, no.2 (2008): 99.
4. P. Ward, *Out of Thin Air: Dinosaurs, Birds, and Earth's Ancient Atmosphere* (Washington, D.C.: Joseph Henry Press, 2006).
5. P. Ward, *Out of Thin Air*. Also see a magnificent summary by our colleague and coauthor on extinction, C. R. Marshall, "Explaining the Cambrian 'Explosion' of Animals," *Annual Review of Earth and Planetary Sciences* 34 (2006): 355–84.
6. J. Valentine, "How Many Marine Invertebrate Fossils?" *Journal of Paleontology* 44 (1970): 410–15; N. Newell, "Adequacy of the Fossil Record," *Journal of Paleontology* 33 (1959): 488–99.
7. D. M. Raup, "Taxonomic Diversity During the Phanerozoic," *Science* 177 (1972): 1065–71; D. Raup, "Species Diversity in the Phanerozoic: An Interpretation," *Paleobiology* 2 (1976): 289–97.
8. J. J. Sepkoski, Jr., "Ten Years in the Library: New Data Confirm Paleontological Patterns," *Paleobiology* 19 (1993): 246–57; J. J. Sepkoski, Jr., "A Compendium of Fossil Marine Animal Genera," *Bulletins of American Paleontology* 363: 1–560.
9. J. Alroy et al., "Effects of Sampling Standardization on Estimates of Phanerozoic Marine Diversification," *Proceedings of the National Academy of Sciences* 98 (2001): 6261–66.
10. J. Sepkoski, "Alpha, Beta, or Gamma; Where Does All the Diversity Go?" *Paleobiology* 14 (1988): 221–34.
11. J. Alroy et al., "Phanerozoic Diversity Trends," *Science* 321 (2008): 97.
12. A. B. Smith, "Large-Scale Heterogeneity of the Fossil Record: Implications for Phanerozoic Biodiversity Studies," *Philosophical Transactions of the Royal Society of London* 356, no. 1407 (2001): 351–67; A. B. Smith, "Phanerozoic Marine Diversity: Problems and Prospects," *Journal of the Geological Society, London* 164 (2007): 731–45; A. B. Smith and A. J. McGowan, "Cyclicity in the Fossil Record Mirrors Rock Outcrop Area," *Biology Letters* 1, no. 4 (2005): 443–45; A. B. Smith, "The Shape of the Marine Palaeodiversity Curve Using the Phanerozoic Sedimentary Rock Record of Western Europe," *Paleontology* 50 (2007): 765–74; A. McGowan and A. Smith. "Are Global Phanerozoic Marine Diversity Curves Truly Global? A Study of the Relationship between Regional Rock Records and Global Phanerozoic Marine Diversity," *Paleobiology*, 34, no. 1 (2008): 80–103.
13. M. J. Benton and B. C. Emerson, "How Did Life Become So Diverse? The Dynamics of Diversification According to the Fossil Record and Molecular Phylogenetics," *Palaeontology* 50 (2007): 23–40.

14. S. E. Peters, "Geological Constraints on the Macroevolutionary History of Marine Animals," *Proceedings of the National Academy of Sciences* 102 (2005): 12326–31.
15. This is one of our favorite "Emperor Has No Clothes" moments in paleontology. A team from University of Kansas hypothesized that the Ordovician could have been caused by in intense gamma-ray burst from deep space. Such events, in which enormous energy pours out of small but energetic stars such as a pulsar or magnetar at galactic distances, are real enough. But the suggestions that one such gamma-ray burst (GRB) fried the Earth, causing the Ordovician mass extinction, is just fanciful. There is not a shred of evidence connecting a GRB to the Ordovician mass extinction. It could as easily have been caused by Vulcans or Darth Vader on a bad day (but were there any other kinds for poor Vader?). See A. L. Melott and B. C. Thomas, "Late Ordovician Geographic Patterns of Extinction Compared with Simulations of Astrophysical Ionizing Radiation Damage," *Paleobiology* 35 (2009): 311–20. Also see www.nasa.gov-vision-universe-starsgalaxies-gammaray_extinction.html.
16. R. K. Bambach et al., "Origination, Extinction, and Mass Depletions of Marine Diversity," *Paleobiology* 30, no. 4 (2004): 522–42.
17. S. A. Young et al., "A Major Drop in Seawater 87Sr-86Sr during the Middle Ordovician (Darriwilian): Links to Volcanism and Climate?" *Geology* 37, 10 (2009): 951–54.
18. S. Finnegan et al., "The Magnitude and Duration of Late Ordovician-Early Silurian Glaciation," *Science* 331, no. 6019 (2011): 903–906.
19. S. Finnegan et al., "Climate Change and the Selective Signature of the Late Ordovician Mass Extinction," *Proceedings of the National Academy of Sciences* 109, no. 18 (2012): 6829–34.

CHAPTER X: *TIKTAALIK* AND THE INVASION OF THE LAND: 475–300 MA

1. For a nice summary of these early tetrapods and their evolutionary positions, try this website: www.devoniantimes.org-opportunity-tetrapodsAnswer.html, and S. E. Pierce et al., "Three-Dimensional Limb Joint Mobility in the Early Tetrapod *Ichthyostega*," *Nature* 486 (2012): 524–27, and P. E. Ahlberg et al., "The Axial Skeleton of the Devonian Tetrapod *Ichthyostega*," *Nature* 437, no. 1 (2005): 137–40.
2. J. A. Clack, *Gaining Ground: The Origin and Early Evolution of Tetrapods*, 2nd ed. (Bloomington: Indiana University Press, 2012).
3. E. B. Daeschler et al., "A Devonian Tetrapod-Like Fish and the Evolution of the Tetrapod Body Plan," *Nature* 440, no. 7085 (2006): 757–63; J. P. Downs et al., "The Cranial Endoskeleton of *Tiktaalik roseae*," *Nature* 455 (2008): 925–29; and a summary: P. E. Ahlberg and J. A. Clack, "A Firm Step from Water to Land," *Nature* 440 (2006): 747–49.
4. N. Shubin, *Your Inner Fish: A Journey into the 3.5-Billion-Year History of the Human Body* (Chicago: University of Chicago Press, 2008); B. Holmes, "Meet Your Ancestor, the Fish That Crawled," *New Scientist*, September 9, 2006.

5. A. K. Behrensmeyer et al., eds., *Terrestrial Ecosystems Through Time: Evolutionary Paleoecology of Terrestrial Plants and Animals* (Chicago and London: University of Chicago Press, 1992); P. Kenrick and P. R. Crane, *The Origin and Early Diversification of Land Plants. A Cladistic Study* (Washington: Smithsonian Institution Press, 1997).
6. S. B. Hedges, "Molecular Evidence for Early Colonization of Land by Fungi and Plants," *Science* 293 (2001): 1129–33.
7. C. V. Rubenstein et al., "Early Middle Ordovician Evidence for Land Plants in Argentina (Eastern Gondwana)," *New Phytologist* 188, no. 2 (2010): 365–69. The press report can be found at www.dailymail.co.uk-sciencetech-article-1319904-Fossils-worlds-oldest-plants-unearthed-Argentina.html.
8. J. T. Clarke et al., "Establishing a Time-Scale for Plant Evolution," *New Phytologist* 192, no. 1 (2011): 266–30; M. E. Kotyk et al., "Morphologically Complex Plant Macrofossils from the Late Silurian of Arctic Canada," *American Journal of Botany* 89 (2002): 1004–1013.
9. Our own work on the insect and vertebrate invasions can be found in P. Ward et al., "Confirmation of Romer's Gap as a Low Oxygen Interval Constraining the Timing of Initial Arthropod and Vertebrate Terrestrialization," *Proceedings of the National Academy of Sciences* 10, no. 45 (2006): 16818–22.

CHAPTER XI: THE AGE OF ARTHROPODS: 350–300 MA

1. Our own work on the insect and vertebrate invasions can be found in P. Ward et al., "Confirmation of Romer's Gap as a Low Oxygen Interval Constraining the Timing of Initial Arthropod and Vertebrate Terrestrialization," *Proceedings of the National Academy of Sciences* 10, no. 45 (2006): 16818–22.
2. R. Dudley, "Atmospheric Oxygen, Giant Paleozoic Insects and the Evolution of Aerial Locomotor Performance," *The Journal of Experimental Biology* 201 (1988): 1043–50; R. Dudley, *The Biomechanics of Insect Flight: Form, Function, Evolution* (Princeton: Princeton University Press, 2000); R. Dudley and P. Chai, "Animal Flight Mechanics in Physically Variable Gas Mixtures," *The Journal of Experimental Biology* 199 (1996): 1881–85; also C. Gans et al., "Late Paleozoic Atmospheres and Biotic Evolution," *Historical Biology* 13 (1991): 199–219l; J. Graham et al., "Implications of the Late Palaeozoic Oxygen Pulse for Physiology and Evolution," *Nature* 375 (1995): 117–20; J. F. Harrison et al., "Atmospheric Oxygen Level and the Evolution of Insect Body Size," *Proceedings of the Royal Society B-Biological Sciences* 277 (2010): 1937–46.
3. D. Flouday et al., "The Paleozoic Origin of Enzymatic Lignin Decomposition Reconstructed from 31 Fungal Genomes," *Science* 336, no. 6089 (2012): 1715-19.
4. Ibid.
5. J. A. Raven, "Plant Responses to High O_2 Concentrations: Relevance to Previous High O_2 Episodes," *Global and Planetary Change* 97 (1991): 19–38; and J. A. Raven et al., "The Influence of Natural and Experimental High O_2 Concentrations on O_2-Evolving Phototrophs," *Biological Reviews* 69 (1994): 61–94.

6. J. S. Clark et al., *Sediment Records of Biomass Burning and Global Change* (Berlin: Springer-Verlag, 1997); M. J. Cope et al., "Fossil Charcoals as Evidence of Past Atmospheric Composition," *Nature* 283 (1980): 647–49; C. M. Belcher et al., "Baseline Intrinsic Flammability of Earth's Ecosystems Estimated from Paleoatmospheric Oxygen over the Past 350 Million Years," *Proceedings of the National Academy of Sciences* 107, no. 52 (2010): 22448–53. Our own take on these experiments is that they are flawed by their failing to test using higher ignition temperatures. Even in low oxygen, a lightning strike causes initial ignition temperatures far higher than those used in this study.
7. D. Beerling, *The Emerald Planet: How Plants Changed Earth's History* (New York: Oxford University Press, 2007).
8. Q. Cai et al., "The Genome Sequence of the Ground Tit *Pseudopodoces humilis* Provides Insights into Its Adaptation to High Altitude," *Genome Biology* 14, no. 3 (2013); www.geo.umass.edu-climate-quelccaya-diuca.html, and P. Ward, *Out of Thin Air: Dinosaurs, Birds, and Earth's Ancient Atmosphere* (Washington, D.C.: Joseph Henry Press, 2006), with references therein to high altitude nesting.
9. P. Ward, *Out of Thin Air*.
10. M. Laurin and R. R. Reisz, "A Reevaluation of Early Amniote Phylogeny," *Zoological Journal of the Linnean Society* 113, no. 2 (1995): 165–223.
11. P. Ward, *Out of Thin Air*.

CHAPTER XII: THE GREAT DYING—ANOXIA AND GLOBAL STAGNATION: 252–250 MA

1. C. Sidor et al., "Permian Tetrapods from the Sahara Show Climate-Controlled Endemism in Pangaea," *Nature* 434 (2012): 886–89; S. Sahney and M. J. Benton, "Recovery from the Most Profound Mass Extinction of All Time," *Proceedings of the Royal Society, Series B* 275 (2008): 759–65.
2. The invertebrate fauna from Meishan, China, is proving to be the best-studied marine fossil record of this catastrophic event. There is now a large literature on this: S.-Z. Shen et al., "Calibrating the End-Permian Mass Extinction," *Science* 334, no. 6061 (2011): 1367–72; Y. G. Jin et al., "Pattern of Marine Mass Extinction Near the Permian–Triassic Boundary in South China," *Science* 289, no. 5478 (2000): 432–36.
3. C. R. Marshall, "Confidence Limits in Stratigraphy," in D. E. G. Briggs and P. R. Crowther, eds., *Paleobiology II* (Oxford: Blackwell Scientific, 2001), 542–45; see also the newer work by our Adelaide colleagues, C. J. A. Bradshaw et al., "Robust Estimates of Extinction Time in the Geological Record," *Quaternary Science Reviews* 33 (2011): 14–19.
4. "End-Permian Extinction Happened in 60,000 Years—Much Faster than Earlier Estimates, Study Says," Phys.org, February 10, 2014. S. D. Burgess et al., "High-Precision Timeline for Earth's Most Severe Extinction," *Proceedings of the National Academy of Sciences* 111, no. 9 (2014): 3316–21.
5. L. Becker et al., "Impact Event at the Permian–Triassic Boundary: Evidence from Extraterrestrial Noble Gases in Fullerenes," *Science* 291 (2001): 1530–33.

6. L. Becker et al., "Bedout: A Possible End-Permian Impact Crater Offshore of Northwestern Australia," *Science* 304 (2004): 1469–76.
7. K. Grice et al., "Photic Zone Euxinia During the Permian-Triassic Superanoxic Event," *Science* 307 (2005): 706–09.
8. C. Cao et al., "Biogeochemical Evidence for Euxinic Oceans and Ecological Disturbance Presaging the End-Permian Mass Extinction Event," *Earth and Planetary Science Letters* 281 (2009): 188–201.
9. L. R. Kump and M. A. Arthur, "Interpreting Carbon-Isotope Excursions: Carbonates and Organic Matter," *Chemical Geology* 161 (1999): 181–98.
10. K. M. Meyer and L. R. Kump, "Oceanic Euxinia in Earth History: Causes and Consequences," *Annual Review of Earth and Planetary Sciences* 36 (2008): 251–88.
11. T. J. Algeo and E. D. Ingall, "Sedimentary C_{org}:P Ratios, Paleoceanography, Ventilation, and Phanerozoic Atmospheric pO_2," *Palaeogeography, Palaeoclimatology, Palaeoecology* 256 (2007): 130–55; C. Winguth and A. M. E. Winguth, "Simulating Permian-Triassic Oceanic Anoxia Distribution: Implications for Species Extinction and Recovery," *Geology* 40 (2012): 127–30; S. Xie et al., "Changes in the Global Carbon Cycle Occurred as Two Episodes during the Permian-Triassic Crisis," *Geology* 35 (2007): 1083–86; S. Xie et al., "Two Episodes of Microbial Change Coupled with Permo-Triassic Faunal Mass Extinction," *Nature* 434 (2005): 494–97; G. Luo et al., "Stepwise and Large-Magnitude Negative Shift in $d^{13}C_{carb}$ Preceded the Main Marine Mass Extinction of the Permian-Triassic Crisis Interval," *Palaeogeography, Palaeoclimatology, Palaeoecology* 299 (2011): 70–82; G. A. Brennecka et al., "Rapid Expansion of Oceanic Anoxia Immediately before the End-Permian Mass Extinction," *Proceedings of the National Academy of Sciences* 108 (2011): 17631–34.
12. P. Ward et al., "Abrupt and Gradual Extinction Among Late Permian Land Vertebrates in the Karoo Basin, South Africa," *Science* 307 (2005): 709–14; C. Sidor et al., "Permian Tetrapods from the Sahara Show Climate-Controlled Endemism in Pangaea"; and S. Sahney and M. J. Benton, "Recovery from the Most Profound Mass Extinction of All Time."
13. R. B. Huey and P. D. Ward, "Hypoxia, Global Warming, and Terrestrial Late Permian Extinctions," *Science*, 308, no. 5720 (2005): 398–401.
14. P. Ward et al., "Abrupt and Gradual Extinction Among Late Permian Land Vertebrates in the Karoo Basin, South Africa."

CHAPTER XIII: THE TRIASSIC EXPLOSION: 252–200 MA

1. The high heat in the lowest Triassic strata is a major confirmation of the greenhouse extinction model.
2. S. Schoepfer et al., "Cessation of a Productive Coastal Upwelling System in the Panthalassic Ocean at the Permian–Triassic Boundary," *Palaeogeography, Palaeoclimatology, Palaeoecology* 313–14 (2012): 181–88.
3. The history of reefs was looked at in our chapter on the Ordovician. George Stanley remains the primary expertise. G. D. Stanley Jr., ed., *Paleobiology and Biology of Corals*, Paleontological Society Papers, vol. 1 (Boulder, CO: The Paleontological Society, 1996), and a very accessible work on many aspects of

modern as well as ancient reefs: G. Stanley Jr., "Corals and Reefs: Crises, Collapse and Change," presented as a Paleontological Society short course at the annual meeting of the Geological Society of America, Minneapolis, MN, October 8, 2011.

4. P. C. Sereno, "The Origin and Evolution of Dinosaurs," *Annual Review of Earth and Planetary Sciences* 25 (1997): 435–89; P. C. Sereno et al., "Primitive Dinosaur Skeleton from Argentina and the Early Evolution of Dinosauria," *Nature* 361 (1993): 64–66; P. C. Sereno and A. B. Arcucci, "Dinosaurian Precursors from the Middle Triassic of Argentina: *Lagerpeton chanarensis*," *Journal of Vertebrate Paleontology* 13 (1994): 385–99. Other important works on early dinosaur and other vertebrate evolution: M. J. Benton, "Dinosaur Success in the Triassic: A Noncompetitive Ecological Model," *Quarterly Review of Biology* 58 (1983): 29–55; M. J. Benton, "The Origin of the Dinosaurs," in C. A.-P. Salense, ed., *III Jornadas Internacionales sobre Paleontología de Dinosaurios y su Entorno* (Burgos, Spain: Salas de los Infantes, 2006), 11–19; A. P. Hunt et al., "Late Triassic Dinosaurs from the Western United States," *Geobios* 31 (1998): 511–31; R. B. Irmis et al., "A Late Triassic Dinosauromorph Assemblage from New Mexico and the Rise of Dinosaurs," *Science* 317 (2007): 358–61; R. B. Irmis et al., "Early Ornithischian Dinosaurs: The Triassic Record," *Historical Biology* 19 (2007):, 3–22; S. J. Nesbitt et al., "A Critical Re-evaluation of the Late Triassic Dinosaur Taxa of North America," *Journal of Systematic Palaeontology* 5 (2007): 209–43; S. J. Nesbitt et al., "Ecologically Distinct Dinosaurian Sister Group Shows Early Diversification of Ornithodira," *Nature* 464 (2010): 95–98.
5. D. R. Carrier, "The Evolution of Locomotor Stamina in Tetrapods: Circumventing a Mechanical Constraint," *Paleobiology* 13 (1987): 326–41.
6. E. Schachner, R. Cieri, J. Butler, G. Farmer, "Unidirectional Pulmonary Airflow Patterns in the Savannah Monitor Lizard," *Nature* 506, no. 7488 (2013): 367–70.
7. A. F. Bennett, "Exercise Performance of Reptiles," in J. H. Jones et al., eds., *Comparative Vertebrate Exercise Physiology: Phyletic Adaptations*, Advances in Veterinary Science and Comparative Medicine, vol. 3 (New York: Academic Press, 1994), 113–38.
8. N. Bardet, "Stratigraphic Evidence for the Extinction of the Ichthyosaurs," *Terra Nova* 4 (1992): 649–56. See also C. W. A. Andrews, *A Descriptive Catalogue of the Marine Reptiles of the Oxford Clay. Based on the Leeds Collection in the British Museum (Natural History), London.* Part II (London: 1910): 1–205, as well as the wonderful new summary by R. Motani, "The Evolution of Marine Reptiles," *Evolution: Education and Outreach* 2, no. 2 (2009): 224–35.
9. P. Ward et al., "Sudden Productivity Collapse Associated with the Triassic-Jurassic Boundary Mass Extinction," *Science* 292 (2001): 115–19; P. Ward et al., "Isotopic Evidence Bearing on Late Triassic Extinction Events, Queen Charlotte Islands, British Columbia, and Implications for the Duration and Cause of the Triassic-Jurassic Mass Extinction," *Earth and Planetary Science Letters* 224, nos. 3–4: 589–600. Our later work in Nevada and back in the Queen Charlottes expanded on this isotopic anomaly. K. H. Williford et al., "An Extended Stable Organic Carbon Isotope Record Across the Triassic-Jurassic Boundary in the Queen Charlotte Islands, British Columbia, Canada," *Palaeogeography, Palaeoclimatology, Palaeoecology* 244, nos. 1–4 (2006): 290–96.

10. P. E. Olsen et al., "Ascent of Dinosaurs Linked to an Iridium Anomaly at the Triassic-Jurassic Boundary," *Science* 296, no. 5571 (2002): 1305–07.
11. J. P. Hodych and G. R. Dunning, "Did the Manicougan Impact Trigger End-of-Triassic Mass Extinction?" *Geology* 20, no. 1 (1992): 51–54; L. H. Tanner et al., "Assessing the Record and Causes of Late Triassic Extinctions," *Earth-Science Reviews* 65, nos. 1–2 (2004): 103–39; J. H. Whiteside et al., "Compound-Specific Carbon Isotopes from Earth's Largest Flood Basalt Eruptions Directly Linked to the End-Triassic Mass Extinction," *Proceedings of the National Academy of Sciences* 107, no. 15 (2010): 6721–25; M. H. L. Deenen et al., "A New Chronology for the End-Triassic Mass Extinction," *Earth and Planetary Science Letters* 291, no. 1–4 (2010): 113–25.

CHAPTER XIV: DINOSAUR HEGEMONY IN A LOW-OXYGEN WORLD: 230–180 MA

1. And just as we pay homage to Bob Bakker, no student of the dinosaurs can do without the magnificent *The Dinosauria* by D. B. Weishampel et al., (Oakland: University of California Press, 2004). Heavy, hefty, and expensive, it is the definitive treatise still in 2014.
2. There is now an extensive literature on air sacs in dinosaurs. Bob Bakker was the first to point it out, and the work of Gregory Paul greatly expanded on this hypothesis.
3. D. Fastovsky and D. Weishampel, *The Evolution and Extinction of the Dinosaurs* (Cambridge: Cambridge University Press: 2005).
4. P. O'Connor and L. Claessens, "Basic Avian Pulmonary Design and Flow-Through Ventilation in Non-Avian Theropod Dinosaurs," *Nature* 436, no. 7048 (2005): 253–56, but see the contrary view of J. A. Ruben et al., "Pulmonary Function and Metabolic Physiology of Theropod Dinosaurs," *Science* 283, no. 5401 (1999): 514–16.
5. W. J. Hillenius and J. A. Ruben, "The Evolution of Endothermy in Terrestrial Vertebrates: Who? When? Why?" *Physiological and Biochemical Zoology* 77, no. 6 (2004): 1019–1042. The work of Greg Erickson is also essential: G. M. Erickson et al., "Tyrannosaur Life Tables: An Example of Nonavian Dinosaur Population Biology," *Science* 313, no. 5784 (2006): 213–17; whereas the important career work of de Ricqlès is summarized in A. de Ricqlès et al., "On the Origin of High Growth Rates in Archosaurs and their Ancient Relatives: Complementary Histological Studies on Triassic Archosauriforms and the Problem of a 'Phylogenetic Signal' in Bone Histology," *Annales de Paléontologie* 94, no. 2 (2008): 57.
6. K. Carpenter, *Eggs, Nests, and Baby Dinosaurs: A Look at Dinosaur Reproduction* (Bloomington: Indiana University Press, 2000).

CHAPTER XV: THE GREENHOUSE OCEANS: 200–65 MA

1. R. Takashima, "Greenhouse World and the Mesozoic Ocean," *Oceanography* 19, no. 4 (2006): 82–92.
2. A. S. Gale, "The Cretaceous World," in S. J. Culver and P. F. Raqson, eds., *Biotic Response to Global Change: The Last 145 Million Years* (Cambridge: Cambridge University Press, 2006), 4–19.

3. T. J. Bralower et al., "Dysoxic-Anoxic Episodes in the Aptian-Albian (Early Cretaceous)," in *The Mesozoic Pacific: Geology, Tectonics and Volcanism*, M. S. Pringle et al., eds. (Washington, D.C.: American Geophysical Union, 1993), 5–37.
4. B. T. Huber et al., "Deep-Sea Paleotemperature Record of Extreme Warmth During the Cretaceous," *Geology* 30 (2002): 123–26; A. H. Jahren, "The Biogeochemical Consequences of the Mid-Cretaceous Superplume," *Journal of Geodynamics* 34 (2002): 177–91; I. Jarvis et al., "Microfossil Assemblages and the Cenomanian-Turonian (Late Cretaceous) Oceanic Anoxic Event," *Cretaceous Research* 9 (1988): 3–103. The work on heteromorphic ammonites including buoyancy has been conducted by Ward and many colleagues around the world. *Ammonoid Paleobiology*, Neil Landman et al., eds. (Springer, 1996), is an excellent introduction. The orientation of *Baculites* was ascertained using scale wax models, in P. Ward, Ph.D. thesis, McMaster University, Ontario Canada, 1976.
5. The wonderful study (one of many!) by Neil Landman and his colleagues was discussed in N. H. Landman et al., "Methane Seeps as Ammonite Habitats in the U.S. Western Interior Seaway Revealed by Isotopic Analyses of Well-preserved Shell Material," *Geology* 40, no. 6 (2012): 507. Other new findings by this group were reported in N. H. Landman et al., "The Role of Ammonites in the Mesozoic Marine Food Web Revealed by Jaw Preservation," *Science* 331, no. 6013 (2011): 70–72, showing for the first time the feeding mechanisms of baculitid ammonites as well as insight into their food sources.
6. Ibid.
7. G. J. Vermeij, "The Mesozoic Marine Revolution: Evidence from Snails, Predators and Grazers," *Palaeobiology* 3 (1977): 245–58.
8. S. M. Stanley, "Predation Defeats Competition on the Seafloor," *Paleobiology* 34, no. 1 (2008): 1–21.
9. T. Baumiller et al., "Post-Paleozoic Crinoid Radiation in Response to Benthic Predation Preceded the Mesozoic Marine Revolution," *Proceedings of the National Academy of Sciences of the United States of America* 107, no. 13 (2010): 5893–96.
10. T. Oji, "Is Predation Intensity Reduced with Increasing Depth? Evidence from the West Atlantic Stalked Crinoid Endoxocrinus parrae (Gervais) and Implications for the Mesozoic Marine Revolution," *Palaeobiology* 22 (1996): 339–51.

CHAPTER XVI: DEATH OF THE DINOSAURS: 65 MA

1. L. W. Alvarez et al., "Extraterrestrial Cause for the Cretaceous-Tertiary Extinction," *Science* 208, no. 4448 (1980): 1095. This was later followed by the discovery of the crater itself: A. R. Hildebrand et al., "Chicxulub Crater: A Possible Cretaceous-Tertiary Boundary Impact Crater on the Yucatán Peninsula, Mexico," *Geology* 19 (1991): 867–71.
2. P. Schulte et al. "The Chicxulub Asteroid Impact and Mass Extinction at the Cretaceous-Paleogene Boundary," *Science* 327, no. 5970 (2005): 1214–18.
3. J. Vellekoop et al., "Rapid Short-Term Cooling Following the Chicxulub Impact at the Cretaceous-Paleogene Boundary," *Proceedings of the National Academy of Sciences* 111, no 21 (2014): 7537–7541.

4. Discussions of this site and the extinction pattern recorded there are in many references, but we rather presumptuously suggest P. Ward, *Under a Green Sky: Global Warming, the Mass Extinctions of the Past, and What They Can Tell Us About Our Future* (Washington, D.C.: Smithsonian, 2007).
5. See also the excellent review by our colleague David Jablonski: D. Jablonski, "Extinctions in the Fossil Record (and Discussion)," *Philosophical Transactions of the Royal Society of London, Series B* 344, 1307 (1994): 11–17.
6. D. M. Raup and D. Jablonski, "Geography of End-Cretaceous Marine Bivalve Extinctions," *Science* 260, 5110 (1993): 971–73. P. M. Sheehan and D. E. Fastovsky, "Major Extinctions of Land-Dwelling Vertebrates at the Cretaceous-Tertiary Boundary, Eastern Montana," *Geology* 20 (1992): 556–60; R. K. Bambach et al., "Origination, Extinction, and Mass Depletions of Marine Diversity," *Paleobiology* 30, no. 4 (2004): 522–42. D. J. Nichols and K. R. Johnson, *Plants and the K–T Boundary* (Cambridge: Cambridge University Press, 2008); P. Ward et al., "Ammonite and Inoceramid Bivalve Extinction Patterns in Cretaceous-Tertiary Boundary Sections of the Biscay Region (Southwestern France, Northern Spain)," *Geology* 19, no. 12 (1991): 1181–84; but see the dissenting N. MacLeod et al., "The Cretaceous-Tertiary Biotic Transition," *Journal of the Geological Society* 154, no. 2 (1997): 265–92.

   Also see P. Shulte et al., "The Chicxulub Asteroid Impact and Mass Extinction at the Cretaceous-Paleogene Boundary," *Science* 327, no. 5970 (2010): 1214–18.
7. V. Courtillot et al., "Deccan Flood Basalts at the Cretaceous-Tertiary Boundary?" *Earth and Planetary Science Letters* 80, nos. 3–4 (1986): 361–74; C. Moskowitz, "New Dino-Destroying Theory Fuels Hot Debate," space.com, October 18, 2009.
8. T. S. Tobin et al., "Extinction Patterns, $\delta^{18}O$ Trends, and Magnetostratigraphy from a Southern High-Latitude Cretaceous-Paleogene Section: Links with Deccan Volcanism," *Palaeogeography, Palaeoclimatology, Palaeoecology* 350–52 (2012): 180–88.

CHAPTER XVII: THE LONG-DELAYED THIRD AGE OF MAMMALS: 65–50 MA

1. The gold standard for vertebrate paleontology has long been Robert L. Carroll, *Vertebrate Paleontology and Evolution* (New York: W. H. Freeman and Company, 1988). New work on the evolution of what we call the third age of mammals in this book can be found in O. R. P. Bininda-Emonds et al. "The Delayed Rise of Present-Day Mammals," *Nature* 446, no. 7135 (2007): 507–11; Z.-X. Luo et al., "A New Mammaliaform from the Early Jurassic and Evolution of Mammalian Characteristics," *Science* 292, 5521 (2001): 1535–40.
2. J. R. Wible et al., "Cretaceous Eutherians and Laurasian Origin for Placental Mammals Near the K-T Boundary," *Nature* 447, no. 7147 (2007): 1003–6; M. S. Springer et al., "Placental Mammal Diversification and the Cretaceous–Tertiary Boundary," *Proceedings of the National Academy of Sciences* 100, no. 3 (2002): 1056–61.
3. K. Helgen, "The Mammal Family Tree," *Science* 334, no. 6055 (2011): 458–59.

4. Q. Ji et al., "The Earliest Known Eutherian Mammal," *Nature* 416, no. 6883 (2002): 816–22.
5. Z.-X. Luo et al., "A Jurassic Eutherian Mammal and Divergence of Marsupials and Placentals," *Nature* 476, no. 7361 (2011): 442–45.
6. K. Jiang, "Fossil Indicates Hairy, Squirrel-sized Creature Was Not Quite a Mammal," UChicagoNews, August 7, 2013; C-F. Zhou, "A Jurassic Mammaliaform and the Earliest Mammalian Evolutionary Adaptations," *Nature* 500 (2013): 163–67.
7. Z.-X. Luo, "Transformation and Diversification in Early Mammal Evolution," *Nature* 450, no. 7172 (2007): 1011–19.
8. J. P. Kennett and L. D. Stott, "Abrupt Deep-Sea Warming, Palaeoceanographic Changes and Benthic Extinctions at the End of the Paleocene," *Nature* 353 (1991): 225–29.
9. U. Röhl et al., "New Chronology for the Late Paleocene Thermal Maximum and Its Environmental Implications," *Geology* 28, no. 10 (2000): 927–30; T. Westerhold et al., "New Chronology for the Late Paleocene Thermal Maximum and Its Environmental Implications," *Palaeogeography, Paleoclimatology, Palaeoecology* 257 (2008): 377–74.
10. P. L. Koch et al., "Correlation Between Isotope Records in Marine and Continental Carbon Reservoirs Near the Palaeocene-Eocene Boundary," *Nature* 358 (1992): 319–22.
11. M. D. Hatch, "C(4) Photosynthesis: Discovery and Resolution," *Photosynthesis Research* 73, nos. 1–3 (2002): 251–56.
12. E. J. Edwards and S. A. Smith, "Phylogenetic Analyses Reveal the Shady History of C_4 Grasses," *Proceedings of the National Academy of Sciences* 107, nos. 6 (2010): 2532–37; C. P. Osborne and R. P. Freckleton, "Ecological Selection Pressures for C_4 Photosynthesis in the Grasses," *Proceedings of the Royal Society B-Biological Sciences* 276, no. 1663 (2009): 1753–60.

CHAPTER XVIII: THE AGE OF BIRDS: 50–2.5 MA

1. A personal note to this chapter. One of us (Ward) has had two parrots as "pets," although it is unclear who was the pet in the relationship between bird and human. What was clear, however, was the level of intelligence. And this is true not just of parrots. Anyone watching crows or other flocking birds can readily see a great and potentially evolving intelligence at work. We dismiss them as "bird brains." Compare the size of a brain of an African gray parrot to that of our own, and then consider that these birds can speak in complete sentences, do math, are complex in behavior. We all want to hope the chickens we eat every day are stupid. Perhaps not.
2. K. Padian and L. M. Chiappe, "Bird Origins," in P. J. Currie and K.Padian, eds., *Encyclopedia of Dinosaurs* (San Diego: Academic Press, 1997), 41–96; J. Gauthier, "Saurischian Monophyly and the Origin of Birds," in K. Padian, *Memoirs of the California Academy of Sciences* 8 (1986): 1–55; L. M. Chiappe, "Downsized Dinosaurs: The Evolutionary Transition to Modern Birds," *Evolution: Education and Outreach* 2, no. 2 (2009): 248–56.
3. J. H. Ostrom, "The Ancestry of Birds," *Nature* 242, no. 5393 (1973): 136;

J. Gauthier, "Saurischian Monophyly and the Origin of Birds," in K. Padian, *Memoirs of the California Academy of Sciences* 8 (1986): 1–55; J. Cracraft, "The Major Clades of Birds," in M. J. Benton, ed., *The Phylogeny and Classification of the Tetrapods, Volume I: Amphibians, Reptiles, Birds* (Oxford: Clarendon Press, 1988), 339–61.

4. A. Feduccia, "On Why the Dinosaur Lacked Feathers," in M. K. Hecht et al., eds. *The Beginnings of Birds: Proceedings of the International* Archaeopteryx *Conference Eichstatt 1984* (Eichstatt: Freunde des Jura-Museums Eichstatt, 1985), 75–79; A. Feduccia et al., "Do Feathered Dinosaurs Exist? Testing the Hypothesis on Neontological and Paleontological Evidence," *Journal of Morphology* 266, no. 2 (2005): 125–66.
5. J. O'Connor, "A Revised Look at Liaoningornis Longidigitris (Aves)." *Vertebrata PalAsiatica* 50 (2012): 25–37.
6. A. Feduccia, "Explosive Evolution in Tertiary Birds and Mammals," *Science* 267, no. 5198 (1995): 637–38; A. Feduccia, "Big Bang for Tertiary Birds?" *Trends in Ecology and Evolution* 18, no. 4 (2003): 172–76.
7. M. Norell and M. Ellison, *Unearthing the Dragon: The Great Feathered Dinosaur Discovery* (New York: Pi Press, 2005); R. Prum, "Are Current Critiques of the Theropod Origin of Birds Science? Rebuttal to Feduccia 2002," *Auk* 120, no. 2(2003): 550–61; S. Hope, "The Mesozoic Radiation of Neornithes," in L. M. Chiappe et al., *Mesozoic Birds: Above the Heads of Dinosaurs* (Oakland: University of California Press, 2002), 339–88; P. Ericson et al., "Diversification of Neoaves: Integration of Molecular Sequence Data and Fossils," *Biology Letters* 2, no. 4 (2006): 543–47; K. Padian, "*The Origin and Evolution of Birds* by Alan Feduccia (Yale University Press, 1996)," *American Scientist* 85: 178–81; M. A. Norell et al., "Flight from Reason. Review of: *The Origin and Evolution of Birds* by Alan Feduccia (Yale University Press, 1996)," *Nature* 384, no. 6606 (1997): 230; L. M. Witmer, "The Debate on Avian Ancestry: Phylogeny, Function, and Fossils," in L. M. Chiappe and L. M. Witmer, eds., *Mesozoic Birds: Above the Heads of Dinosaurs* (Berkeley: University of California Press, 2002), 3–30.
8. C. Pei-ji et al., "An Exceptionally Preserved Theropod Dinosaur from the Yixian Formation of China," *Nature* 391, no. 6663 (1998): 147–52; G. S. Paul, *Dinosaurs of the Air: The Evolution and Loss of Flight in Dinosaurs and Birds* (Baltimore: Johns Hopkins University Press, 2002), 472; X. Xu et al., "An *Archaeopteryx*-like Theropod from China and the Origin of Avialae," *Nature* 475 (2011): 465–70.
9. D. Hu et al., "A Pre-*Archaeopteryx* Troodontid Theropod from China with Long Feathers on the Metatarsus," *Nature* 461, no. 7264 (2009): 640–43; A. H. Turner et al., "A Basal Dromaeosaurid and Size Evolution Preceding Avian Flight," *Science* 317, no. 5843 (2007): 1378–81; X. Xu et al., "Basal Tyrannosauroids from China and Evidence for Protofeathers in Tyrannosauroids," *Nature* 431, 7009 (2004): 680–84; C. Foth, "On the Identification of Feather Structures in Stem-Line Representatives of Birds: Evidence from Fossils and Actuopalaeontology," *Paläontologische Zeitschrift* 86, no. 1 (2012): 91–102; R. Prum and A. H. Brush, "The Evolutionary Origin and Diversification of Feathers," *Quarterly Review of Biology* 77, no. 3 (2002): 261–95.

10. M. H. Schweitzer et al., "Soft-Tissue Vessels and Cellular Preservation in *Tyrannosaurus rex*," *Science* 307, no. 5717 (2005); C. Dal Sasso and M. Signore, "Exceptional Soft-Tissue Preservation in a Theropod Dinosaur from Italy," *Nature* 392, no. 6674 (1998): 383–87; M. H. Schweitzer et al., "Heme Compounds in Dinosaur Trabecular Bone," *Proceedings of the National Academy of Sciences of the United States of America* 94, no. 12 (1997): 6291–96.
11. Dr. Paul Willis, "Dinosaurs and Birds: The Story," The Slab, http://www.abc.net.au/science/slab/dinobird/story.htm.
12. J. A. Clarke et al., "Insight into the Evolution of Avian Flight from a New Clade of Early Cretaceous Ornithurines from China and the Morphology of *Yixianornis grabaui*," *Journal of Anatomy* 208 (3 (2006): 287–308.
13. N. Brocklehurst et al., "The Completeness of the Fossil Record of Mesozoic Birds: Implications for Early Avian Evolution," *PLOS One* (2012); J. A. Clarke et al., "Definitive Fossil Evidence for the Extant Avian Radiation in the Cretaceous," *Nature* 433 (2005): 305–8.
14. L. Witmer, "The Debate on Avian Ancestry: Phylogeny, Function and Fossils," in L. Chiappe et al., eds., *Mesozoic Birds: Above the Heads of Dinosaurs* (Berkeley, California: University of California Press, 2002), 3–30; L. M. Chiappe and G. J. Dyke, "The Mesozoic Radiation of Birds," *Annual Review of Ecology and Systematics* 33 (2002): 91–124; J. W. Brown et al., "Strong Mitochondrial DNA Support for a Cretaceous Origin of Modern Avian Lineages," *BMC Biology* 6 (2008): 1–18; J. Cracraft, "Avian Evolution, Gondwana Biogeography and the Cretaceous-Tertiary Mass Extinction Event," *Proceedings of the Royal Society B-Biological Sciences* 268 (2001): 459–69; S. Hope, "The Mesozoic Radiation of Neornithes," in L. M. Chiappe et al., eds., *Mesozoic Birds: Above the Heads of Dinosaurs* (Berkeley: University of California Press, 2002), 339–88; Z. Zhang et al., "A Primitive Confuciusornithid Bird from China and Its Implications for Early Avian Flight," *Science in China Series D* 51, no. 5 (2008): 625–39.
15. N. R. Longrich et al., "Mass Extinction of Birds at the Cretaceous-Paleogene (K-Pg) Boundary," *Proceedings of the National Academy of Sciences* 108 (2011): 15253–57; G. Mayr, *Paleogene Fossil Birds* (Berlin: Springer, 2009), 262; J. A. Clarke et al., "Definitive Fossil Evidence for the Extant Avian Radiation in the Cretaceous," *Nature* 433 (2005): 305–8; T. Fountaine, et al., "The Quality of the Fossil Record of Mesozoic Birds," *Proceedings of the Royal Academy of Sciences B-Biological Science* 272 (2005): 289–94.
16. P. Ericson et al. "Diversification of Neoaves: Integration of Molecular Sequence Data and Fossils," *Biology Letters* 2, no.4 (2006): 543–47; but see J. W. Brown et al., "Nuclear DNA Does Not Reconcile 'Rocks' and 'Clocks' in Neoaves: A Comment on Ericson et al.," *Biology Letters* 3, no. 3 (2007): 257–20; A. Suh et al., "Mesozoic Retroposons Reveal Parrots as the Closest Living Relatives of Passerine Birds," *Nature Communications* 2, no.8 (2011).
17. K. J. Mitchell et al., "Ancient DNA Reveals Elephant Birds and Kiwi Are Sister Taxa and Clarifies Ratite Bird Evolution," *Science* 344, no. 6186 (2014): 898–900.

CHAPTER XIX: HUMANITY AND THE TENTH EXTINCTION: 2.5 MA TO PRESENT

1. P. Ward, *Rivers in Time* (New York: Columbia University Press, 2000).
2. R. Leakey and R. Lewin, *The Sixth Extinction* (Norwell, MA: Anchor Press, 1996).
3. "Lucy's Legacy: The Hidden Treasures of Ethiopia," Houston Museum of Natural Science, 2009.
4. D. Johanson and M. Edey, *Lucy, the Beginnings of Humankind* (Granada: St Albans, 1981); W. L. Jungers, "Lucy's Length: Stature Reconstruction in *Australopithecus afarensis* (A.L.288-1) with Implications for Other Small-Bodied Hominids," *American Journal of Physical Anthropology* 76, no. 2 (1988): 227–31.
5. B. Yirka, "Anthropologist Finds Large Differences in Gait of Early Human Ancestors," Phys.org, November 12, 2012; P. A. Kramer, "Brief Communication: Could Kadanuumuu and Lucy Have Walked Together Comfortably?" *American Journal of Physical Anthropology* 149 (2012): 616–2; P. A. Kramer and D. Sylvester, "The Energetic Cost of Walking: A Comparison of Predictive Methods," *PLoS One* (2011).
6. D. J. Green and Z. Alemseged, "*Australopithecus afarensis* Scapular Ontogeny, Function, and the Role of Climbing in Human Evolution," *Science* 338, no. 6106 (2012): 514–17.
7. J. P. Noonan, "Neanderthal Genomics and the Evolution of Modern Humans," *Genome Res.* 20, no. 5 (2010): 547–53.
8. K. Prufer et al., "The Complete Genome Sequence of a Neanderthal from the Althai Mountains," *Nature* 505, no. 7481 (2014): 43–49.
9. P. Mellars, "Why Did Modern Human Populations Disperse from Africa ca. 60,000 Years Ago?" *Proceedings of the National Academy of Sciences* 103, no. 25 (2006): 9381–86.
10. P. Ward, *The Call of Distant Mammoths: What Killed the Ice Age Mammals* (Copernicus, Springer-Verlag, 1997).

Index

A Note on the Authors

Peter Ward is a Professor of Biology and Professor of Earth and Space Sciences at the University of Washington in Seattle, and has appeared in numerous television documentaries and his eight-hour series, *Animal Armageddon*, was televised in 2009. He is the author of many books, including the bestselling *Rare Earth* and *On Methuselah's Trail*, which was nominated for a *Los Angeles Times* Book Award.

Joe Kirschvink received his PhD from Princeton University and is the Nico and Marilyn Van Wingen Professor of Geobiology at the California Institute of Technology. He originated the 'Snowball Earth' concept to explain weird features of Earth's oldest glaciations, discovered the tiny magnets that animals use for navigation, and has recognised several major shifts in Earth's spin axis that have driven biological evolution.